Introduction to Plasmonics

Introduction to Plasmonics
Advances and Applications

edited by

Sabine Szunerits
Rabah Boukherroub

Pan Stanford Publishing

Published by

Pan Stanford Publishing Pte. Ltd.
Penthouse Level, Suntec Tower 3
8 Temasek Boulevard
Singapore 038988

Email: editorial@panstanford.com
Web: www.panstanford.com

British Library Cataloguing-in-Publication Data
A catalogue record for this book is available from the British Library.

Introduction to Plasmonics: Advances and Applications

Copyright © 2015 Pan Stanford Publishing Pte. Ltd.

All rights reserved. This book, or parts thereof, may not be reproduced in any form or by any means, electronic or mechanical, including photocopying, recording or any information storage and retrieval system now known or to be invented, without written permission from the publisher.

For photocopying of material in this volume, please pay a copying fee through the Copyright Clearance Center, Inc., 222 Rosewood Drive, Danvers, MA 01923, USA. In this case permission to photocopy is not required from the publisher.

Cover image: Courtesy of Alexandre Barras, Interdisciplinary Research Institute, Villeneuve d'Ascq, France

ISBN 978-981-4613-12-5 (Hardcover)
ISBN 978-981-4613-13-2 (eBook)

Printed in the USA

Contents

Foreword   xiii
Preface   xvii

1. **Propagating Surface Plasmon Polaritons**   1
   Atef Shalabney
   1.1 Introduction   1
   1.2 Surface Plasmons on Smooth Surfaces   3
      1.2.1 Surface Plasmon at Single Interface   4
      1.2.2 Surface Plasmon in Multilayer Systems   11
      1.2.3 Electromagnetic Energy Confinement and Field Enhancement   14
      1.2.4 Excitation of Surface Plasmon Polaritons   19
   1.3 Applications   22
      1.3.1 Surface Plasmon Resonance–Based Sensors   22
      1.3.2 Enhanced Spectroscopy and Emissive Processes   29
   1.4 Concluding Remarks   32

2. **Different Strategies for Glycan Immobilization onto Plasmonic Interfaces**   35
   Sabine Szunerits and Rabah Boukherroub
   2.1 Introduction   35
   2.2 Carboxymethylated Dextran Layers: The BiAcore Chip   37
   2.3 Self-Assembled Monolayers Based on Thiolated Functional Groups   40
   2.4 Polymer Films   43

2.5	Lamellar SPR Structures		47
2.5	Conclusion		55

3. Biophysics of DNA: DNA Melting Curve Analysis with Surface Plasmon Resonance Imaging — 61

Arnaud Buhot, Julia Pingel, Jean-Bernard Fiche, Roberto Calemczuk, and Thierry Livache

3.1	Introduction		61
3.2	Temperature Regulation of SPRi for DNA Melting Curves Analysis		63
	3.2.1	SPRi Apparatus with Temperature Regulation	63
	3.2.2	Equilibrium versus Out-of-Equilibrium Melting Curves	65
	3.2.3	Stability of Grafting Chemistries at High Temperatures	67
		3.2.3.1 Electro-copolymerization of poly-pyrrole	67
		3.2.3.2 Thiol self-assembling monolayer	68
3.3	Physico-Chemistry of DNA Melting at a Surface		69
	3.3.1	Effects of Denaturant Molecules	70
	3.3.2	Effects of Salt Concentration	73
3.4	Detection of Single Point Mutation from Melting Curve Analysis		76
	3.4.1	Detection with Oligonucleotides Targets	77
	3.4.2	Detection Limit of Somatic Mutations	78
	3.4.3	Homozygous and Heterozygous Detection of PCR Products	80
3.5	Conclusion		85

4. Plasmon Waveguide Resonance Spectroscopy: Principles and Applications in Studies of Molecular Interactions within Membranes — 89

Isabel D. Alves

4.1	Introduction	89
4.2	Plasmon Spectroscopy	90

		4.2.1	Description of Surface Plasmons	90
		4.2.2	Types of Surface Plasmon Resonances	92
			4.2.2.1 Conventional surface plasmon resonance	92
			4.2.2.2 Plasmon-waveguide resonance	94
		4.2.3	PWR Spectral Analysis	96
	4.3	PWR Applications		100
		4.3.1	Lipid Bilayers	100
			4.3.1.1 Solid-supported lipid bilayers	101
			4.3.1.2 Membranes composed of cellular membrane fragments	102
		4.3.2	GPCR Insertion into Membranes, Activation and Signaling	104
		4.3.3	Role of Lipids in GPCR Activation, Signaling, and Partition into Membrane Microdomains	108
		4.3.4	Interaction of Membrane Active Peptides with Lipid Membranes	111
	4.4	PWR Ongoing Developments		113

5. Surface-Wave Enhanced Biosensing — 119

Wolfgang Knoll, Amal Kasry, Chun-Jen Huang, Yi Wang, and Jakub Dostalek

	5.1	Introduction	119
	5.2	Surface Plasmon Field-Enhanced Fluorescence Detection	121
	5.3	Long-Range Surface Plasmon Fluorescence Spectroscopy	126
	5.4	Optical Waveguide Fluorescence Spectroscopy	132
	5.5	Conclusions	136

6. Infrared Surface Plasmon Resonance — 143

Stefan Franzen, Mark Losego, Misun Kang, Edward Sachet, and Jon-Paul Maria

	6.1	Introduction	143
	6.2	The Hypothesis That Surface Plasmon Resonance Will Be Observed in Free Electron Conductors	149

6.3	Confirmation of the Hypothesis That Conducting Metal Oxides Can Support Surface Plasmon Resonance	151
6.4	The Effect of Carrier Concentration	153
6.5	The Effect of Mobility	155
6.6	Hybrid Plasmons: Understanding the Relationship between Localized LSPR and SPR	157
6.7	The Effect of Materials Properties on the Observed Surface Plasmon Polaritons	159
6.8	Detection of Mid-Infrared Surface Plasmon Polaritons	159
6.9	The Search for High Mobility Conducting Metal Oxides	161
6.10	Conclusion	163

7. The Unique Characteristics of Localized Surface Plasmon Resonance — 169

Gaëtan Lévêque and Abdellatif Akjouj

7.1	Localized Surface Plasmon Resonance of a Single Particle		170
	7.1.1	Single Particle in the Quasi-Static Approximation	172
		7.1.1.1 Case of the spherical particle	174
		7.1.1.2 Case of the spheroidal particle	176
	7.1.2	Beyond the Quasi-Static Approximation	178
7.2	Examples of Coupled Plasmonic Systems		183
	7.2.1	Chain of Identical Particles	183
	7.2.2	Chain of Different Particles	184
7.3	Localized Surface Plasmon for a Periodic Nanostructure		186
	7.3.1	Model and Simulation Method	186
	7.3.2	Absorption Spectra for Au Nanostructures Array	188
	7.3.3	Influence of the Thickness of a Diamond Dielectric Overlayer on the LSPR	191
	7.3.4	Conclusion	196

8. Advances in the Fabrication of Plasmonic Nanostructures: Plasmonics Going Down to the Nanoscale — 199
Thomas Maurer

- 8.1 Introduction — 199
- 8.2 Top-Down Techniques: A Mask-Based Process — 201
  - 8.2.1 Conventional Lithography Techniques: Photolithography and Particle Beam Lithography — 201
    - 8.2.1.1 Photolithography — 202
    - 8.2.1.2 Particle beam lithography — 204
  - 8.2.2 Advanced Lithography Techniques: Masks Coming from Researcher Imagination — 205
    - 8.2.2.1 Multilevel laser interference lithography — 205
    - 8.2.2.2 Nanostencil lithography — 206
    - 8.2.2.3 Self-assembly techniques for mask fabrication: nanosphere lithorgaphy and block copolymer lithography — 208
  - 8.2.3 Direct Writing — 210
    - 8.2.3.1 Particle beam–induced etching and particle beam–induced deposition — 211
    - 8.2.3.2 Laser ablation — 212
    - 8.2.3.3 3D laser lithography — 212
  - 8.2.4 Printing, Replica Molding and Embossing — 214
    - 8.2.4.1 Printing — 214
    - 8.2.4.2 Replica molding — 215
    - 8.2.4.3 Embossing — 217
  - 8.2.5 Conclusion about the Top-Down Strategy — 218
- 8.3 Bottom-Up Techniques: Atom by Atom Building — 219
  - 8.3.1 The Bottom-Up Strategy — 219
    - 8.3.1.1 Physical route — 219

		8.3.1.2	Electrochemical route	221
		8.3.1.3	Chemical route	221
	8.3.2	Self-Organization, the Next Challenge of Plasmonics		224
		8.3.2.1	Laboratory self-assembly techniques	225
	8.3.3	Mass Production Using Wet Coating Processes		232
8.4	Mixing Top-Down and Bottom-Up Routes			232
	8.4.1	Porous Membranes for Ordered Nanowires Growth		232
	8.4.2	Copolymer Template Control of Plasmonic Nanoparticle Synthesis via Thermal Annealing		232
	8.4.3	Let's Play Your Imagination		233
8.5	Conclusion: First, Choose Materials			234

9. Colorimetric Sensing Based on Metallic Nanostructures — 251

Daniel Aili and Borja Sepulveda

9.1	Introduction and Historical Perspective		251
9.2	Synthesis of Gold Nanoparticles		254
9.3	Optical Properties of Gold Nanoparticles		254
9.4	Colloidal Stability and Surface Chemistry of Gold Nanoparticles		255
	9.4.1	Surface Functionalization	256
9.5	Molecular Recognition for Modulation of Nanoparticle Stability		257
	9.5.1	Cross-Linking Assays	259
	9.5.2	Redispersion Assays	263
	9.5.3	Non-Cross-Linking Assays	266
9.6	Outlook and Challenges		267
	9.6.1	Assays with Reversed Sensitivity and Plasmonic ELISA	268
	9.6.2	Assays for the Future	269

10. Surface-Enhanced Raman Scattering: Principles and Applications for Single-Molecule Detection — 275

Diego P. dos Santos, Marcia L. A. Temperini, and Alexandre G. Brolo

10.1	Introduction	275
10.2	Raman Scattering	276
10.3	SERS	283
10.4	SERS Substrates	295
10.5	Single-Molecule SERS	297
10.6	Conclusion	313

11. Graphene-Based Plasmonics — 319

Sinan Balci, Emre Ozan Polat, and Coskun Kocabas

11.1	Introduction: Plasmons in Reduced Dimensions	319
11.2	Optical Properties of Graphene	322
11.3	Synthesis of Graphene	325
11.4	Plasma Oscillations on Graphene–Metal Surface	328
11.5	Graphene Functionalized SPR Sensors	332
11.6	Graphene Passivation for SPR Sensors	334
11.7	Biomolecular Detection Using Graphene Functionalized SPR Sensors	336
11.8	Graphene Oxide Functionalization	337
11.9	Gate-Tunable Graphene Plasmonics	338
11.10	Conclusion	341

12. SPR: An Industrial Point of View — 347

Iban Larroulet

12.1	Introduction	347
12.2	Companies	348
12.3	Future Trends	351

Index — 353

Foreword

The phenomenon of surface plasmon polaritons (or surface plasmons for short) propagating as a bound electromagnetic wave, as surface light, along an interface between a (noble) metal and a dielectric medium has been known for a long time. Already in 1902, R. W. Wood, while monitoring the spectrum of (white) light after reflection by an optical (metallic) diffraction grating, noticed, "I was astounded to find that under certain conditions, the drop from maximum illumination to minimum, a drop certainly of from 10 to 1, occurred within a range of wavelengths not greater than the distance between the sodium lines" (taken from: "On a remarkable case of uneven distribution of light in a diffraction grating spectrum" *Philos. Mag.* **4**: 396). Later, this observation was understood as the first example for the optical excitation of a surface plasmon mode by light, i.e., by plane waves, reflected off the surface of a metallic grating. Various other coupling schemes using prisms to fulfill the required matching conditions between energy and momentum of surface plasmons and plane waves demonstrated a broad range of experimental configurations that allowed for the excitation of this surface light. The wavelength-dispersed direct observation of surface plasmons, excited by white light upon reflection at the metallized base of a right-angle prism as first reported by J. D. Swalen et al. ("Plasmon surface polariton dispersion by direct optical observation," *Am. J. Phys.* (1980) **48**: 670)) is shown in Figure 1. The dark curve—the "Black Rainbow"—demonstrates directly the dispersion of surface plasmons, i.e., their momentum given as the angle of resonance as a function of the excitation wavelength.

These and other experiments marked the early days of surface plasmon research aiming at elucidating the basics and potential applications, e.g., in biosensing, of these evanescent waves propagating at the interface between a mere dielectric medium and a metal that shows collective excitations of its conduction electron cloud, the plasmons. This surface light interacts with spatial (refractive index) heterogeneities at the interface in much the same way as plane waves do, thus giving rise to the full set of optical features known from normal photons interacting with refractive index variations, like refraction, diffraction, or scattering (elastic and inelastic), and can be used for imaging purposes or for the excitation of fluorescence emission provided the chromophores are located near the interface, i.e., within the decay length of the evanescent field. Many of these aspects of surface plasmons are described in the first chapters of this book dealing with propagating surface plasmon modes.

Figure 1 The Black Rainbow.

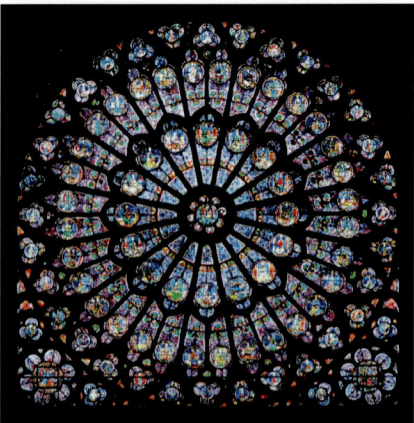

Figure 2 Gothic stained glass rose window of Notre-Dame de Paris. The colors originate from localized surface plasmons, excited in Au colloids, nanoparticles, that are embedded in the glass matrix.

More recently, another very old form of matter, i.e., colloidal gold, reinvented and renamed "Au nanoparticles," attracted a lot of interest in the context of (localized) surface plasmon excitation. Although used in stained glass for centuries (cf., e.g., the colored glass window of the cathedral Notre-Dame in Paris, shown in Figure 2) it was not until the discovery of their enormous enhancement factors seen in Raman spectroscopy or in non-linear optical spectroscopy that these nanoscopic objects became so prominent in the emerging communities of nano-scientists and nano-engineers. The obtainable optical field enhancements are well-understood as given by the resonant excitation of localized surface plasmons in particles, shells, rods, triangles, cubes, beau-ties, and a dozen other shapes of the corresponding nanoscopic resonators made from different (noble) metals. Together with a smart surface functionalization optimized for specific chemical or biological sensing platforms, this allowed for the development of spectroscopic tools, e.g., for the vibrational characterization of even single molecules or for the ultra-sensitive detection of bio-molecules from an analyte solution with an unprecedented limit of detection.

This novel class of surface plasmons in nanostructures is in the focus of the second set of chapters of the current book.

Enjoy the reading of this state-of-the-art summary of the basics and some applications of surface plasmon optics.

Wolfgang Knoll
AIT Austrian Institute of Technology
Vienna, Austria

Preface

It was in the autumn of 2002, working at the CEA (commissariat à l'énergie atomique) in Grenoble, when I became familiar with surface plasmon resonance (SPR). I had no idea that this topic would fascinate me to a point that it would become one of several common research activities with Rabah Boukherroub and finally result in the acceptance to co-edit a book on this interdisciplinary topic. The fascination around the field of plasmonics is that this topic is continuously changing and many researchers and scientists from different fields have joined the field in one way or the other.

This book presents the most widely employed plasmonic approaches and the numerous applications associated with these optical readouts. It seems that several elements underlie plasmonics research today. Advances made in nanoscience and nanotechnology have made the fabrication of plasmonic nanostructures, deposition of thin films, and development of highly sensitive optical characterization techniques possible. The different approaches to nanostructuring metals has led to a wealth of interesting optical properties and functionality via the manipulation of the plasmon modes that such structures support. The sensitivity of plasmonic structures to changes in their local dielectric environment has led to the development of new sensing strategies and systems for chemical analysis and identification.

The first part of the book deals with propagating surface plasmon resonance. Chapter 1 (by Shalabney) explores the properties, excitation, and some applications of surface plasmon polaritons on smooth and planar surfaces. Since SPR biosensors combine two building blocks, the SPR interface where the plasmonic wave is generated and an appropriate surface functionalization, it is clear that the overall performance of an SPR biosensor depends on both the intrinsic optical performance of the SPR sensor and the characteristics of the surface functionalization. Chapter 2 (by

Szunerits and Boukherroub) reviews several approaches used to impart biofunctionality to SPR interfaces and structures and hence transforming them into biosensors; a special focus was put on glycan-functionalized SPR interfaces. More recently, a third advantage was brought by the SPR imaging (SPRi) performances, which permit to follow multiple interactions in parallel. Chapter 3 (by Buhot, Pingel, Fiche, Calemczuk, and Livache) presents the SPRi apparatus and its interests for the DNA sensing and analysis.

Plasmon waveguide resonance (PWR) spectroscopy, a relatively new plasmonics-based biophysical method is presented in Chapter 4 (by Alves). Its use for the study of supported proteolipid membranes and interaction of membrane active peptides with lipid membranes is presented. Chapter 5 (by Knoll, Kasry, Huang, Wang, and Dostalek) gives an excellent insight into the race for the most sensitive platforms for biosensing applications. The chapter, entitled "Surface-Wave Enhanced Biosensing," discusses surface-plasmon field-enhanced fluorescence, long-range surface plasmon fluorescence spectroscopy, and optical waveguide fluorescence spectroscopy. This first part will be concluded by Chapter 6 (by Franzen, Losego, Kang, Sachet, and Maria) on infrared surface plasmon resonance. The state of the art of mid-infrared instrumentation and materials is discussed.

The second part of the book considers the interest of localized surface plasmons. Chapter 7 (Akjouj and Lévêque) summarizes the physical concept behind localized surface plasmon resonance. Chapter 8 (by Maurer) outlines, in a systematic manner, the different fabrication methods employed to generate plasmonic nanostructures. The use of metallic nanostructures in solution for colorimetric sensing is presented in Chapter 9 (by Aili and Sepulveda). Chapter 10 (by dos Santos, Temperini, and Brolo) follows up and describes surface-enhanced Raman scattering as means to obtain chemical information and as high sensitive detection tool. Chapter 11 (by Kocabas, Balci, and Polat) is focused on a rather novel aspect of plasmonics, graphene plasmonics. The last chapter, Chapter 12 (by Larroulet), entitled "SPR: An Industrial Point of View," rounds up the book and gives an overall conclusion.

The authors and editors of this book hope that the content will be of interest for researchers, students, and anybody interested in the diverse aspects of plasmonics. We wish you a good time reading the chapters.

Sabine Szunerits
Rabah Boukherroub

Chapter 1

Propagating Surface Plasmon Polaritons

Atef Shalabney
ISIS, Université de Strasbourg and CNRS (UMR 7006),
8 allèe Gaspard Monge, 67000 Strasbourg, France
shalabney@unistra.fr

1.1 Introduction

One of the most common approaches to describe the electronic properties of the solid state is the plasma concept. In this approach, the free electrons of a metal are treated as an electron liquid of high density of about 10^{23} cm^{-3}. If one ignores the lattice in a first approximation, plasma oscillations will propagate through the metal bulk with energy quanta, called volume plasmons, of $\hbar\omega_p = \hbar\sqrt{4\pi n_e q_e^2 / m_e}$, where n_e, q_e, and m_e are the electron density, charge, and mass, respectively. This energy quanta, $\hbar\omega_p$, is of the order of few electron volts and can be excited by electron-loss spectroscopy means.

In the presence of planar boundary, there is a new mode, called surface plasmon (SP), with broad spectrum of eigen frequencies from $\omega = 0$ to $\omega = \omega_p/\sqrt{1 + \varepsilon_d}$ (ε_d—the dielectric constant of the medium adjacent to the metal) depending on the wave vector $\vec{k}$. Introducing the concept "surface plasma" in addition to the plasma

---

Introduction to Plasmonics: Advances and Applications
Edited by Sabine Szunerits and Rabah Boukherroub
Copyright © 2015 Pan Stanford Publishing Pte. Ltd.
ISBN 978-981-4613-12-5 (Hardcover), 978-981-4613-13-2 (eBook)
www.panstanford.com

concept extended the plasmon physics and opened the door for huge amount of studies in the last few decades. In an ideal semi-infinite medium, the dispersion relation of surface plasmon waves (SPW) lies to the right of the light line, which means that the SPs have longer wave vector than light waves of the same energy $h\omega$. Therefore, they are called "non-radiative" surface plasmons, i.e., they cannot decay by emitting a photon and, conversely, light incident on an ideal surface cannot excite SP.

However, under certain circumstances, SPs can couple to light-producing surface plasmon polaritons (SPPs), which their electromagnetic fields decay exponentially into the space perpendicular to the surface with maximum at the interface, as is characteristic for surface waves. From the basic dispersion relation of the SP, one can see that the increase in the momentum with respect to the light line is associated with the binding of the SP to the surface. The resulting momentum mismatch between light and SPs of the same frequency must be bridged if light is to be used to generate SPs.

There are three main techniques by which the missing momentum can be provided. The first is using a prism coupling to enhance the momentum of the incident light. The second makes use of a periodic corrugation in the metal's surface. The third involves scattering from a topological defects on the surface, such as sub-wavelength hole. The diffraction (scattering) of light by a diffraction metallic grating allows incident light to be momentum matched and thus coupled to SPs. The reverse process, importantly, allows the otherwise non-radiative SP mode to couple with light in a controlled way with good efficiency, which is vital if SP-based photonic circuits are to be developed.

This chapter is designated to explore the properties, excitation, and some applications of SPP on smooth and planar surfaces. In this case, most of the SPWs physics can be entirely understood as the picture is well mathematically defined. The second section presents the basic electrodynamics description of SPWs on planar interface in the ideal case between two semi-infinite metallic and dielectric media. In this subsection, the role of the metal optical properties will be extensively addressed with appropriate modeling of the dispersion. Although this case cannot describe the practical uses and excitations of SPP, it may provide the reader with basic physical insights into the main properties of these waves. In the following subsections, the full dispersion relation for multilayer systems will be derived with the electromagnetic fields distribution through the layers. Two major methods of SPP excitation will be discussed in Section 1.2.4 and few applications of SPPs are discussed in Section 1.3. Our concluding remarks are briefly given in Section 1.4.

1.2 Surface Plasmons on Smooth Surfaces

Treating the optics of SPs can be done by different approaches. From the electrodynamics point of view, SPs are a particular case of surface waves, a topic that has large importance in the realm of radio-wave propagation, such as propagation along the surface of the earth. From the optics point of view, SPs are modes of an interface and from the solid-state physics point of view; SPs are collective excitation of electrons.

In this chapter, the macroscopic electrodynamics point of view will be adopted by deriving the dispersion relation of the surface wave. Within this approach, the material properties are accounted for by using a dielectric constant without any microscopic model. In order to focus on the case of SPPs, the basic Drude model will be introduced into the dispersion relation of the SP. However, the limitations of this model will be addressed, and the Lorenz–Drude model will be discussed as a good alternative in some spectral regions [1].

First, the dispersion relation of single interface bounded by semi-infinite metal and dielectric will be derived. This is the ideal case, from which one can easily understand the principles, conditions, and limitations of the SP excitation. In the next part, the practical case of multi-layer system consisting of metal slab will be addressed and different variations of this structure will be outlined. The main feature of SP waves, namely the sub wavelength confinement of the electromagnetic energy to the surface will be emphasized with its importance to different applications. The most prominent techniques of exciting SPs on smooth surfaces will be also explained.

1.2.1 Surface Plasmon at Single Interface

SP waves are a longitudinal charge density distribution generated on the metal side interface when light propagates through the metal. Both the metal and the dielectric sample can be with complex refractive indices $n_m = \sqrt{\varepsilon_m}$ and $n_d = \sqrt{\varepsilon_d}$, respectively, as shown in the inset of Fig. 1.1. Within the electrodynamics approach, the surface plasmon propagates as an electromagnetic wave parallel to the x direction with magnetic field oriented parallel to the y direction, that is, transverse magnetic polarization (TM or P) state. The condition of (TM) polarization state is needed to generate the charge distribution on the metal interface, which is considered as the first condition for SP excitation. The SP phenomenon can be easily understood, and its main characteristics can be determined by solving Maxwell's equation to the boundary-value problem.

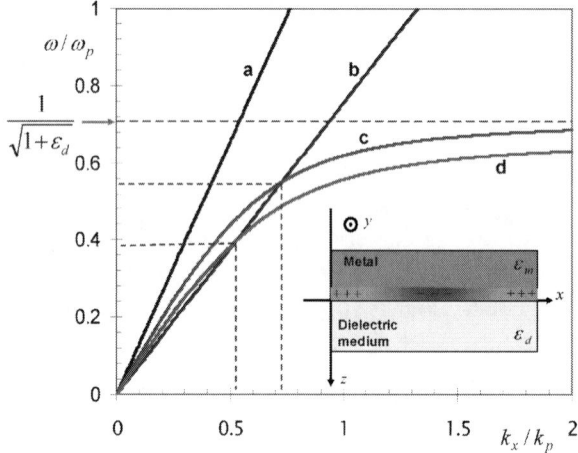

Figure 1.1 a and b show typical dispersion relation of a photon propagating in free space and inside a dielectric, respectively in order to compare with the SP dispersion. Curves c and d show the dispersion of SP at the interface between metal and dielectric. Line c corresponds to higher RI for the dielectric medium than case d. The frequency is normalized to the plasma frequency ω_p, and the x component of the light wave vector is normalized to the plasma wave number, which is defined in the text. The inset at the right bottom shows metal/dielectric interface with the coordinates system as described in the text.

In this section, we will treat the single interface case only to introduce the main features of SP waves. Hence, the full solution of the electrodynamics problem will not be detailed. In the next section, when the multi-layer system is being addressed, full description of the electromagnetic field involved in the boundary-problem will be done.

By assuming TM plane and harmonic electromagnetic fields in the metal and the dielectric sample with the appropriate waves:

$$E_i(r,t) = (E_{ix}, 0, E_{iz})e^{-k_{zi}|z|}e^{i(k_{xi}x-\omega t)}$$
$$H_i(r,t) = (0, H_{iy}, 0)e^{-k_{zi}|z|}e^{i(k_{xi}x-\omega t)}, \qquad (1.1)$$

where, (i is metal, dielectric sample), k_{zi} is the z component of the wave vector in the layer i, and k_{xi} is the x component of the wave vector (or propagation constant). Substituting the fields from Eq. 1.1 into Maxwell's equations and applying the continuity conditions and the relation $k_{zi} = \sqrt{\varepsilon_i(\omega^2/c^2) - k_i^2}$ leads to

$$\frac{k_{z1}}{\varepsilon_1}H_{1y} + \frac{k_{z2}}{\varepsilon_2}H_{2y} = 0$$
$$H_{1y} - H_{2y} = 0. \qquad (1.2)$$

A solution could be obtained for the last equations set by equating the determinant to zero leading to $k_{z1}\varepsilon_2 + k_{z2}\varepsilon_1 = 0$. Together with the phase-matching condition ($k_{x1} = k_{x2} = k_x$) gives an explicit expression for k_x (here $\varepsilon_m = \varepsilon_{mr} + i\varepsilon_{mi}$ and ε_d stand for ε_1 and ε_2, respectively).

$$k_x = \frac{\omega}{c}\sqrt{\frac{\varepsilon_m \varepsilon_d}{\varepsilon_m + \varepsilon_d}} \qquad (1.3)$$

The value in the left hand side of Eq. (1.3) represents the surface plasmon wave vector $k_x = k_{sp}$; hence, this equation defines the dispersion relation for the propagating SP.

The requirement for surface wave imposes k_z to be pure imaginary; therefore, the term under the root of Eq. 1.3 should be larger than one, and this can be satisfied only if the two dielectric constants have opposite signs. Since in the range of optical frequencies, the real part of the metal dielectric constant is negative ($\varepsilon_{mr} < 0$) and assuming that the dielectric medium adjacent to the metal does not or weakly absorbs in this region, the condition for SP wave (Eq. 1.3) can be fulfilled only if metallic substance possesses one of the interface sides.

In order to obtain the dispersion relation in terms of $\omega(k)$, we introduce the simplified Drude model for the metal dispersion $\varepsilon_m = 1 - \omega_p^2/\omega^2$ (ω_p is the plasma frequency given in the introduction) in Eq. (1.3) and the dispersion of the SP can be written as follows:

$$\omega^2 = \frac{1}{2}\omega_p^2 + \frac{1}{2}k_x^2 c^2 \left(1 + \frac{1}{\varepsilon_d}\right) - \frac{1}{2}\sqrt{\omega_p^4 + 2\omega_p^2 k_x^2 c^2 \left(1 - \frac{1}{\varepsilon_a}\right) + k_x^4 c^4 \left(1 + \frac{1}{\varepsilon_a}\right)^2} \quad (1.4)$$

One can see by the curves in Fig. 1.1 that the SP dispersion always lies to the right of the light cone (line a in the figure). However, the dispersion curve of the SP approaches the light line $\sqrt{\varepsilon_d}\omega/c$ at small k_x, it remains larger than this line so that the SPs cannot transform into light; thus, it is a nonradiative SP. At large k_x (or in case that $\varepsilon_{mr} \to -\varepsilon_d$) the value of the plasmon frequency approaches $\omega_p/\sqrt{1 + \varepsilon_d}$. In this limit, the group velocity of the SP goes to zero as well as the phase velocity, so that the SP resembles localized fluctuations of the electron plasma.

It is important to emphasize that the Drude model is a crude approximation for noble metals when frequency approaches the plasma frequency. Nevertheless, this model was used in our discussion to derive the SP dispersion relation in the ideal case and to introduce few key issues for the sake of simplicity. In fact, the Drude model describes the metal optical response in the limit of zero losses where there is no damping to the free electrons oscillations inside the metal due to the electromagnetic field.

Before proceeding with the main features of SPs, it is necessary to discuss the optical properties of metals more extensively. It is important to keep in mind that although the Drude model can be very useful tool, its accuracy is much better in the IR than in the visible range. When the frequency increases, photons can excite electrons in electronic bands of lower energies so that new absorption channels are available. In addition, the Drude model cannot describe spatial dispersion and size-dependent dielectric constant. Therefore, simple phenomenological models such as the

Lorenz–Drude oscillator model based on the damped harmonic oscillator approximation can be used to describe the optical properties of metals. Here, we briefly discuss the Lorenz–Drude model often used for parameterization of the optical constants of metals.

It has been shown that the complex dielectric function $\varepsilon_m = \varepsilon_{mr} + i\varepsilon_{mi}$ can be expressed by combining the intra-band contribution, also referred to as free-electrons effect and the inter-band transitions, usually referred to as bound-electrons effect.

$$\varepsilon_m = \varepsilon_{mr} + i\varepsilon_{mi} = \varepsilon_{\text{free-electrons}} + \varepsilon_{\text{bound-electrons}} \quad (1.5)$$

The intra-band contribution (free electrons effect), $\varepsilon_{\text{free-electrons}}$, of the dielectric function is described by the same Drude model as before with introducing the damping term Γ_0, representing the losses that were not considered in the former discussion.

$$\varepsilon_{\text{free-electrons}} = 1 - \frac{f_0 \omega_p^2}{\omega(\omega - i \cdot \Gamma_0)} \quad (1.6)$$

The inter-band contribution of the dielectric constant can be described, similar to the Lorenz result for insulators, as follows:

$$\varepsilon_{\text{bound-electrons}} = \sum_{j=1}^{K} \frac{f_j \omega_p^2}{(\omega_j^2 - \omega^2) + i \cdot \Gamma_j \omega}, \quad (1.7)$$

where in the last two equations ω_p is the plasma frequency and K is the number of oscillators involved with frequency ω_j strength f_j and lifetime $1/\Gamma_j$ for each oscillator

In practical use, the parameters involved in the Lorenz–Drude model are calculated by fitting to the available experimental data in the literature. The appropriate parameters for noble metals (Ag, Au, and Cu) usually used in plasmonic together with some transition metals (Ni, Ti, and Pt) can be found in Table 1.1. The parameters in Table 1.1 were mainly taken from [2].

In Fig. 1.2, the real and imaginary parts of gold and silver refractive (RI) index are shown in the range 0.5–10 μm. Silver and gold RIs are presented since these two metals are commonly used in plasmonic nanostructures due to several properties that will be discussed latter in this section and the following sections.

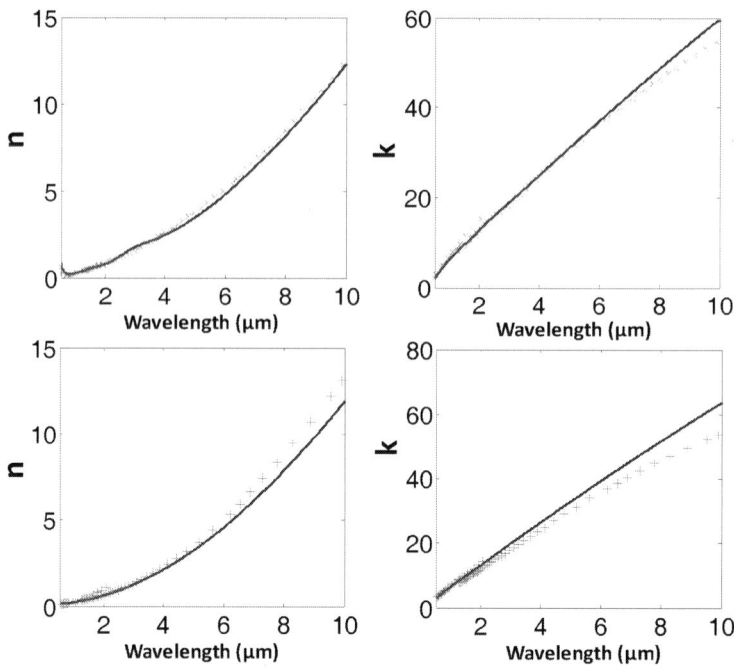

Figure 1.2 Calculated real (n) and imaginary (k) parts in solid blue lines versus experimental data in red crosses for gold (top panel) and silver (bottom panel). The parameters in table 1 were used in the calculations.

Table 1.1 Lorenz–Drude model parameters that give the best fit with the experimental data of some metals.

	Ag	Au	Cu	Ni	Ti	Pt
ω_p	9.01	9.03	10.83	15.92	7.29	9.59
f_0	0.845	0.760	0.575	0.096	0.148	0.333
Γ_0	0.048	0.053	0.030	0.048	0.082	0.080
f_1	0.065	0.024	0.061	0.1	0.899	0.191
Γ_1	3.886	0.241	0.378	4.511	2.276	0.517
ω_1	0.816	0.415	0.291	0.174	0.777	0.780
f_2	0.124	0.010	0.104	0.135	0.393	0.659
Γ_2	0.452	0.345	1.056	1.334	2.518	1.838
ω_2	4.481	0.830	2.957	0.582	1.545	1.314
f_3	0.011	0.071	0.723	0.106	0.187	0.547

Γ_3	0.065	0.870	3.213	2.178	1.663	3.668
ω_3	8.185	2.969	5.300	1.597	2.509	3.141
f_4	0.840	0.601	0.638	1.729	0.001	3.576
Γ_4	0.916	2.494	4.305	6.292	1.762	8.517
ω_4	9.083	4.304	11.18	6.089	19.43	9.249
f_5	5.646	4.384				
Γ_5	2.419	2.214				
ω_5	20.29	13.32				

Note: All values of ω_j and Γ_j are given in eV units.

Back to Eq. 1.3, one can see that in practice, the propagation constant of the SP is not purely real. The real and imaginary part of the k_{sp} can be easily derived from Eq. 1.3 by introducing the complex representation of the metal dielectric constant from Eqs. 1.5–1.7.

$$k_{x\text{-real}} = \frac{\omega}{c}\sqrt{\frac{\varepsilon_{mr} \cdot \varepsilon_d}{\varepsilon_{mr} + \varepsilon_d}}; \quad k_{x\text{-imag}} = \frac{\omega}{c}\left(\frac{\varepsilon_{mr} \cdot \varepsilon_d}{\varepsilon_{mr} + \varepsilon_d}\right)^{3/2}\left(\frac{\varepsilon_{mi}}{2\varepsilon_{mr}^2}\right) \quad (1.8)$$

The imaginary part of the SP wave vector induces, in fact, additional damping factor parallel to the interface that originates solely from the dispersion when assuming that the dielectric medium adjacent to the metal does not absorb.

This damping factor can be quantified through the *propagation distance* L_x of the SP along the interface before damping to the half of its initial intensity; this propagation length can be expressed from the last equation as follows:

$$L_x = \frac{1}{2|k_x''|} = \frac{\lambda}{2\pi} \cdot \frac{\varepsilon_{mr}^2}{\varepsilon_{mi}} \cdot \left[\frac{\varepsilon_a + \varepsilon_{mr}}{\varepsilon_a \cdot \varepsilon_{mr}}\right]^{3/2} \quad (1.9)$$

The damping along the interface is also combined with decay in the direction perpendicular to the interface due to the imaginary character of the z component of the wave vector (see Eq. 1.3 together with the k_z expression). The simultaneous decay of the electromagnetic fields in parallel and perpendicular to the surface extremely confines the SP to the interface as will be shown in Section 1.2.3. The decay distance, often called the *penetration depth* of the field, can be calculated in the same manner as the propagation distance using $1/|k_z|$. The penetration inside the metal δ_m and the dielectric δ_d sides are

$$\delta_m = \frac{\lambda}{2\pi} \cdot \sqrt{\frac{\varepsilon_d + \varepsilon_{mr}}{-\varepsilon_{mr}^2}}; \quad \delta_d = \frac{\lambda}{2\pi} \cdot \sqrt{\frac{\varepsilon_d + \varepsilon_{mr}}{-\varepsilon_d^2}} \quad (1.10)$$

The penetration depths presented in Eq. 1.10 correspond to the field as can be seen in Fig. 1.3 and not to the intensity; factor 1/2 should be introduced to account for the intensity damping distance.

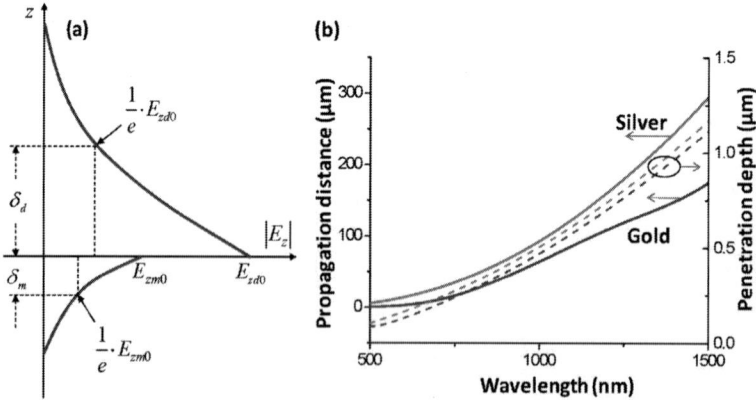

Figure 1.3 (a) Enhancing the field component perpendicular to the interface due to SPP. δ_m, δ_d, are the decay lengths of the field perpendicular to the surface in the metal and the dielectric regions, respectively (b) propagation distances (solid curves) and penetration depths (dotted curves) for gold (blue) and silver (red) for gold (silver)/air interface. The penetration depths correspond to the field intensity.

One expects that the field penetrates to larger extent inside the dielectric medium due to the smaller dielectric constant. While the penetration depths perpendicular to the interface can be sub wavelength values (tens of nanometers), the propagation distance can reach few hundreds of microns in the standard SP configurations (see Fig. 1.3).

1.2.2 Surface Plasmon in Multilayer Systems

In the previous section, the single metal/dielectric interface was only considered in order to introduce the main characteristics of SP waves generated at the interface. In real uses and experimental studies, this case is not indeed practical. In this chapter, the modes generated in a multi-layer configuration will be extensively addressed. We exploit this opportunity to introduce the full rigorous calculations, based on the electrodynamics theory to calculate the dispersion of SP supported by multi-interfaces. This rigorous treatment is arranged in top-to-bottom hierarchy. Namely, the dispersion relation will be derived for the case of two thin layers bounded by two semi-infinite media, which constitute the general case of our discussion. Simpler and more particular study cases will be then derived from the general dispersion relation.

Derivation of the dispersion relation for SPs for two adjacent thin layers with asymmetric environment:

Assume two thin layers with thicknesses a and b bounded between two semi-infinite dielectric, homogenous, and isotropic media (see Fig. 1.4). Layer (2) with thickness a and permittivity ε_2 is in the range $0 < z < a$ and layer (3) with thickness b and permittivity ε_3 extends in the range $-b < z < 0$.

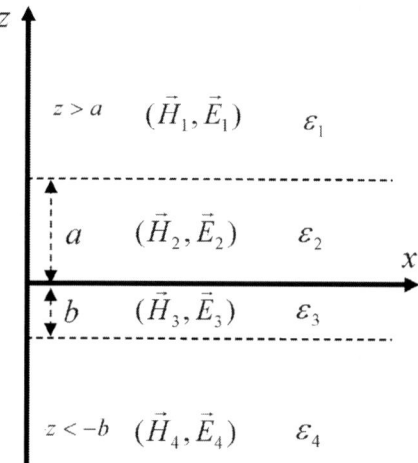

Figure 1.4 Schematic illustration of two thin layers with thicknesses a and b, and permittivities ε_2 and ε_3 bounded by two semi-infinite media with permittivities ε_1 and ε_4.

The first semi-infinite medium (medium 1) locates on the top of layer 2 with permittivity ε_1, whereas the second semi-infinite medium (medium 4) locates beneath layer 3 with permittivity ε_4.

In order to derive the dispersion relation of the entire system, p-polarized (TM) wave that propagates in the x direction is considered. The electromagnetic fields inside media i (i = 1, 2, 3, 4) that have no y dependence are of the form $\vec{H}_i$ = (0, H_{yi}, 0) and $\vec{E}_i$ = (E_{xi}, 0, E_{zi}).

Using the relations $E_{xi} = (-i/\omega\varepsilon_0\varepsilon_i)\cdot(\partial H_{yi}/\partial z)$ and $E_{zi} = (-k_x/\omega\varepsilon_0\varepsilon_i)\cdot H_{yi}$, from Maxwell equations, the fields in all regions can be written as follows:

$z > a$
$$H_{y1} = A\cdot \exp i(k_x x - \omega t)\cdot \exp(ik_1 z)$$
$$E_{x1} = \frac{A}{\omega\varepsilon_0}\left(\frac{k_1}{\varepsilon_1}\right)\cdot \exp i(k_x x - \omega t)\cdot \exp(ik_1 z)$$
$$E_{z1} = \frac{-A\cdot k_x}{\omega\varepsilon_0}\left(\frac{1}{\varepsilon_1}\right)\cdot \exp i(k_x x - \omega t)\cdot \exp(ik_1 z)$$

(1.11)

$0 < z < a$
$$H_{y2} = \exp i(k_x x - \omega t)\cdot\left[B\cdot \exp(ik_2 z) + C\cdot \exp(-ik_2 z)\right]$$
$$E_{x2} = \frac{1}{\omega\varepsilon_0}\cdot\left(\frac{k_2}{\varepsilon_2}\right)\exp i(k_x x - \omega t)\cdot\left[B\cdot \exp(ik_2 z) - C\cdot \exp(-ik_2 z)\right]$$
$$E_{z2} = \frac{-k_x}{\omega\varepsilon_0}\cdot\left(\frac{1}{\varepsilon_2}\right)\exp i(k_x x - \omega t)\cdot\left[B\cdot \exp(ik_2 z) + C\cdot \exp(-ik_2 z)\right]$$

$-b < z < 0$
$$H_{y3} = \exp i(k_x x - \omega t)\cdot\left[D\cdot \exp(ik_3 z) + E\cdot \exp(-ik_3 z)\right]$$
$$E_{x3} = \frac{1}{\omega\varepsilon_0}\cdot\left(\frac{k_3}{\varepsilon_3}\right)\exp i(k_x x - \omega t)\cdot\left[D\cdot \exp(ik_3 z) - E\cdot \exp(-ik_3 z)\right]$$
$$E_{z3} = \frac{-k_x}{\omega\varepsilon_0}\cdot\left(\frac{1}{\varepsilon_3}\right)\exp i(k_x x - \omega t)\cdot\left[B\cdot \exp(ik_3 z) + C\cdot \exp(-ik_3 z)\right]$$

$z < -b$
$$H_{y4} = F\cdot \exp i(k_x x - \omega t)\cdot \exp(-ik_4 z)$$
$$E_{x4} = \frac{F}{\omega\varepsilon_0}\left(\frac{k_4}{\varepsilon_1}\right)\cdot \exp i(k_x x - \omega t)\cdot \exp(-ik_4 z)$$
$$E_{z4} = \frac{-F\cdot k_x}{\omega\varepsilon_0}\left(\frac{1}{\varepsilon_4}\right)\cdot \exp i(k_x x - \omega t)\cdot \exp(-ik_4 z)$$

The wave vector in layer i has the components k_x and k_i, which satisfy the relation $k_x^2 + k_i^2 = \varepsilon_i(\omega/c)^2$. The component k_z in region i is indicated as k_i for brevity. The continuity relations of the tangential field's components E_{xi} and H_{yi} at the three interfaces (1/2), (2/3), and (3/4) gives two linear homogenous equations for the coefficients B and C:

$$\begin{bmatrix} m_{11} & m_{12} \\ m_{21} & m_{22} \end{bmatrix} \begin{bmatrix} B \\ C \end{bmatrix} = \begin{bmatrix} 0 \\ 0 \end{bmatrix}, \quad M \equiv \begin{bmatrix} m_{11} & m_{12} \\ m_{21} & m_{22} \end{bmatrix} \quad (1.12)$$

The last equation, in fact, has the same form of Eq. 1.2 for the ideal interface but with new m_{ij} elements, which are as follows:

$$m_{11} = \exp(ik_2 a) \cdot [\varepsilon_2 k_1 - \varepsilon_1 k_2]$$
$$m_{12} = \exp(-ik_2 a) \cdot [\varepsilon_2 k_1 + \varepsilon_1 k_2]$$
$$m_{21} = \exp(-ik_3 b) \cdot [\varepsilon_3 k_4 + \varepsilon_4 k_3] \cdot [\varepsilon_3 k_2 + \varepsilon_2 k_3] - \exp(ik_3 b) \cdot [\varepsilon_3 k_4 - \varepsilon_4 k_3] \cdot [\varepsilon_3 k_2 - \varepsilon_2 k_3]$$
$$m_{22} = \exp(-ik_3 b) \cdot [\varepsilon_3 k_4 + \varepsilon_4 k_3] \cdot [\varepsilon_2 k_3 - \varepsilon_3 k_2] + \exp(ik_3 b) \cdot [\varepsilon_3 k_4 - \varepsilon_4 k_3] \cdot [\varepsilon_3 k_2 + \varepsilon_2 k_3]$$

$$(1.13)$$

Using the same analogy with Eq. 1.2, in order to obtain a solution to the equation system (1.12), the determinant $|M|$ has to be zero. The later condition leads to the dispersion relation of the entire system described in Fig. 1.4, which is

$$\exp(2ik_2 a) = \frac{-[1 + r_{23} \cdot r_{34} \cdot \exp(2ik_3 b)]}{r_{12} \cdot [r_{34} \cdot \exp(2ik_3 b) + r_{23}]}; \quad r_{ij} = \left(\frac{k_i}{\varepsilon_i} - \frac{k_j}{\varepsilon_j}\right) \bigg/ \left(\frac{k_i}{\varepsilon_i} + \frac{k_j}{\varepsilon_j}\right) \quad (1.14)$$

One can see that the coefficients r_{ij} are, indeed, the reflection coefficients of the interface (i/j) obtained from the Fresnel equations for p-polarized light. The dispersion relation in Eq. 1.14 is the central result of this section, and will be used through all the discussion below. At this point, one should remember that dispersion relation is defined as the relation $\omega(k_x)$, which can be obtained by introducing $k_i^2 = \varepsilon_i(\omega/c)^2 - k_x^2$ into Eq. 1.14 and ending with only two dependent variables relation.

The first particular case to be discussed is the case of a single thin layer between two homogeneous semi-infinite media, which can be easily realized by setting $b = 0$ in Eq. 1.14. Actually, this gives the dispersion relation for a thin film, which was addressed in several works [3].

$$\exp(2ik_2 a) = \frac{-1}{r_{12} \cdot r_{24}} = \frac{-\left(\dfrac{k_1}{\varepsilon_1} + \dfrac{k_2}{\varepsilon_2}\right) \cdot \left(\dfrac{k_2}{\varepsilon_2} + \dfrac{k_4}{\varepsilon_4}\right)}{\left(\dfrac{k_1}{\varepsilon_1} - \dfrac{k_2}{\varepsilon_2}\right) \cdot \left(\dfrac{k_2}{\varepsilon_2} - \dfrac{k_4}{\varepsilon_4}\right)} \tag{1.15}$$

One should note that the layers are numbered according to Fig. 1.4. Assuming layer 2 (with thickness a) is a metal layer, and considering the thick metal layer limit, the exponent in the left side of Eq. (1.15) vanishes and two solutions of k_x are obtained, describing two distinguished plasmons at the two metal interfaces, which are given as follows:

$$k_{x(1/2)} = \frac{\omega}{c} \cdot \sqrt{\frac{\varepsilon_1 \cdot \varepsilon_2}{\varepsilon_1 + \varepsilon_2}}; \quad k_{x(2/4)} = \frac{\omega}{c} \cdot \sqrt{\frac{\varepsilon_2 \cdot \varepsilon_4}{\varepsilon_2 + \varepsilon_4}} \tag{1.16}$$

Due to the large imaginary part in the k_z inside the metal layer, the thick metal layer limit can be satisfied only by approaching the thickness a to the wavelength or even sub wavelength values. Furthermore, one can easily obtain the single interface dispersion curve that was discussed in the previous section from the last equations.

In the derivation of the dispersion relation, no restrictions were assumed on k_z. However, when SPP modes are to be derived, the lowest-order bound modes are of interest. These bounded modes only decay in the z direction and do not oscillate; therefore, the wave vectors in the z direction should be considered as pure imaginary numbers. In this case, Eq. 1.15 is valid when k_i are replaced by $i|k_i|$.

1.2.3 Electromagnetic Energy Confinement and Field Enhancement

One of the most prominent characteristic of SP waves is the confinement nature of the electromagnetic fields associated with the SP to the interface at which they are generated. Once the SP wave vector is determined by Eq. 1.14, the normalized distribution of the fields can be calculated using the equation set 1.11. Let us for simplicity, without losing the generality, assume that the magnetic field amplitude A at the 1/2 interface ($z = a$ in Fig. 1.4) is known. In this case, one can easily express the fields' amplitudes in Eq. 1.11 as follows:

$$B = \frac{1}{2} \cdot A \cdot \exp\{i(k_1 - k_2)a\} \cdot \left[1 + \frac{k_1 \varepsilon_2}{k_2 \varepsilon_1}\right]$$

$$C = \frac{1}{2} \cdot A \cdot \exp\{i(k_1 + k_2)a\} \cdot \left[1 - \frac{k_1 \varepsilon_2}{k_2 \varepsilon_1}\right]$$

$$\begin{bmatrix} D \\ E \end{bmatrix} = \frac{1}{2} \begin{bmatrix} 1 + \dfrac{k_2 \varepsilon_3}{k_3 \varepsilon_2} & 1 - \dfrac{k_2 \varepsilon_3}{k_3 \varepsilon_2} \\ 1 - \dfrac{k_2 \varepsilon_3}{k_3 \varepsilon_2} & 1 + \dfrac{k_2 \varepsilon_3}{k_3 \varepsilon_2} \end{bmatrix} \begin{bmatrix} B \\ C \end{bmatrix}$$

$$F = D \cdot \exp\{-i(k_3 + k_4)b\} + E \cdot \exp\{i(k_3 - k_4)b\} \tag{1.17}$$

In the same manner, one can normalize all the fields in a specific system to any other component of the electromagnetic wave. As a classical example for the behavior of the fields, we consider a case in which we have two layers embedded between two semi-infinite media as depicted in Fig. 1.4. The first layer is dielectric with refractive index 1.5 and lies in the region 0 < z < 300 nm. The second layer is made of silver with 47 nm thickness, namely located in the region 300 < z < 347 nm. The two layers are bounded between two infinite dielectric media, with ε_1 = 3.16, ε_4 = 1 as can be seen in Fig. 1.5a. Solving the dispersion relation (1.14) for the system yields two plasmons that can be generated on the 2/3 and 3/4 interfaces: the two interfaces of the metal with the dielectric materials. Numerically, one can obtain that the plasmon at the air side (3/4 interface) propagates with k_{sp4} = 1.02 × 10^7 m^{-1}, while the Plasmon at the 2/3 interface propagates with k_{sp3} = 1.619 × 10^7 m^{-1}. The last values of the propagation constant of the waves are the real values of K_{sp}. From Fig. 1.5, one can see that once the plasmon is excited at the metal/air interface, the fields become to be tightly confined to this surface and rapidly decay into both the metal and the air sides. In both Figs. 1.5 and 1.6, the intensities of the electric and magnetic fields were plotted in the two thin layers and the air when z = 0 is set to be the first interface (1/2 interface). Figure 1.5 was obtained by substituting the first solution k_{sp4} into the fields' distribution (Eq. 1.17).

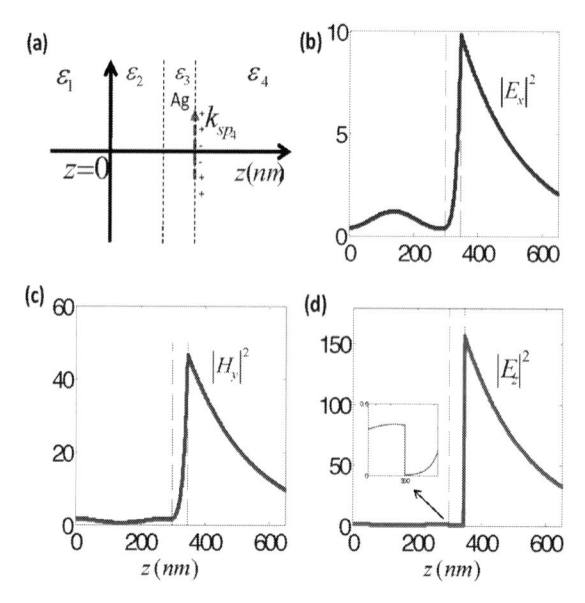

Figure 1.5 (a) Two thin layers with $\varepsilon_2 = 2.25$, $\varepsilon_3 = \varepsilon_{Ag}$ embedded between two semi-infinite media with $\varepsilon_1 = 3.16$, $\varepsilon_4 = 1$. The profiles in b, c, and d describe the intensity distributions of E_x, H_y, and E_z, respectively, for Plasmon on Ag/air interface (K_{sp4} in the text). The inset in d shows the discontinuity in the Z component of the electric field at the 2/3 interface. The wavelength is 632.8 nm and $z = 0$ is set to be the first interface (1/2 interface).

Substituting the second solution of the dispersion relation in Eq. 1.17 gives the distribution of the fields when the plasmon at the 2/3 interface is excited. These fields can be seen in Fig. 1.6.

This example was brought in order to show the fields enhancement associated with the SPP wave created on the surface. This enhancement in the fields is the key feature for the majority of plasmonics applications as will be further discussed in the following sections. However, the general structure of two layers was considered; in most of the common applications, only one thin film is located between the two media and then Eq. 1.15 should be used to calculate the propagation constant of the plasmon. In this context, a considerably interesting special case in which the film is embedded between two symmetric media should be further discussed.

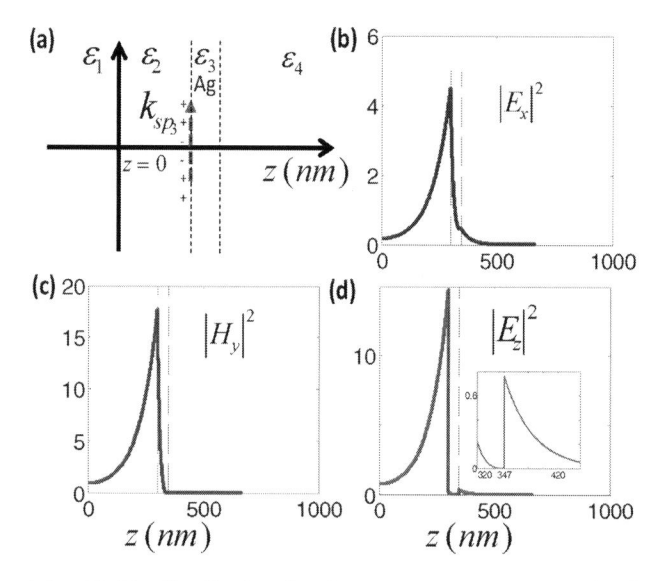

Figure 1.6 Fields distribution for the same case as in Fig. 1.5, when the second solution k_{sp3} is considered. The inset in d shows the discontinuity in E_z intensity at the 3/4 interface.

When $b = 0$ and $\varepsilon_1 = \varepsilon_4 = \varepsilon_d$ in Fig. 1.4, the structure becomes symmetric by means of the dielectric response of the surrounding media and $k_1 = k_4 = k_d$. Due to the new conditions, Eq. 1.15 splits into two equations:

$$i \cdot \tan\left(\frac{k_2 \cdot a}{2}\right) = \frac{k_d \cdot \varepsilon_2}{k_2 \cdot \varepsilon_d}$$
$$i \cdot \tan\left(\frac{k_2 \cdot a}{2}\right) = \frac{k_2 \cdot \varepsilon_d}{k_d \cdot \varepsilon_2} \tag{1.18}$$

It can be easily shown that these two dispersion relations describe two different modes that can be supported by the new system. It can be shown that the two equations in 1.18 give two different modes with odd and even parity of Ex. these two modes are of significant interest when the insulator-metal-insulator (IMI) and metal-insulator-metal (MIM) structures are considered. In the first structure, the thin layer constitutes the metallic core of the structure and the surrounding media are dielectric, whereas in MIM structure the thin core layer is dielectric bounded between two metallic thick shells. Considering the IMI structure, one can see that the confinement of the fields to the metal film decreases upon decreasing the metallic film thickness.

This decrease causes that the mode evolves into a plane wave supported by the infinite dielectric boundaries. The higher concentration of the mode energy inside the dielectric and not into the metal extremely decreases the losses and consequently increases the propagation distance of the mode. Due to their relatively long propagation distance at the surface, these modes are called long-range SPP (LRSPP) modes [4].

In Fig. 1.7, E_x and H_y intensities distribution in a simple IMI configuration is shown. The metallic core is thin gold film with

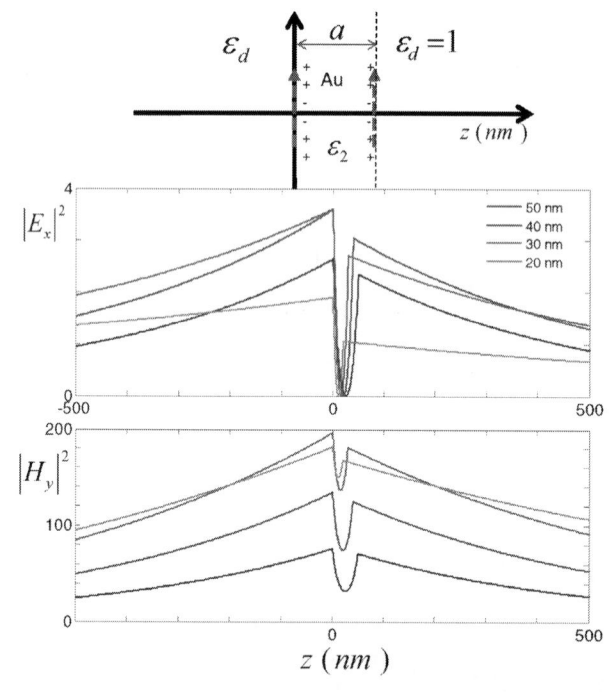

Figure 1.7 E_x and H_y intensities distribution versus the metallic core thickness in IMI structure. The metallic core is gold and the wavelength is 850 nm. Here $z = 0$ is the gold/air interface. The inset on top shows the symmetric IMI structure.

varying thickness bounded by the free space from both sides. When the gold thickness decreases, the relative energy portion inside the gold decreases and the intensities are more concentrated into the free space sides until the cut-off thickness is reached. The case of the MIM can be interesting for plasmonics waveguides as well. More details about the wide uses of IMI and MIM configurations for several applications can be found in [3].

1.2.4 Excitation of Surface Plasmon Polaritons

As was early shown in Fig. 1.1, the wave vector of the SP generated at the metal\dielectric interface is greater than the wave vector inside the dielectric, which, in fact, causes the evanescent decay on both sides of the interface.

Moreover, it was clearly demonstrated in Fig. 1.1 that the wave vector of light falling on the metallic film satisfies the dispersion line of a photon in free space ($\omega = ck_0$), which lies on the left of the dispersion line of the plasmon for any frequency. Hence, a light beam that hits the metal surface from air will never excite the plasmon at the metal interface unless the photon momentum is enhanced in order to match the SP momentum, and an intersection between the two lines is obtained. The intersection represents the resonance phenomenon and defines an operating point in which the frequency and the wave vector of both the exciting light and the plasmon are determined.

Since the early stages of SP field, the interaction between volume and surface plasmons in thin metallic films were studied using low-energy electron beams [5]. Electrons hit on a thin metallic film transfer momentum and energy to the electrons of the metal.

If the energy loss of the scattered electrons together with the angle of diffraction were observed, the momentum transferred to the metal electrons can be easily determined. In the works of Ritchie [6], an additional loss, apart from the volume plasmon loss, was observed in lower energies. The physics of SPs has been studied intensively with electrons and the fundamental properties of SPs have been found by using the loss-energy spectroscopy techniques. However, in this section we will highlight the optical techniques rather than the traditional electronic techniques to excite SPs. In most of the modern uses and applications of SP, optical approaches are commonly implemented to overcome the momentum-matching problem associated with SP excitation.

Back to Fig. 1.1, once an optical beam emerges from the free space is being used to excite the SP, the wave vector k_{inc} of the incident beam should be enhanced to match the wave vector of the SP on the metal/dielectric interface. For this purpose, two major methods are used. The first is by using a prism coupler and the second is using diffraction gratings to achieve the momentum matching condition. Although many techniques, perhaps with different names, are often referred to when SPs excitation are considered, we want to stress that most of these techniques can be easily included into the prism coupling and diffraction gratings methods. In fact, they are not different in substance from the tow methods that we are going to highlight, even if they are presented by different configurations in the literature.

The *prism coupling* method enhances the wave vector of the incident beam by the high refractive index of the prism. Therefore, the momentum matching condition becomes $k_{inc} = k_0 n_p \sin\theta_p = k_{sp}$ and can be now fulfilled (k_{sp} as was defined in Eq. 1.3). If the incidence angle inside the prism is larger than the critical angle for the total reflection from the dielectric interface ($\theta_{inc} = \theta_p > \theta_c \sin^{-1}(n_d/n_p)$), the momentum-matching condition can be achieved and the SP at the metal/dielectric interface can be excited (see Fig. 1.8). This configuration is called the Kretschmann configuration [5] and the matching between the incident photon and the SPP referred to as surface plasmon resonance (SPR). The excitation of the SP is accompanied with a minimum in the reflectivity due to the enhanced absorption by which one can be used as a probe for the SP excitation. The SPs in the Kretschmann configuration is a leaky mode thus the absorption is due to leakage of radiation back into the prism and not only due to the inherent losses in the metal. The minimum in the reflection as can be seen in Fig. 1.8d is highly sensitive to the optical properties of the adjacent dielectric medium as will be further discussed in the applications section.

In the *grating coupling* method, on the other hand, the wave vector of the incident light is enhanced due to the normal diffraction modes of the gratings as can be seen from Fig. 1.8. Unlike the Kretschmann configuration, by using the grating coupling method the incident angle is modified and not the effective refractive index that the incident beam experiences. Since the diffracted angle can be modified by 2D grating as well, the general momentum-matching condition becomes:

$$k_{SP} = (k_x + iG_x) \cdot \hat{x} + (k_y + jG_y) \cdot \hat{y}$$
$$G_x = \frac{2\pi}{a_x}, \; G_y = \frac{2\pi}{a_y}, \quad (1.19)$$

where G_x and G_y are the reciprocal lattice vectors of the grating in x and y, respectively, a_x and a_y the corresponding grating periods, and i and j are integers.

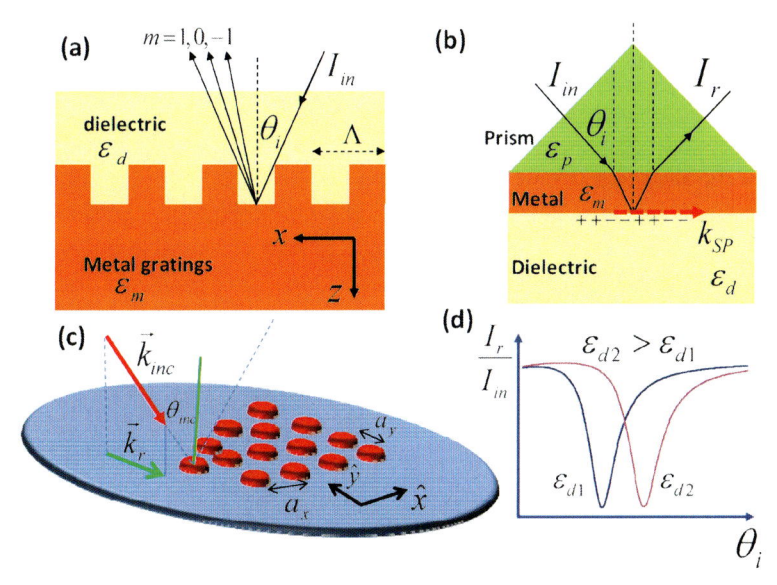

Figure 1.8 (a) 1D grating configuration for exciting SPP. The incident beam diffracts into several modes, which can enhance the wave vector of the incident light and couple to the SP at the interface between the metallic gratings and the adjacent dielectric medium. (b) Kretschmann configuration for exciting SPP at the metal/dielectric interface by increasing the wave-vector using the high refractive index of the prism. (c) 2D grating configuration (see the text). (d) Typical reflectivity from the Kretschmann configuration, minimum in the reflectivity is observed once the SP is excited. The minimum is sensitive to the refractive index of the dielectric medium.

It should be noted that $\vec{k}_r = k_x \cdot \hat{x} + k_y \cdot \hat{y}$ is the projection of the incident wave vector $\vec{k}_{\text{inc}}$ in the grating plane as shown in Fig. 1.8c. The case of the 1D grating can be easily derived from Eq. 1.19 by cancelling the y contribution. Using other techniques such as excitation using highly focused optical beams [7] and near field excitation [8] are not different methods than those we have mentioned above. One can see that those configurations are basically based on the same two concepts that we discussed with different optical set-ups, configurations, and measurement techniques.

1.3 Applications

The property of the SPP in confining the electromagnetic fields to the interface between metal and dielectric enables many applications of SPP in photonics, biology, and materials science. The wide spectrum of applications evolved since the early stages of the field and is being developed progressively. Since this chapter is dedicated for the creation, characterization, and manipulation of SPP on smooth surfaces, only some of the relevant applications would be addressed. Even if our discussion is restricted to the applications of SPP on smooth surfaces, one can find a huge number of works utilizing these waves in several areas of research and industry. However, in this section two major realms of applications will be extensively presented. In the first part, we will show how SPWs can be used for sensing in general and biosensing in particular. In the second part, on the other hand, the enhancement of radiative processes due to SPPs excitations is considered. We still want to stress that there are many other uses and interesting applications for SPWs on smooth surfaces such as plasmonic waveguides, nonlinear plasmonics, and optical imaging. These applications will not be considered in the present chapter due to the limited space and the apriority in our opinion. Nevertheless, many publications, books, and book chapters can be found on these applications [9, 10, 3].

1.3.1 Surface Plasmon Resonance–Based Sensors

Optical biosensors are powerful detection and analysis tools that have vast applications in biomedical research, healthcare, pharmaceuticals, environmental monitoring, homeland security, and the battlefield. They can be used to provide qualitative information, such as whether two molecules interact, and quantitative information, such as kinetic and equilibrium constants for complex formation, for a wide range of biological systems. In general, there are two detection protocols that can be implemented in optical biosensing: fluorescence-based detection and label-free detection. In fluorescence-based detection, either target molecules or bio-recognition molecules are labeled with fluorescent tags, such as dyes; the intensity of the fluorescence indicates the presence of the target molecules and the strength of the interaction between target and bio-recognition molecules. Although fluorescence-based detection methods are usually extremely sensitive, they are accompanied by laborious labeling processes. In label-free detection methods, on the other hand, no labeling processes are needed, and molecules are detected in their natural forms. This is relatively easy and cheap to perform, and enables dynamic monitoring as well. Furthermore, some label-free procedures measure the refractive index (RI) change induced by molecular interactions, which is often related to the concentration of the sample or the surface density, and therefore there is no need for large volumes of samples. This characteristic makes the label-free detection methods advantageous when ultra small (femto-liters to nano-liters) volumes of the sample are involved (see Fig. 1.9)

However, both protocols are being widely used in optical biosensors, providing complementary information on bimolecular interactions, and making optical biosensors more versatile than other sensing technologies.

The central concept in the sensing process, when one talks about SPP-based optical biosensors, is that the sample that is being sensed (called analyte through this section) comes into proximity with the interface on which the SPP is excited. Since the excitation properties are strongly dependent on the refractive index of the surrounding medium, any variations in the analyte can be probed by SPW at the interface.

To demonstrate the basic mechanism on which SPR sensors are based, we will consider the KR configuration that was used in the previous section to excite SPP. In Fig. 1.10, the dispersion of the SPP at the interface between the metal and an analyte is shown with the cross section points with the light line when it emerges from the high-RI prism as was discussed in the previous section. When the momentum matching condition is satisfied, the intersection point between the plasmon dispersion and the photon dispersion determines the resonance operating point, namely the resonance frequency ω and momentum k. This resonance can be optically probed by monitoring the dip in the reflectance R.

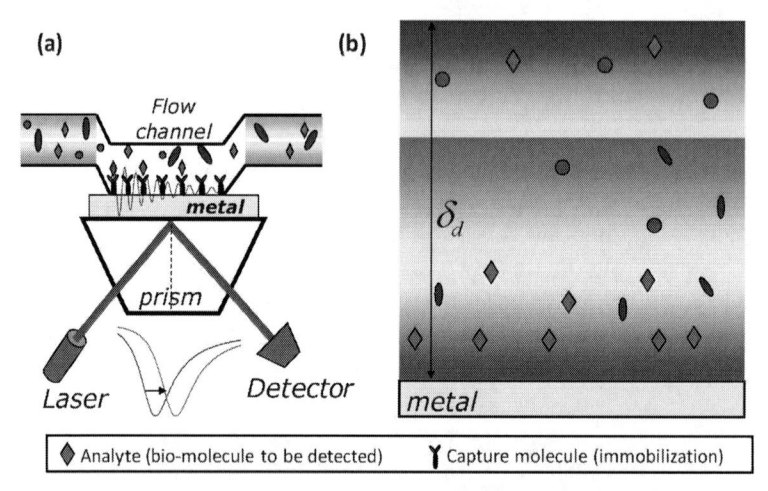

Figure 1.9 Schematic representation of SPR based label free detection method. (a) In the KR configuration, the SPP propagates parallel to the surface of the metal film (often Ag or Au films are used). The lower inset shows the change in the optical response of the structure upon the binding of the specific molecule to be sensed, and the shift demonstrates the change in the signal after binding. (b) The figure illustrates the penetration depth of the SPP field inside the analyte region. The color gradient in each magnified illustration represents the field intensity distribution (the enhanced field at the surface (red) decreases towards the dielectric medium (blue)).

Each of the straight lines in Fig. 1.10 represents a fixed incident angle light line, and by moving along this line one can change the wavelength continuously. By changing the analyte RI, the dispersion of the plasmon will change and consequently changes the resonance point as can be seen in Fig. 1.10C. This route, in which the incidence angle is fixed and the spectrum of the reflectance is probed, is called the spectral interrogation configuration. Alternatively, one can keep the wavelength of the incident light constant and change the incidence angle (the slope of the light line) in order to meet the new resonance point as shown in Fig. 1.10D. This route, in which the reflectance versus the incidence angle is probed, is called the angular interrogation configuration.

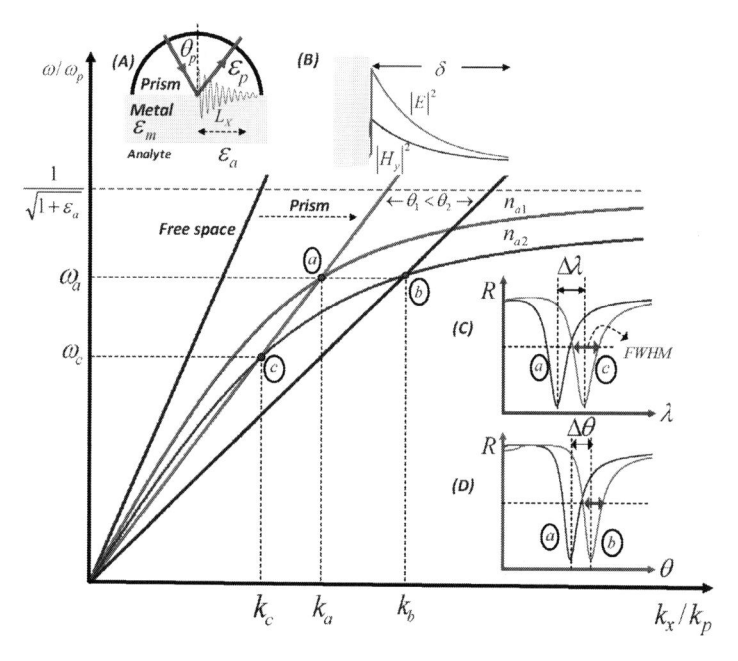

Figure 1.10 Dispersion relation plot of the plasmon wave vector k_{sp} at the interface between metal and dielectric semi-infinite medium, for two different analyte RIs. The lines from left to right represent the prism dispersion line in KR configuration when the light comes from free space and when it emerges from a prism in two different angles. Insets A and B show the propagation distance and the penetration depth of the EM fields, respectively. Insets C and D show the reflectivity in different resonance points, demonstrating spectral (C) and angular (D) interrogation.

By using the simple dispersion relation that was derived in Section 2.1, using Eq. 1.3 for the plasmon dispersion and considering the matching condition in the KR configuration $k_{inc} = k_0 n_p \sin\theta_p = k_{sp}$ one can easily derive the sensitivity of the two configurations to variations in the analyte RI.

The angular and spectral sensitivity for the angular and spectral interrogation in the KR configuration are as follows:

$$S_\theta \equiv \frac{d\theta}{dn_a} = \frac{\varepsilon_{mr} \cdot \sqrt{-\varepsilon_{mr}}}{(\varepsilon_{mr} + n_a^2) \cdot \sqrt{\varepsilon_{mr}(n_a^2 - n_p^2) - n_a^2 \cdot n_p^2}}$$

$$S_\lambda \equiv \frac{d\lambda}{dn_a} = \frac{\varepsilon_{mr}^2}{\frac{1}{2} \cdot \left|\frac{d\varepsilon_{mr}}{d\lambda}\right| \cdot n_a^3 + \frac{\varepsilon_{mr} \cdot n_a}{n_p} \cdot \frac{dn_p}{d\lambda} \cdot (\varepsilon_{mr}^2 + n_a^2)} \quad (1.20)$$

The complex dielectric constant of the metal $\varepsilon_m = \varepsilon_{mr} + i\varepsilon_{mi}$, the prism RI is n_p, and the analyte RI n_a.

In addition to possibilities to track the angular and spectral movements in the reflectance response, one can also follow the change in phase or intensity once the analyte changes as can be seen from Fig. 1.11. Due to the steep change in the intensity of the reflectance in the vicinity of the resonance, the intensity can be used

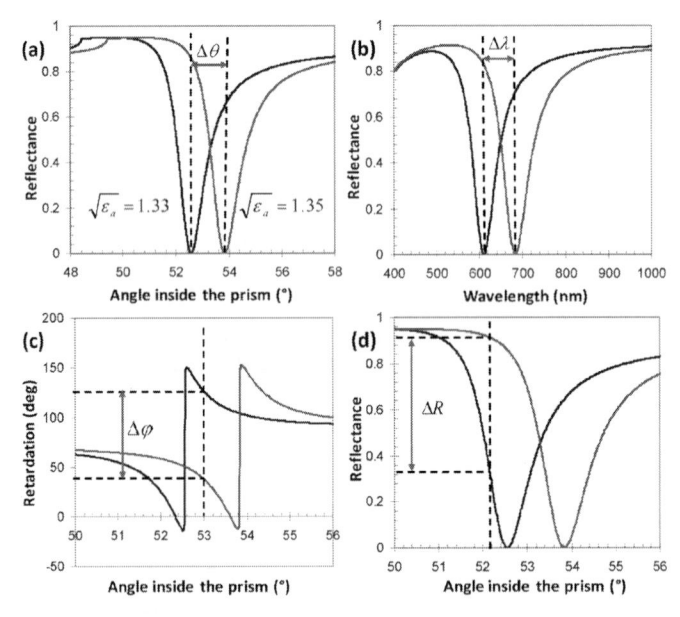

Figure 1.11 Demonstration of the four different routes using the KR for sensing: (a) angular interrogation, (b) spectral interrogation, (c) phase interrogation, and (d) intensity interrogation. The blue curves in all the panels correspond to analyte RI $n_a = 1.33$ and the pink curves correspond to RI $n_a = 1.35$.

to probe the analyte RI change with high sensitivity. Furthermore, upon exciting the SPWs, strong retardation to the reflected light occurs because the high contrast between the P and S polarization components in the incident light. This abrupt change in phase in resonance can be used for sensing as well. For extensive review to the methods of enhancing the sensitivity of SPR sensors, see Shalabney and Abdulhalim [11].

The mechanism of sensing on planar plasmonic surfaces is based on the exceptional high electromagnetic fields at the interface. The sensitivity of SPR-based sensors scales with the intensity of the field, which penetrates and overlaps with the analyte substance [12]. This was the origin of the motivation using the long-range SPR (LR-SPR) structures, whose physical properties were presented in Section 1.2.3. By exciting the LR-SPP mode at the surface of very thin metallic layers, the field is significantly enhanced and the propagation distance increases, which together enhance the overlap between the SPP field and the analyte. This configuration of using LR-SPR was intensively investigated recently, for more details see [13, 14]. In order to obtain the symmetrical claddings with respect to the metal film, a buffer layer with RI close to the analyte RI is used (see inset in Fig. 1.12c). Using this configuration, two distinguishable dips can be seen in the reflectance. The first is associated with the short range SPR with broad dip and the second corresponds to the LR-SPR with extremely narrow dip and very large propagation distance along the surface. Within a good approximation, the propagation distance of the plasmon is proportional to the reciprocal of the full width half maximum (FWHM) of the dip. Furthermore, using the LR-SPR structure enhances the fields at the analyte interface and the penetration depth inside the analyte, enhancing with that the overlap with the analyte region. The short range SPR and LR-SPR dips are shown in Fig. 1.12c and the comparison between the fields of the LR-SPR and conventional SPR are shown in Fig. 1.12d.

The interest in modifying the conventional KR configuration to improve SPR sensor performances created wide range of planar multilayer structures in which the SPP field is enhanced and the sensitivity is increased.

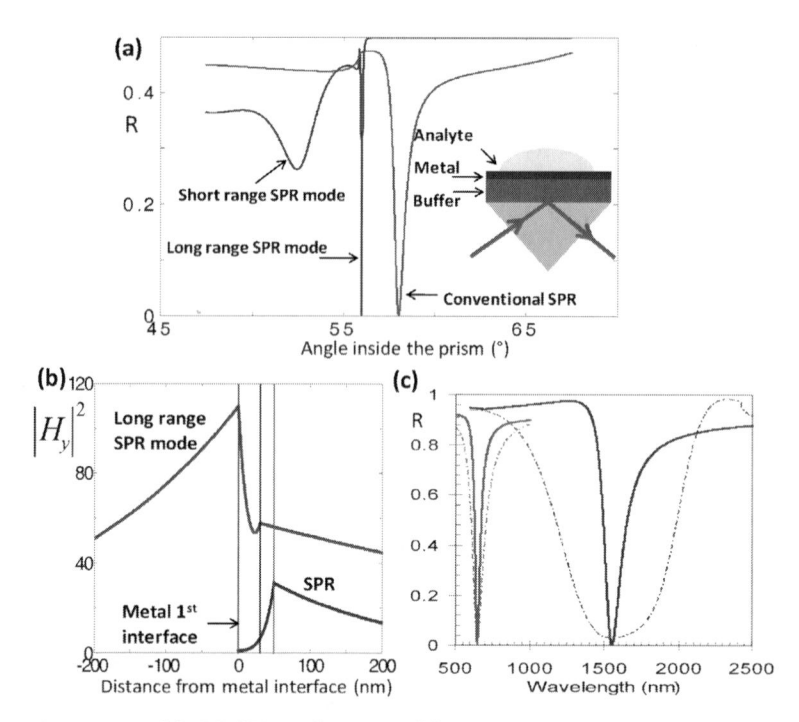

Figure 1.12 (a) LR-SPR reflectivity (blue curve) versus the incidence angle inside the prism for the structure in the inset. The buffer layer is Teflon with RI 1.31, and the metal layer is gold with 30 nm thickness. The reflectivity is measured for 830 nm wavelength, and the analyte RI is 1.33. The green curve shows the reflectivity of the conventional SPR configuration for the same wavelength but the gold film thickness is 50 nm, (b) the magnetic field intensity distribution for both cases in part a showing the enhancement in the field for the LR-SPR case with respect to the conventional SPR, (c) SPR spectra (dashed curves) and NGWSPR (solid curves) in two different incident angles showing the narrowing effect caused by the additional dielectric film. The responses of SPR and NGWSPR were designed to be resonant at the same wavelengths to facilitate the comparison between the two configurations.

Dielectric-coated plasmonic interfaces and their role in enhancing the sensitivity of analyte–ligand interactions are extensively examined due to the advanced deposition techniques that were recently developed [15]. Unlike the LR-SPR case in which the buffer layer is added between the prism and the metal layer, adding a thin dielectric film on top of the metallic layer showed prominent effects on SPR sensors performances. As well as enhancing both the angular and the spectral sensitivities of the sensor, these dielectric coatings may improve the resolution of the sensor by considerably narrowing the dip widths. Since the additional top layer thickness is below the cutoff thickness of the first guided TM mode, this configuration was initially called near guided wave SPR (NGWSPR) [16].

Adding 10 nm of Si on top of a silver layer in the conventional KR set-up enhances the sensitivity and narrows the dips in the spectral interrogation as can be seen in Fig. 1.12a. In this figure, part (a) shows the experimental set-up using the NGWSPR configuration with the additional layer, and part (b) shows differences between the dip widths for both conventional SPR and NGWSPR. The dashed curves in Fig. 11.12c. This figure shows the differences in dips widths between conventional SPR and NGWSR. The comparison was done in two different spectral regions by tuning the incidence angle appropriately. The incidence angles were chosen to obtain the dips for both SPR and NGWSPR at the same wavelength. The narrowing in the dips significantly improves the resolution of sensors based on the NGWSPR configuration with respect to the conventional SPR. For extensive details on the NGWSPR configuration, see [15].

In the present section, we present the basic mechanism of SPR-based biosensors and some of the recent modifications for the conventional KR configuration. Various configurations were used in the context of SPR-based optical sensors on flat and planar surfaces. Again, due to the limitation of space, we cannot address all these configurations here. For more details on the variety of SPR sensors, see [11].

1.3.2 Enhanced Spectroscopy and Emissive Processes

Within the wide perspective, using the basic KR configuration and the spectral/angular interrogation of the reflectance to probe changes in the adjacent medium can be included in SPR-based spectroscopy as well. However, we intended in this short section to emphasize radiative processes on planar surfaces and their dependence on the SPP that can be simultaneously generated on these surfaces. In order to make the discussion more concrete, we will address two radiative optical phenomena in which SPP have very prominent role. Surface enhanced Raman scattering (SERS) and surface enhanced fluorescence (SEF) are of great interest for spectroscopy as well as for biosensing communities.

In the following paragraphs, we will briefly describe the effect of SPP on enhancing these two optical phenomena by manifesting the main SPP-mediated mechanisms of enhancing these emission processes.

Several mechanisms of enhancement were proposed in the early days of SERS investigations to account for the experimental facts observed. These mechanisms converged with the years into two classes, which were called the electromagnetic mechanism (EM) and the chemical mechanism (CM). The EM mechanism is basically caused by enhancing the electric field at the vicinity of the molecule, which yields Raman scattering. The key result is that surface plasmons can be excited, resulting in an enhanced electromagnetic field close to the surface. In Raman scattering, the intensity depends on the square of the incident field strength, as a result, the intensity will be enhanced relative to what it would be in the absence of the surface. The Raman emitted field may also be enhanced, generally by a different amount than the incident field, because the frequency is different. The overall enhancement associated with the incident and the emitted field enhancement is called the EM contribution to SERS. The total averaged Raman enhancement (EN) following Schatz [17] is

$$\text{EN} = \Re(\lambda_{ex}) \cdot \Re(\lambda_{sc}) \tag{1.21}$$

The components $\Re(\lambda_{ex})$ and $\Re(\lambda_{sc})$ refer to the enhancements in the incident and the scattered fields, respectively. When doing so, we also make the simplifying approximation that the field enhancement for the Raman emitted photon with wavelength λ_{sc} can be calculated by using the same expression for the field as the incidence photon with the excitation wavelength λ_{ex}. This assumption can be done since the two processes are very close spectrally and spatially. The SERS enhancement factor is then proportional to $|E(\lambda_{ex})|^2 \cdot |E(\lambda_{sc})|^2$, which was indicated by $\Re(\lambda_{ex})$ and $\Re(\lambda_{sc})$ in Eq. 1.21.

The chemical mechanism (CM) of enhancement comes; in particular, to explain the difference between the intensities obtained from different molecules despite the fact that they have the same polarizability and hence demonstrate the same EM effect. Although this mechanism is of interest for SERS community, it still out of the scope of this section and only the EM mechanism will be addressed.

Combining the properties of enhancing the electromagnetic field at the surfaces caused by SPP and Eq. 1.21, one can immediately notice the contribution of SPP to SERS. Under the assumption that the incident photon and scattered photon are very close spectrally (with respect to the spectral line width of the SP) brings us to the result that the enhancement due to the EM mechanism is proportional to the fourth power of the field enhancement. To this end, we attempted to explain the enhancement in SERS within the fundamental picture of fields' enhancement due to SPP excitations. For more details on the origins of SERS enhancement, the reader is referred to appropriate review articles [18, 19]. Although SPP on planar surfaces enhance the fields, SERS is often observed from molecules on rough surfaces and nanostructures. Since localized surface plasmon (LSPR) resonance is not within the scope of this chapter, we leave this discussion to other chapters.

Unlike SERS, fluorescence is a linear emission process that depends on the field intensity in which the fluorescent molecule is embedded. Hence, molecules that are located in the near-field of SPP can significantly enhance their fluorescence. However, care should be taken when the molecule is placed near metallic surface in order not to quench the fluorescence by non-radiative relaxations. When the incident field in resonance with the SP at the surface where the molecule is located, significant enhancement in the emission could occur as long as non-radiative energy transfer from the molecule to the surface cannot take place. These non-radiative relaxations have lower probability when the molecule is not too close to the surface. Therefore, the distance between the emitter and the surface supporting SPP should be appropriately optimized when fluorescence enhancement is needed. For more details on the fluorescence rate variations from single emitter near metallic surfaces, see [20].

When the spectral and spatial overlap between the emitter absorption band and the photonic mode are further enhanced, the field–matter interaction may enter to the strong coupling regime. In this limit, the emission is split into two new hybrid states that are not observed before the strong coupling. The classical example of enhancing fluorescence in a cavity, known as the Purcell effect [21], in which possible emission channels are modified due to the interaction between the molecule and the cavity. In the strong coupling regime, the photonic mode and the emitter exchange energy with rate larger than the dissipation rate of both the cavity photonic mode and the exaction fluorescence decoherence. For more details on the strong coupling effect, see [22, 23].

1.4 Concluding Remarks

Plasmonics is vastly expanding field and plasmonics on smooth and planar surfaces provide the basic steps to understand the phenomenon. Although in this chapter we attempted to present the most basic theory of plasmonics on planar surfaces, we still believe that for graduate students and newcomers to the field, this introduction is sufficient to build the basic understanding of plasmonics. The mathematical platform given in the first sections together with the study cases presented in this chapter may provide the reader with the elementary tools to explore this area. Although in the applications section we presented only two applications for SPP on smooth surfaces, we still want to stress that there is numerous of applications that were not addressed and can be found in the relevant literature.

References

1. Enoch, S., Bonod, N. (2012). *Plasmonics: From Basics to Advanced Topics*, Springer.
2. Rakic A. D., Djurisic A. D., Elzar J. M., Majewski M. L. (1998). Optical properties for metallic films for vertical-cavity optoelectronic devices, *Appl. Opt.*, **37**, 22, 5271–5283.
3. Maier, S. A. (2007). *Plasmonics, Fundamentals and Applications*, Springer.
4. Sarid, D. (1981). Long-range surface-plasma waves on very thin metal films, *Phys. Rev. Lett.*, **47**, 1927–1930.
5. Raether, H. (1988). "Surface plasmons on smooth surfaces", Springer.
6. Ritchie, R. H. (1957). Plasma Losses by fast electron in thin films, *Phys. Rev.*, **106**, 874–881.

7. Bouhelier, A., Wiederrecht, G. P. (2005). Excitation of surface plasmon polaritons: plasmonic continuum spectroscopy, *Phys. Rev. B*, **71**(19), 195406–195412.
8. Hecht, B., Bielefeldt, H., Novotny, L., Inouye, Y., Pohl, D. W. (1996). Local excitation, scattering, and interference of surface plasmons, *Phys. Rev. Lett.*, **77**, 1889–1892.
9. Berini, P. (1999). Plasmon polariton modes guided by a metal film of finite width, *Opt. Lett.*, **24**(15), 1011–1013.
10. Lamprecht, B., Krenn, J. R., Schider, G., Ditlbacher, H., Salerno, M., Felidj, N., Leitner, A., Aussnegg, F. R., Weeber, J. C. (2001). Surface plasmon propagation in microscale metal strips, *Appl. Phys. Lett.*, **79**, 51.
11. Shalabney, A., Abdulhalim, I. (2011). Sensitivity-enhancement methods for surface plasmon, *Sensors*, **5**(4), 571–606.
12. Shalabney, A., Abdulhalim, I. (2010). Electromagnetic fields distribution in multilayer thin film structures and the origin of sensitivity enhancement in surface plasmon resonance sensors, *Sens. Actuators A: Phys.*, **159**(1), 24–32.
13. Nenninger, G. G., Tobiska, P., Homola, J., Yee, S. S. (2001). Long-range surface plasmons for high-resolution surface plasmon resonance sensors, *Sens. Actuators B: Chem.*, **74**(1–3), 145–151.
14. Slavik, R., Homola, J. (2007). Ultrahigh resolution long range surface plasmon-based sensor, *Sens. Actuators B: Chem.*, **123**(1), 10–12.
15. Szunerits, S., Shalabney, A., Boukherroub, R., Abdulhalim, I. (2012). Dielectric coated plasmonic interfaces: their interest for sensitive sensing of analyte–ligand interactions, *Rev. Anal. Chem.*, **31**, 15–28.
16. Shalabney, A., Abdulhalim, I. (2012). Figure-of-merit enhancement of surface plasmon resonance sensors in the spectral interrogation, *Opt. Lett.*, **37**(7), 1175–1177.
17. Schatz, G. (1984). Theoretical studies of surface enhanced Raman scattering, *Acc. Chem. Res.*, **17**(10), 370–376.
18. Moskovits, M. (1985). Surface enhanced spectroscopy, *Rev. Mod. Phys.* **57**(3), 783–828.
19. Kerker, M., Wang, D. S., Chew, H. (1980). Surface enhanced Raman scattering (SERS) by molecules adsorbed at spherical particles, *Appl. Optics*, **19**(19), 3373–3388.
20. Anger, P., Bharadwaj, P., Novotny, L. (2006). Enhancement and Quenching of single-molecule fluorescence, *Phys. Rev. Lett.*, **96**, 113002–113005.
21. Purcell, E. M. (1946). Spontaneous emission probabilities at radio frequencies, *Phys. Rev.*, **69**, 681.

22. Houdrè, R. (2005). Early stages of continuous wave experiments on cavity-polaritons, *Phys. Stat. Sol. B*, **242**(11), 2167–2196.
23. Schwartz, T., Hutchison J. A., Genet C., Ebbesen T. W. (2011). Reversible switching of ultrastrong light-molecule coupling, *Phys. Rev. Lett.*, **106**, 196405.

Chapter 2

Different Strategies for Glycan Immobilization onto Plasmonic Interfaces

Sabine Szunerits and Rabah Boukherroub

Interdisciplinary Research Institute, Parc de la Haute Borne,
50 avenue de Halley, BP 70478, 59658 Villeneuve d'Ascq, France

sabine.szunerits@iri.univ-lille1.fr

2.1 Introduction

Surface plasmon resonance (SPR) spectroscopy has become an accepted bioanalytical technique for routine characterization of molecular recognition events at a solid interface. It has proven to be a powerful approach for the determination of kinetic parameters (association and dissociation rate constants) as well as thermodynamic parameters, including affinity constants of interacting partners such as antigen/antibody, DNA/DNA, glycoconjugate/lectin, etc., without the requirement of tagging one of the two partners. With such SPR-based affinity sensors, it is now easy and quite fast to acquire the relevant data that serve as a basis to understand biological specificity and function. The basis for the ability of SPR to determine binding affinities and kinetic parameters

---

Introduction to Plasmonics: Advances and Applications
Edited by Sabine Szunerits and Rabah Boukherroub
Copyright © 2015 Pan Stanford Publishing Pte. Ltd.
ISBN 978-981-4613-12-5 (Hardcover), 978-981-4613-13-2 (eBook)
www.panstanford.com

lies in the high sensitivity of surface plasmons excited at the metal–dielectric interface to changes in the refractive index of the adjacent medium. When a biomolecular interaction (e.g., specific binding of analyte to ligand on SPR interface) takes place, the refractive index near the surface is altered resulting in a change of the SPR signal. It has, however, to be kept in mind that refractive index changes are also initiated by non-specific surface interactions. One important aspect when constructing an SPR sensor is the suppression of non-specific adsorption of the bioreceptor on the sensing surface. In most cases, the degree of non-specific adsorption determines the sensitivity and specificity of a biosensor. Suitable surface functionalization schemes resulting in the binding of specific biorecognition elements onto the SPR surface is one of the first steps in SPR-based protocols. SPR-based biosensors combine two building blocks: the SPR interface where the plasmonic wave is generated and an appropriate surface functionalization (Fig. 2.1). It is thus obvious that the overall performance of an SPR biosensor depends on both the intrinsic optical performance of the SPR sensor and the characteristics of the surface functionalization.

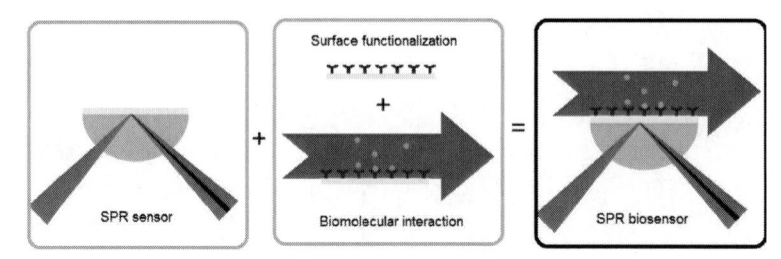

Figure 2.1 Formation of an SPR affinity sensor through suitable surface functionalization. Biological analytes, represented as green dots, interact with the biorecognition elements, represented as brown Y. The large blue arrow indicates the flow of the solution to be analyzed; practically this flow is ensured by a microfluidic system.

This chapter reviews the various approaches used to impart biofunctionality to SPR interfaces and structures and hence transforming them into biosensors. In fact, the technique of SPR becomes only interesting when the metallic film supporting the propagation of the plasmons is chemically modified. While bioreceptors can be physisorbed onto the sensor surface, covalent attachment is often preferred, as it provides strong and stable binding of the receptor to the SPR surface. This allows consequently easy regeneration of the sensor interface using conditions, which can remove the analyte from the surface, but not the attached ligand itself.

The literature is rich in examples on the use of SPR. Advances in biology continue to reveal that cell-surface oligosaccharides play an essential role in the development and maintenance of all living systems and that they confer an exquisite level of structural and functional diversity characteristic of higher organisms [1, 2]. Despite the ubiquity and importance of carbohydrates in biology, difficulties in the study of carbohydrate interactions have hindered the development of a mechanistic understanding of carbohydrate structure and function. The structural complexity of carbohydrates is one of the major obstacles. In addition, the binding affinities are typically weak in the 10^{-3} to 10^{-6} M range of dissociation constants, compared with antibody-antigen interactions (10^{-8} to 10^{-12} M). SPR technology has allowed answering several of open questions over the years.

In this chapter, we focus on surface chemistry approaches developed for the fabrication of glycan-modified SPR interfaces.

2.2 Carboxymethylated Dextran Layers: The BiAcore Chip

The two most widely used approaches for the introduction of functional groups onto gold SPR chips are based on carboxylated dextran (CM-dextran) layers (Fig. 2.2a) and on the formation of self-assembled monolayers (SAMs) on gold surfaces using a variety of different thiolated organic compounds (Fig. 2.2b).

Dextrans are hydrophilic and non-charged polymeric carbohydrates, soluble in water in any proportion forming highly hydrated hydrogels. Due to these properties, dextrans display very low non-specific interactions with bioreceptors. Owing to the high concentration of hydroxyl groups in the dextran matrix, chemical modification is possible without significantly altering its hydrophilicity. The essentially non-branched polymer chains are highly flexible and ligands immobilized in dextran matrices are thus well accessible. Introduced by Lofas and Johnsoon [3], the matrix is constructed by self-assembly of 1, ω-hydroxyalkythiol

(16-mercaptohexadecane-1-ol) onto gold, followed by covalently linking of the dextran polymer by activation of the hydroxyl groups with epichlorohydrin under basic condition to yield epoxides, which by further reaction with bromoacetic acid results in the formation of surface linked carboxymethylated dextran films (Fig. 2.2a) [4]. The thickness of the polymer matrix is about 100 nm and the carboxylic acid groups can be further used to covalently link amine-terminated or thiol-terminated bioreceptors [5].

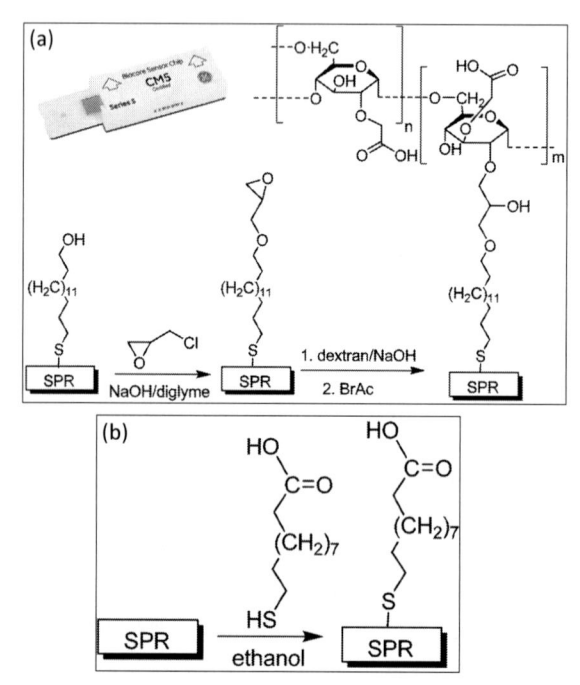

Figure 2.2 Modification of gold SPR interfaces by (a) carboxymethylated dextran polymer films and (b) SAMs of 11-mercaptoundecanoic acid.

The references that can be cited using carboxymethylated dextran-modified SPR chips are countless, as these interfaces are commercialized by Biacore under the name "CM5 chip." One recent report using both surface functionalization strategies, i.e., glycan immobilization on carboxymethylated SPR chip or through SAM formation, concerns the kinetic analysis of inhibition of glucoamylase on acarbose-modified SPR surfaces [6]. Acarbose, a valienamine-containing pseudo-tetrasaccharide clinically in use as a regulator of glucose release in type 2 diabetes patients, is a very good inhibitor of glucoamylase (GA). To study the kinetics of the interaction of wild-type glucoamylase and active site mutants with inhibitors such as acarbose, a fast and efficient method for anchoring native acarbose to SPR chips was developed [6]. The Biacore CM5 gold chip is covered with surface bound dextran polymer that contains $\alpha(1 \rightarrow 6)$ and $\alpha(1 \rightarrow 3)$-linked glycosyl

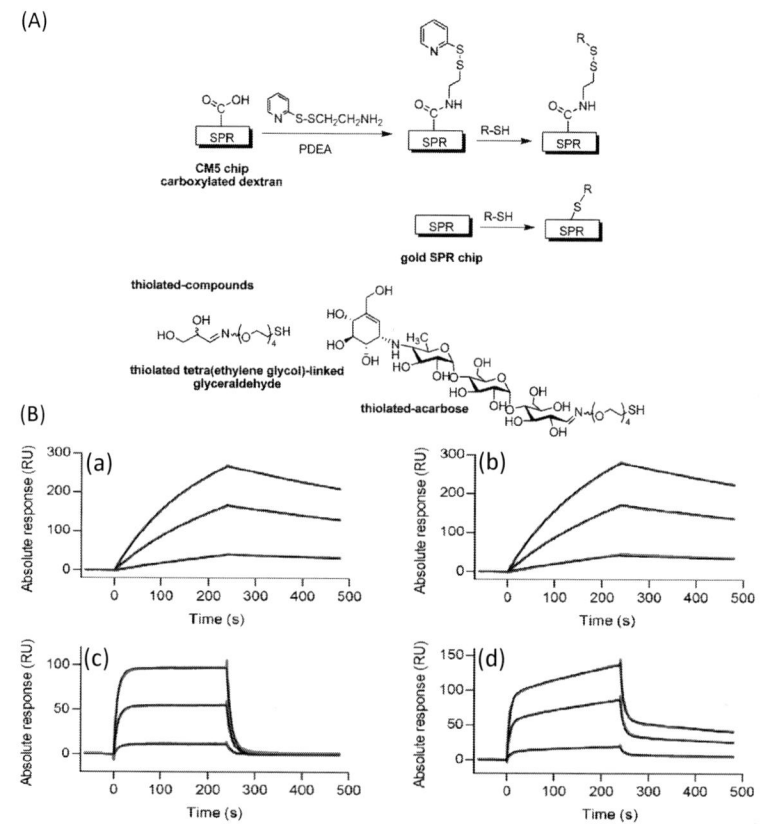

Figure 2.3 (A) Carbohydrate ligand immobilization strategies on CM5-dextran chips and gold-based SPR interfaces; (B) SPR sensorgrams obtained using in situ conjugation formation with GA wild-type (a) and mutant Y175F (b), E180Q (c) and R54L (d) at pH 7; concentrations: 58, 290, 580 nM (a, b) or 580, 2900, 5800 nM (c, d). Sensorgrams were corrected for unspecific interaction to control (glyceraldehydes). Reprinted with permission from Ref [6].

residues and was used as potential substrate for GA inhibition. The performance was compared to a gold surface modified with a thiolated glycan (Fig. 2.3a). Thiol-reactive 2-pyridyldisulfide groups (PDEA) were introduced on the CM5 surface to enable disulfide bond formation with the glycoconjugate and glyceraldehyde control. Alternatively, gold–sulfur bond formation of the thiolated conjugates directly on gold was achieved. Careful examination and comparison with pre-formed oxime glycoconjugates showed that the organization of the acarbose ligands on the commercial dextran matrix chip allows unhindered access to GA even at high ligand densities. A comparison with SAMs of acarbose ligands showed that the presence of the matrix did not affect the binding kinetics of the interaction. At pH 7, the association and dissociation rate constants for the acarbose-glucoamylase interaction are 10^4 $M^{-1}s^{-1}$ and 10^3 s^{-1}, respectively, and that the conformational change to a tight enzyme-inhibition complex affects the dissociation rate constant by a factor of 10^2 s^{-1} (Fig. 2.3b). The acarbose-modified SPR surface could also be used as a glucoamylase sensor for rapid, label-free affinity screening of small carbohydrate-based inhibitors in solution, which is otherwise difficult with immobilized enzymes or other proteins.

2.3 Self-Assembled Monolayers Based on Thiolated Functional Groups

As already mentioned above, the other approach for the introduction of functional groups onto thin gold SPR films is based on thiolated organic compounds. Gold–sulfur (Au–S) bonds are readily formed on gold, enabling in an easy and fast manner the direct attachment of receptors to the gold surface through the formation of self-assembled monolayers [7, 8]. The Au–S chemistry has thus made possible the routine analysis of aqueous binding events to immobilize molecules at nearly neutral pH and moderate temperature. Generally, SAM-modified SPR interfaces are beneficial compared to polymeric layers both when the analytes of interest are larger than molecules, such as cells and viruses, and for kinetic parameter determination, when a low amount of non-specific binding is required and a low level of immobilization is recommended [9]. Otherwise, there is no difference in binding kinetics between monolayers and dextan hydrogels [10].

Two different approaches are used to modify SPR interfaces with thiolated molecules. The first strategy is based on the separate synthesis of the bioreceptor derivative with a pendant alkanethiol group and subsequent formation of the SAM [11–13]. This is often used in the case of DNA, where a thiol (SH) group can be easily incorporated onto the 5′ end. The drawback of this strategy arises from the higher synthetic effort, especially for more complex bioreceptors. Furthermore, as the complexity of tethered bioreceptors increases, there is no guarantee that the molecules will pack to form a structurally well-defined monolayer. To control the surface density, the use of mixed monolayers by diluting with short chain thiols (e.g., mercaptohexanol) is thus often preferred [11]. The second strategy consists on a multistep chemical transformation of a SAM-modified SPR interface [14–18]. As seen, for example, in Fig. 2.4a the carboxylic acid groups of an SPR interface modified with 11-mercaptoundecanoic acid (MUA) is further reacted with amine-terminated DNA and other biomolecules by the carbodiimide-catalyzed amide bond formation reaction with the surface linked carboxyl groups using ethyl(dimethylaminopropyl) carbodiimide (EDC) and *N*-hydroxysuccinimide (NHS).

The intricacy of SAMs formation on gold has also peaked widespread interest for glycobiology. The possibility to form single-component as well as mixed SAMs allowing to control the surface linked glycan density has made SAMs of particular interest for the preparation of carbohydrate interfaces. The effect of surface glycan density on carbohydrate-protein interactions was systematically studied by Dhayal and Ratner through the immobilization of thiolated trimannoside using oligoethylene glycol (OEG) of different chain lengths as diluting compound (Fig. 2.4b) [19]. Mannose-specific Concanavaline A (Con A) was used to optimize the binding reaction with carbohydrate microarrays of different glycan densities. Figure 2.4c shows the SPR imaging array of mixed glycan/OEG SAMs with varying concentration of glycan and the corresponding SPR response after interaction with Con A. Proteins RNase A and B served as negative and positive controls for the Con A binding due to the presence of a large asparagines-like glycan containing mannose residues in RNase B, which is completely absent in RNase A. Quantitative analysis of Con A binding to different surface densities of trimannoside mixed SAMs with short OEG shows a markedly

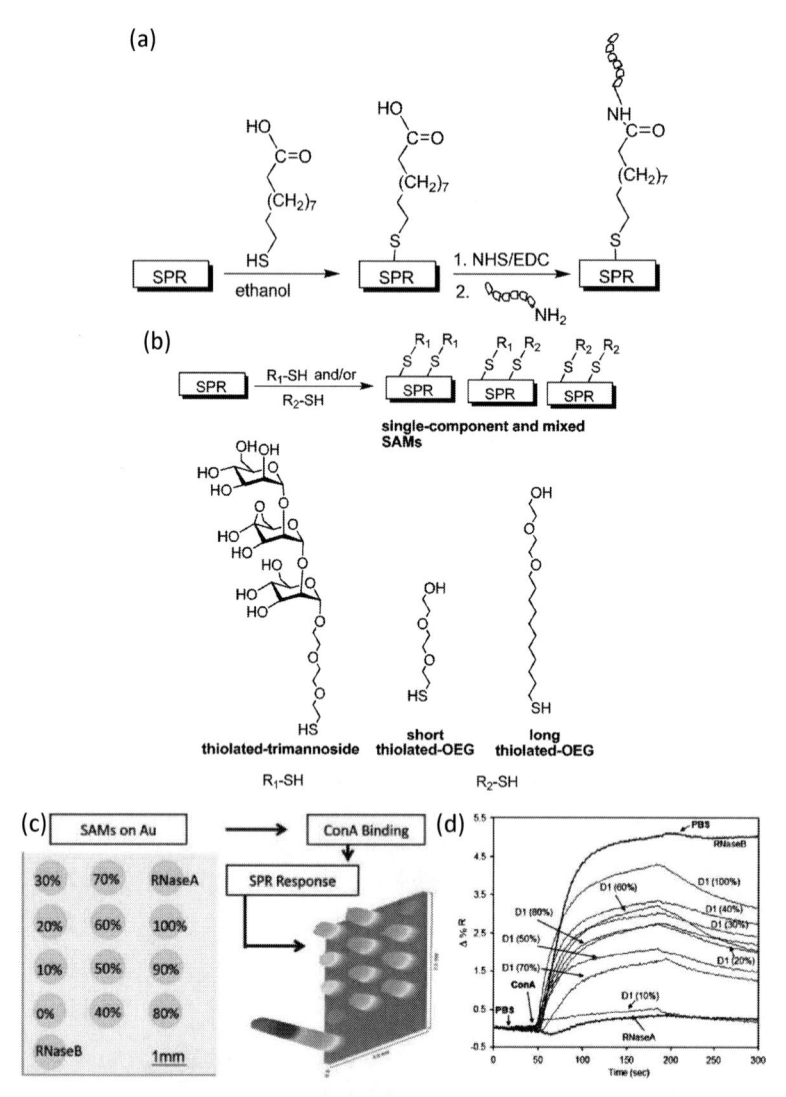

Figure 2.4 (a) Formation of 11-mercaptoundecanoic acid (MUA) SAM on gold followed by linking of amine-terminated DNA molecules. (b) Formation of trimannoside-modified SPR chip using long and short OEG as diluting compounds. (c) Image of glycan array prepared by spotting thiolated molecules onto gold and varying the percentage of trimannoside present together with SPRi response after incubation with Con A. (d) SPR sensorgram of Con A binding to a density profile of trimannoside in mixed SAMS with short OEG-thiol.

different profile from that of Con A binding to mixed sugar/OEG SAMS using the longer alkanethiol. In the case of long OEG, Con A binding is nearly nonexistent when the glycan concentration was less than 60%. In the case of short OEG, the relative binding of Con A rises dramatically with increasing concentration of trimannoside from 0–40%; beyond 40–60%, there is a plateau and slight decrease in overall Con A binding (Fig. 2.4d). At high surface densities, the immobilized glycan appears to become less readily bound by Con A. A number of explanations are possible, including steric hindrance at the surface by neighboring carbohydrates and carbohydrate-carbohydrate interactions between surface-bound glycans as well as multivalency effects.

2.4 Polymer Films

A convenient way to modify the surface of conducting materials such as gold is through electrodeposition of thin oligomeric or polymeric films from monomer solutions [20–23]. The intensive use and interest of such interfaces for anchoring bioreceptors onto biosensor interfaces is driven mainly by the following factors: (i) the polymer films are uniform, (ii) their thickness can be readily controlled and (iii) the surface modification is limited to the surface of the electrode. In addition, electrochemistry can be easily integrated with SPR measurements [21, 24, 25]. Both methods are compatible in the sense that they both rely on a conductive substrate: for electrochemistry the gold film of the SPR interface is the working electrode, without disturbing its use for the generation of surface plasmons.

Polypyrrole is probably the most intensively investigated and widely used conducting polymer for biosensing, mainly owing to its stability, conductivity and biocompatibility [20]. The entrapment of a bioreceptor within the polymer constitutes a simple one-step method during the electrochemical polymerization, but suffers greatly from the poor accessibility of the target molecules due to the polymer hydrophobicity. The use of functional polypyrrole films is thus a more promising approach [20, 22].

The potential of pyrrole chemistry in connection with oligosaccharides (lactose and 3′-silyllactose) (Fig. 2.5a) for the generation of sensors specific to two lectins, *Arachis hypogaea*

(PNA) and *Maackia amurensis* (MAA) agglutinins has been described recently [21]. The electrochemical behavior of pyrrole-lactosyl (I) is seen in Fig. 2.5b. Upon oxidative scanning up to 0.85 V vs. Ag/AgCl, the cyclic voltammogram of the pyrrole derivative displays an irreversible anodic peak at 0.77 V, which is attributed to the oxidation of the pyrrole unit leading to highly reactive radical cations. Such electrochemical polymerization can be performed on SPR interfaces by amperometrically biasing the gold interface at $E = 0.94$ V vs. Ag/AgCl in a solution of the pyrrole monomer (3 mM in 0.1 M LiClO$_4$) until a predefined charge has passed. The maximal surface coverage Γ_{theo} corresponding to 100% polymerization efficiency can be estimated from the electrical charge Q_a passed during the oxidation process according to $\Gamma_{theo} = Q_a/nFA$, where F is the Faraday constant, n the number of electrons exchanged during the oxidation ($n = 2.3$), and A the surface area. The actual surface coverage Γ_{real} can be estimated after the electrochemical characterization of the formed polymer. Figure 2.5c shows the cyclic voltammograms of the gold/polymer interface when cycled in aqueous 0.1 M LiClO$_4$ at a scan rate of 100 mV s^{-1}. An anodic peak at ~0.2 V vs. Ag/AgCl is detected and assigned to a one-electron oxidation of the polypyrrole matrix for three pyrrole monomers polymerized. From the integrated area, the polymerization efficiency on gold was determined being 10%. The surface coverage of polypyrrole–lactosyl on the gold SPR interface deposited by passing $Q_a = 11$ mC cm^{-2} is $\Gamma_{real} = 5 \times 10^{-9}$ mol cm^{-2}. The change of the SPR signal of a polypyrrole-lactosyl-modified SPR interface is seen in Fig. 2.5d. The thickness of the polypyrrole–lactosyl film was estimated to be 40 nm for $Q_a = 11$ mC cm^{-2}. The specificity of gold/polypyrrole–lactosyl to PNA was investigated using SPR (Fig. 2.5e). Addition of the lectin MAA to the gold/polypyrrole–lactosyl interface showed some nonspecific adsorption, which remains even after rinsing with phosphate buffer. Indeed, MAA is sensitive to all polypyrrole-lactosyl surfaces even though more specific to sialyllactosyl films. Addition of specific PNA results in a larger change in SPR intensity. A reverse selectivity was observed on the gold/polypyrrole-3'-sialyllactosyl film, where a strong affinity of MAA was seen.

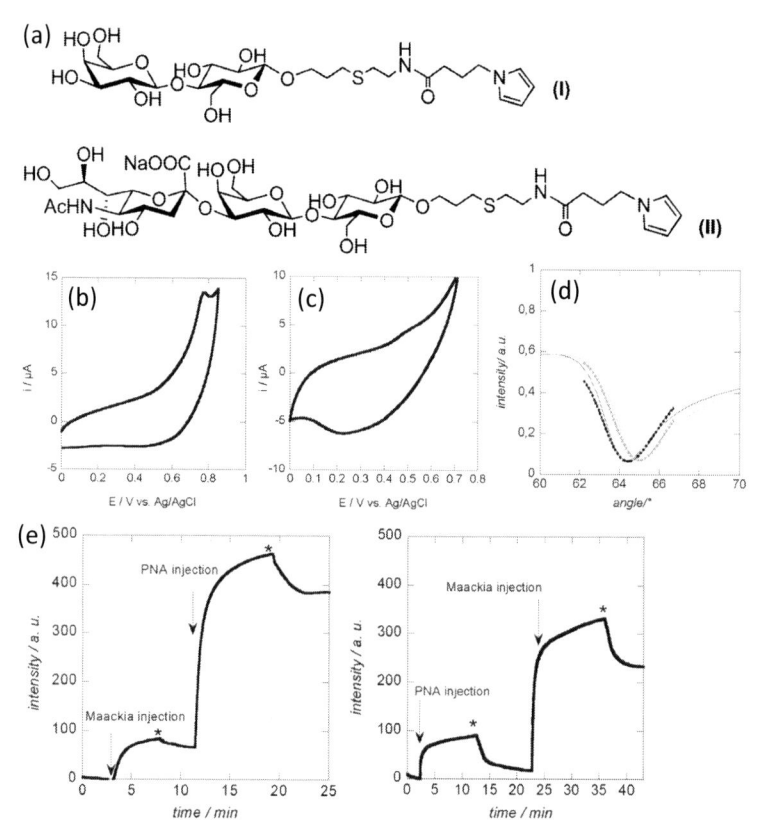

Figure 2.5 (a) Structure of pyrrole-lactosyl (**I**) and pyrrole-3′-sialyllactosyl (**II**); (b) Cyclic voltammogram of pyrrole-lactosyl (3 mM) in LiClO$_4$ (0.1 M/water); v = 100 mV s^{-1}, A = 0.21 cm^2; (c) cyclic voltammogram of polypyrrole–lactosyl film formed at E_{app} = 0.95 V/Ag/AgCl using pyrrole-lactosyl (3 mM) in LiClO$_4$ (0.1 M/water), passed charge: 0.1 mC; (d) Angular scan of the reflected light intensity of the thin gold film (black line) and after the deposition of pyrrole-lactosyl (grey line); Dashed lines are experimental values, full lines correspond to fitted curves (n_{prism} = 1.58, n_{Ti} = 2.40 + i3.13, d = 5 nm, n_{gold} = 0.248 + i3.77, d = 50 nm, $n_{solution}$ = 1.33, $n_{ppy\text{-lactose}}$ = 1.35 + i0.00); polymerization conditions: E_{app} = 0.95 V/Ag/AgCl, Q = 11 mC cm^{-2}, pyrrole-lactosyl (3 mM in 0.1M LiClO$_4$); (e) Specific interactions on different interfaces: (left) polypyrrole–lactosyl (5 nmol cm^{-2}), (right) polypyrrole-3′-sialyllactosyl 2 nmol cm^{-2}: MAA lectin (8.3 µmol L^{-1}), PNA lectin (8.3 µmol L^{-1}).

From the SPR sensorgrams, the association (k_{ass}) and dissociation (k_{diss}) rate constants can be calculated using a linear transformation of the SPR sensorgram data according to

$$k_s = k_{ass} \times c + k_{diss}, \qquad (2.1)$$

where k_s is the slope of a dR/dt versus R plot (R being the change in the SPR response amplitude) [26, 27]. Figure 2.6a displays a plot of the extracted k_s values for the different lectin concentrations from which the association rate constants for PNA and MAA are determined to be k_{ass}^{PNA} = 2.45 × 10^3 M^{-1} s^{-1} and k_{ass}^{MAA} = 2.18 × 10^3 M^{-1} s^{-1}. The value of the dissociation rate constants k_{diss} correspond to the intercept, providing k_{diss}^{PNA} = 1.09 × 10^{-3} s^{-1} and k_{diss}^{MAA} = 1.03 × 10^{-3} s^{-1}. From the association and dissociation rate constants, the equilibrium dissociation constant K_d ($K_d = k_{diss}/k_{ass}$)

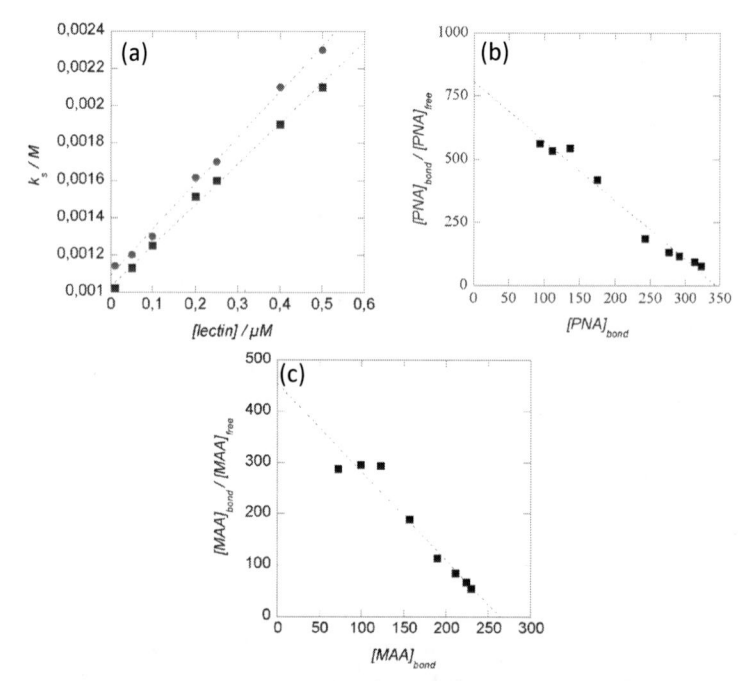

Figure 2.6 (a) Plots of k_s vs. PNA (circles) and MAA (squares) lectin concentration, the values of k_s were obtained from analysis of SPR sensorgrams on polypyrrole–lactosyl film. (b) Scatchard plots on polypyrrole–lactosyl film with increasing PNA lectin concentration. (c) Scatchard plots on polypyrrole-3′-sialyllactosyl film with increasing MAA lectin concentration [21].

can then be determined being K_d^{PNA} = 4.45 × 10^{-7} M and K_d^{MAA} = 4.72 × 10^{-7} M. These values are in good agreement with the equilibrium dissociation constant calculated by Scatchard analysis for the different film thicknesses (Fig. 2.6b) where K_d values of K_d^{PNA} = 5.8 × 10^{-7} M are found for PNA and lactosyl. The interaction between MAA and the 3′-sialylactose film gave a K_d value of the same magnitude (K_d^{MAA} = 5.4 × 10^{-7} M) [21].

Plasma polymerization techniques have shown to be efficient for the deposition of highly reactive and functional thin films with potential applications in the biomedical field. One of the advantages of plasma polymerization techniques is that the density of a particular functional group and the degree of cross-linking within the polymer network can be largely tailored by careful control over the energy input during the deposition. The types of films that have been found to be particularly suitable for biosensing applications are those containing reactive groups such as carbonyl, amine, carboxylic acid and anhydride. For example, amine-terminated or amine bearing bioreceptors such as bovine serum albumin (BSA) can be irreversibly linked to plasma-polymerized films carrying maleic anhydride functions [28]. Until now, these approaches have not been validated for the study of carbohydrate binding events.

2.5 Lamellar SPR Structures

While the surface chemistry developed on gold has been of great value [7, 8], the limitations of working on gold are becoming more noticeable with increasingly complex fabrication requirements for biometric systems and arrays. Alternative routes and improvements were thus sought after. Silicon dioxide-based materials such as glass (silicate) are standard materials for biosensing being inexpensive and benefiting from a rich variety of well-developed attachment schemes based on silane-coupling chemistry. For many years, the use of such lamellar structured SPR chips for biosensing applications was hampered as the gold/SiO$_2$ interface was in most cases not stable upon immersion in water and/or PBS buffer. The thin silica layer peels off the surface within a few minutes [29]. These limitations have been overcome using different approaches (Table 2.1).

Table 2.1   Formation of glass coated SPR interfaces

	Method	Ref.
1	Sol-gel technique based on 3-(mercaptopropyl) trimethoxysilane followed by hydrolysis of the trimethoxysilyl groups to generate surface silanol groups necessary for the condensation reaction of spin-coated tetramethoxysilane. The silica films were 3–100 nm in thickness and further functionalization with amine- and biotin–modified silanes introduced chemical functionalities onto the SPR interface.	[29]
2	Layer-by-layer deposition of poly(allylamine) hydrochloride) and sodium silicate, followed by calcination at high temperature to from stable films of 2–15 nm in thickness.	[30]
3	Spin coating poly(hydroxymethylsiloxane) onto the SPR chip, and post-treatment via plasma oxidation.	[31, 32]
	Plasma-enhanced chemical vapor deposition (PECVD) based on the decomposition of a mixture of silane gas (SiH_4) and nitrous oxide (N_2O) near the substrate surface, enhanced by the use of a vapor containing electrically charged particles or plasma at 300°C.	[33–39]

The classical strategy for adding functionality to such interfaces is based on the use of the silanization reaction as depicted in Fig. 2.7A. Amine-terminated oligonucleotides (ODNs) were in this way grafted on such an interface using a standard procedure developed for glass, including the following sequences: (i) reaction of the silicon oxide layer with 3-aminopropyltri-methoxysilane to produce amine terminal groups on the surface, (ii) transformation of the amine to aldehyde termination by chemical coupling with a bifunctional linker such as glutaraldehyde and immobilization of ODNs bearing a terminal amine group [39].

The formation of glycan-modified glass-based SPR was achieved through the reaction of surface linked amine groups with 4-azidobenzoic acid. Mannose units were integrated onto this surface via the Cu(I) catalyzed 1,3 dipolar cycloaddition between the surface azide groups and the alkyne-terminated glycan reacting partner through the formation of a triazole linkage (Fig. 2.7B). This reaction termed "click" chemistry has been established as an attractive method for the immobilization of carbohydrates onto solid surfaces [40, 41]. Nevertheless, the requirement for prior derivatization of glycans with either thiol or propargyl/azide functions by one- or multi-step sequences presents a significant hurdle for the construction of functional microarrays especially when complex glycans are to be immobilized. The field would undoubtedly benefit greatly from alternate direct conjugation strategies where the need for functionalized glycans, prior to immobilization, would be obviated.

A recently developed strategy involves photo-induced (irradiation at 365 nm) attachment of unmodified glycans directly to SPR interfaces modified with perfluorophenyl azide derivatives (Fig. 2.7C) [42]. It might be argued that such a coupling strategy might result in "uneffective" glycan interfaces. The high reactivity of the perfluorophenylnitrene intermediate, generated through UV irradiation of the azide, is known to give a wide range of potential reaction products, although it is the C–H insertion pathway that is reported to be preferred [43]. Nevertheless, this pathway can be, in principle, used to conjugate sugars to surfaces via centers other than their anomeric carbons and alpha and beta selectivity is not guaranteed. To investigate whether photochemically immobilized sugar ligands do indeed maintain their expected binding affinity and selectivity towards the appropriate lectin partner, the SPR signal obtained on mannose-modified SPR interface formed via the photocoupling strategy (with perfluorophenylazide-functionalized surfaces) was compared to that recorded for the corresponding mannose-decorated interface obtained through the "click" approach. It was found that both modification methods gave identical amounts of mannose on the respective interfaces to allow the direct comparisons of the SPR signals. As can be seen in Fig. 2.7D, an immediate change in the SPR signal is observed upon injection of the specific lectin, as a consequence of the association of the mannopyranoside specific *Lens culinaris* lectin with both mannose-decorated surfaces. The difference of the change in the SPR signal is somewhat larger in the case of the mannose interface formed through the photocoupling approach compared with that fabricated via the "click strategy (Fig. 2.7D). The random photoimmobilization seems to be thus less random than expected, having a strong tendency to take preferential place at the reducing ends [42].

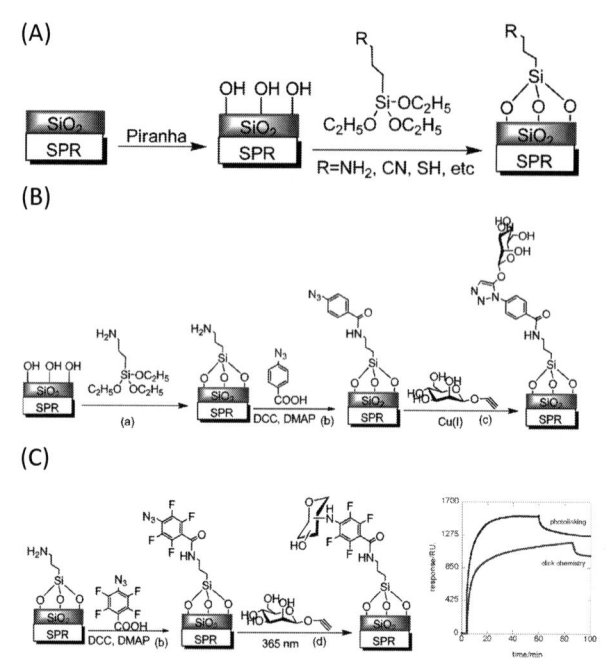

Figure 2.7 (A) Surface modification of glass-based SPR interfaces. (B) Schematic representation of the SPR surface functionalization strategies employed for the covalent linking of mannose (B) "click chemistry" and (C) photochemical linking. (a) 3-aminopropyltrimethoxylsilane, (b) 3-azidobenzoic acid, N,N'-dicyclohexylcarbodiimide (DCC), 4-dimethylaminopyridine (DMAP), (c) "click" chemistry with alkynyl-terminated mannose, (d) photocoupling with perfluoronated azido benzoic acid. (C) SPR binding curves of Lens culinaris (10 µg/mL) to mannose-modified SPR interfaces formed through UV irradiation of perfluorophenylazide-modified interface (black) and linked to phenylazide via "click" (blue) [42].

While stable and reusable devices are formed using silanization, one of the drawbacks of silane chemistry is that it is not entirely well controlled. In addition, the Si–O–Si bonds are prone to hydrolysis when exposed to harsh environments, such as highly saline conditions. Recent studies have demonstrated that carbon-based surfaces could overcome this hurdle [44]. While amorphous carbon coated SPR interfaces can be formed by sputtering, a significant loss in sensitivity was observed due to the unfavorable optical properties of thin amorphous carbon films [44]. Nevertheless, the in situ synthesis of oligonucleotide arrays utilizing photochemically protected oligonucleotide building blocks was possible. We and others have found that the sensitivity can remain the same as on gold by choosing amorphous silicon-carbon alloys (abbreviated as a-$Si_{1-x}C_x$:H) with the right chemical composition [45]. Amorphous silicon-carbon alloys can be deposited as thin films, and changing the carbon content of the film allows for fine-tuning the material properties. Increasing the carbon content leads to the optical band gap enlargement and to a transparent material, and at the same time decreasing the refractive index, which is beneficial for the fabrication of lamellar SPR interfaces. Stable hydrogenated coatings of about 5 nm in thickness of a-$Si_{0.63}C_{0.37}$:H film can be formed on gold using PECVD in the "low-power" regime. Surface hydrogenated a-$Si_{0.63}C_{0.37}$:H can be conveniently functionalized by stable organic layers through robust Si–C covalent bonds in a similar way as crystalline silicon (Fig. 2.8) [45]. The acid function introduced through hydrosilylation of undecylenic acid can be further converted to an activated ester group using EDC/NHS chemistry, to which amine-terminated bioreceptors can be easily anchored.

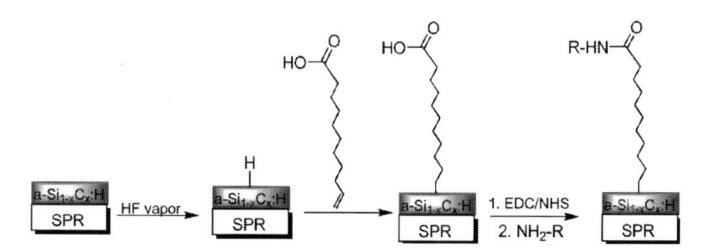

Figure 2.8 Surface hydrogenation of gold-based SPR interfaces coated with 5 nm a-$Si_{1-x}C_x$:H thin film and subsequent functionalization with undecylenic acid.

Lamellar SPR structures provide also an interesting means of working with silver rather gold-based interfaces. While gold is most commonly used as it possesses highly stable optical and chemical properties, silver provides the sharpest SPR signal and is reported to have an enhanced sensitivity to thickness and refractive index variation in comparison to gold. In addition, the penetration length of a 50 nm-thick gold film is about 164 nm with a light source of 630 nm, whereas a 50 nm-thick silver film has an enlarged penetration length of 219 nm. Although the favorable optical properties of silver should favor its use over gold interfaces for

SPR sensing, the main drawback of the silver-based SPR interfaces is its chemical instability. Silver rapidly oxidizes when exposed to air, and the process is accelerated in aqueous solutions, making it difficult to get reliable optical signals and to perform long time measurements needed to detect biological interactions. As a consequence, silver-based SPR interfaces can only be employed when coated with a thin and dense protecting layer, stabilizing the interface while keeping the favorable optical properties of silver. Only a few attempts have been reported until now to protect silver SPR surfaces and to take advantage of the optical properties of silver [46, 47] (Table 2.2).

Table 2.2 Formation of silver-based SPR interfaces

	Method	Ref.
1	Bimetallic silver/gold layers (total thickness of 50 nm with the gold thickness comprised between 12 and 25 nm)	[46]
2	Coating with ITO by magnetron sputtering	[48, 49]
3	Coating with antimony-doped tin oxide hybrid by radio frequency sputtering	[50, 51]
4	Coating with α-$Si_{0.63}C_{0.37}$ by PECVD sputtering	[45, 52]
5	Graphene	[53–56]

More recently, thin graphene layers are considered as an alternative coating for silver [57]. Graphene-functionalized SPR surfaces are believed to have several advantages: (i) Graphene has a very high surface-to-volume ratio, which is expected to be beneficial for efficient adsorption of biomolecules as compared to gold, (ii) the presence of graphene on the top of silver passivates the surface against oxidation, making possible the use of silver as plasmon active interface, (iii) controlling the number of graphene layers transferred onto the metal interface allows controlling the SPR response and the sensitivity of SPR measurements, (iv) graphene should increase the adsorption of organic and biological molecules as their carbon-based ring structure allows π–π stacking interaction with the hexagonal cells of graphene. However, π-stacking interaction is not specific and any organic molecule or biomolecule of appropriate aromatic structure will integrate onto the graphene matrix. This is currently a major limitation of these SPR interfaces.

While for a long time, the formation of reproducible graphene-based SPR interfaces was a considerable hurdle to overcome, chemical and/or mechanical transfer of CVD grown graphene sheets onto gold and silver (Fig. 2.9a) [53, 56] as well as electrophoretic deposition of reduced graphene oxide (Fig. 2.9b) [54, 58] have been successfully applied. We have shown that

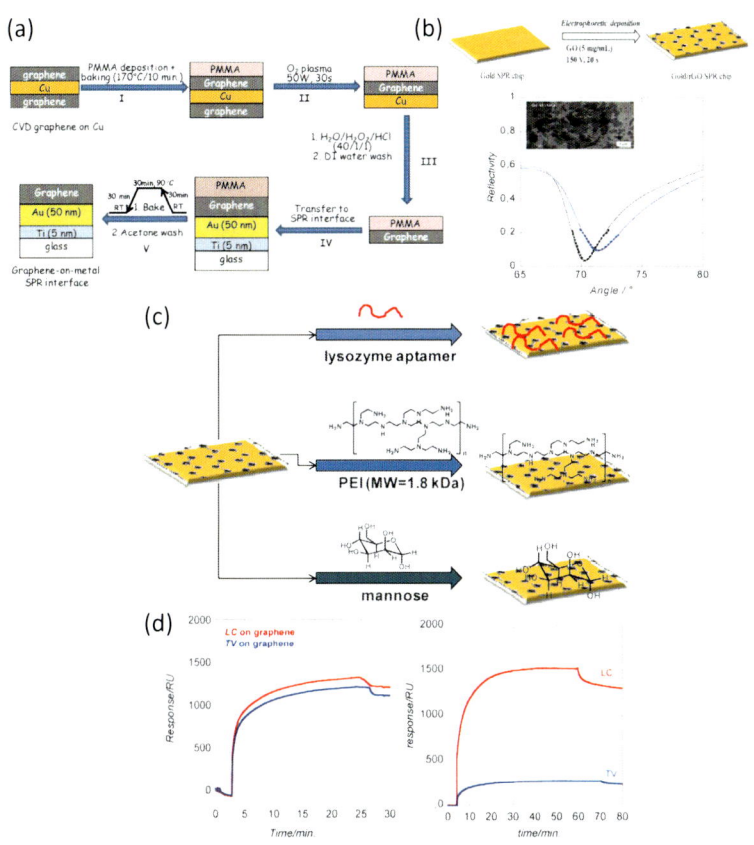

Figure 2.9 (a) Schematic representation of the transfer process of CVD grown graphene on Cu onto gold SPR substrates. (b) Electrophoretic deposition of reduced graphene oxide from a solution of graphene oxide (GO), SPR signal of gold interface before and after electrophoretic deposition; inset: SEM image. (c) Graphene functionalization with aptamers, glycans, and polymers. (d) SPR binding curves of *Lens culinaris* (500 µg/mL) and *Triticum Vulgaris* (500 µg/mL) on unmodified graphene-on-metal (left) and on mannose-modified SPR interfaces.

specificity can be simply added to graphene-based SPR interfaces through immersion of the interface into aptamers [54], polymers such polyethyleneimide (PEI), and sugars [56, 58] (Fig. 2.9c). In the case of aptamer-modified graphene, SPR sensing of lyzosyme was possible, while in the case of sugar and polymer-modified interfaces the interaction of such interfaces with pathogens and lectins was demonstrated (Fig. 2.9c). While unmodified graphene could not differentiate between *Lens culinaris* (*LC*; 46 kDa) and *Triticum vulgaris* (TV, 36 kDa), on mannose-modified graphene (prepared by simple immersion of the SPR interface for 4 h into a 100 mM solution of mannose) an immediate change of the SPR signal was observed upon injection of *Lens culinaris* as a consequence of its association with the mannose. In the case of mannose unspecific *Triticum vulgaris*, the increase of the SPR signal was much smaller. The association constant K_A of the mannose-modified interfaces with *LC* lectin was estimated to be $(3.6 \pm 0.6) \times 10^6$ M^{-1}. This indicates that the presence of non-covalently bound mannose allows tuning the surface selectivity in an easy manner.

Such interfaces were also recently used to study the adhesion strength of different *Escherichia coli* strains. A variety of physical and chemical parameters are of importance for adhesion of bacteria to surfaces. In the colonization of mammalian organisms for example, bacterial fimbriae and their adhesins not only seek particular glycan sequences exposed on diverse epithelial linings, but they also enable the bacteria to overcome electrostatic repulsion exerted by their selected surfaces. We investigated the adhesion behavior of different *E. coli* strains to electrophoretically formed rGO SPR interfaces modified with differently charged ligands such as positively charged polyethyleneimine (PEI) and negatively charged poly(sodium 4-styrenesulfonate) (PSS) as well as with glycans such as mannose or lactose. The summary of adhesion results suggested that local charge variations close to the binding pockets of the bacteria seem to determine bacteria affinity and that chemically based affinity effects have to be considered in addition (Fig. 2.10). Glycan interfaces showed indeed specific binding events with sugar-specific *E. coli* strains such as UTI89 and att25. A detection limit of $\approx 10^2$ cfu/mL was achieved for *E. coli* 107/86 and UTI89 on PSS and PEI-modified Au/rGO SPR interfaces, respectively. The advantage of such an approach is thus multiple: in contrast to bacteria sensing with bacteriophage where the size of the phage has an important implication on the proportion of the target bacteria which can be probed by the evanescent field of the SPR [59], the modified rGO SPR interfaces allow probing the entire bacteria without any constrain. The integration of affinity targets for different *E. coli* strains can be easily achieved through non-covalent interactions between the rGO matrix and various organic ligands. The strategy is thus adaptable to any other bacteria and more specifically if one is interested in examining bacterial affinities to molecules on a surface and quantitative detection of pathogens in solution. It should be possible to design novel experiments based on simplified model systems to answer many fundamental questions using this technique, combining a plasmonic read out and graphene-based surface functionalization schemes.

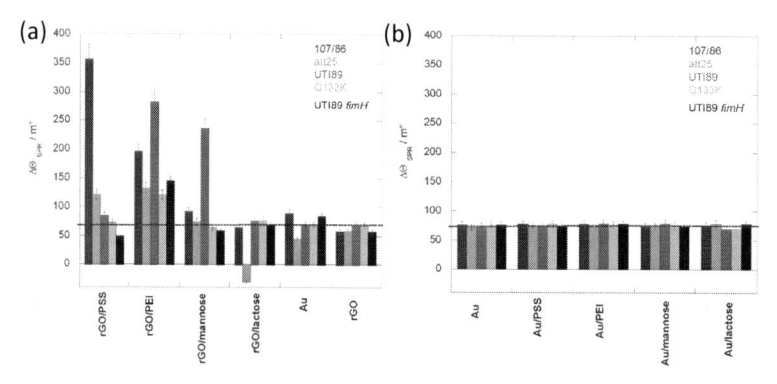

Figure 2.10 Bar diagramm summarizing all the recorded changes of the SPR response upon addition of 1.5×10^9 cfu/mL *E. coli* strains onto differently modified (a) rGO and (b) gold interfaces (Five replicate measurements were performed for each experiment): 107/86 (blue), att25 (green), UTI89 (red), UTI89 Q133K (grey), UTI89 *fimH*⁻ (black).

2.5 Conclusion

Through the use of unique surface functionalization schemes and advanced substrate design, the scope of SPR for the study of affinity binding events has changed considerably over time. As the field matured, SPR has become also an ideal technique for the characterization of low affinity interactions as is the case between glycans and their associated binding partners. Probing carbohydrate-lectin, carbohydrate-bacteria, and other carbohydrate related interactions with sufficient specificity was made possible through the development of surface chemistry strategies displaying the glycan moieties in a fashion that are readily accessible by the analyte.

The ultimate goal for SPR biosensing technology is to effectively compete with other popular labeled techniques in particular fluorescence methods. Unlike fluorescent-based assays, SPR as a label-free method eliminates errors caused by the unlikely but possible binding of the fluorescent label itself to proteins being screened. The increasing support and interest from industry to investigate different aspects of SPR makes the future of SPR biosensing very promising.

Acknowledgments

Financial support from the Centre National de la Recherche Scientifique (CNRS), the University Lille 1, the Nord Pas de Calais region, the Institut Universitaire de France (IUF) and the European Commission through ITN-project MATCON (contract no. 238201) is acknowledged.

References

1. Munoz, F. J., Rumbero, A., Sinisterra, J. V., Santos, J. I., André, S., Gabius, H.-J., Jiménez-Barbero, J., and Hérnaiz, M. J. (2008). *Glycoconjugate J.*, **25**, 633.
2. Ambrosi, M., Cameron, N. R., and Davis, B. G. (2005). *Org. Biomol. Chem.*, **3**, 1593.
3. Lofas, S., and Johnsoon, B. J. (1990). *Chem. Soc. Chem. Commun.*, **21**, 1526.
4. Lofas, S., Johnsson, B., Tegendal, K., and Ronnberg, I. (1993). *Colloids Surf. B*, **1**, 83–89.
5. Jung, S.-H., Jung, J.-W., Suh, I.-B., Yuk, J.-S., Kim, W.-J., Choi, E.-Y., Kim, Y.-M., and Ha, K.-S. (2007). *Anal. Chem.*, **79**, 5703–5710.
6. Sauer, J., Hachem, M. A., Svensson, B., Jensen, K. J., and Thygesen, M. B. (2013). *Carbohydrate Res.*, **375**, 21.
7. Love, J. C., Estroff, L. A., Kriebel, J. K., Nuzzo, R. G., and Whitesides, G. M. (2005). *Chem. Rev.*, **105**, 1103.

8. Ulman, A. (1996). *Chem. Rev.*, **96**, 1533.
9. Kyprianou, D., Guerreiro, A. R., Nirschl, M., Chianella, I., Subrahmanyam, S., Turner, A. P. F., and Piletsky, S. (2010). *Biosens. Bioelectron.*, **25**, 1049–1055.
10. Lahiri, J., Isaacs, L., Tien, J., and Whitesides, G. M. (1999). *Anal. Chem.*, **71**, 777–790.
11. Peeters, S., Stakenborg, T., Reekmans, G., Laureyn, W., Lagae, L., Van Aerschot, A., and Ranst, M. V. (2008). *Biosens. Bioelectron.*, **24**, 72.
12. Malic, L., Veres, T., and Tabrizian, M. (2009). *Biosens. Bioelectron.*, **24**, 2218.
13. Ananthanawat, C., Vilaivan, T., Mekboonsonglarp, W., and Hoven, V. P. (2009). *Biosens. Bioelectron.*, **24**, 3544.
14. Wang, J., Munir, A., Li, Z., and Shou, H. S. (2010). *Biosens. Bioelectron.*, **25**, 124.
15. Frasconi, M., Tel-Vered, R., Riskin, M., and Willner, I. (2010). *Anal. Chem.*, **82**, 2512–2519.
16. Liu, Y., Dong, Y., Jauw, J., Linman, M. J., and Cheng, Q. (2010). *Anal. Chem.*, **82**, 3679–3685.
17. Chen, Y., Nguyen, A., Niu, L., and Corn, R. M. (2009). *Langmuir*, **25**, 5054–5060.
18. Hoa, X. D., Kirk, A. G., and Tabrizian, M. (2009). *Biosens. Bioelectron.*, **24**, 3043.
19. Dhayl, M., and Ratner, D. M. (2009). *Langmuir*, **25**, 2181.
20. Le, H. Q. A., Sauriat-Dorizon, H., and Korri-Youssoufi, H. (2010). *Anal. Chim. Acta*, **674**, 1.
21. Gondran, C., Dubois, M.-P., Fort, S., Cosnier, S., and Szunerits, S. (2008). *Analyst*, **133**, 206.
22. Dong, H., Cao, X., Li, C. M., and Hu, W. (2008). *Biosens. Bioelectron.*, **23**, 1055.
23. Su, X., Teh, H., Aung, K. M. M., Zong, Y., and Gao, Z. (2008). *Biosens. Bioelectron.*, **23**, 1715.
24. Zhang, N., Schweiss, R., Zong, Y., and Knoll, W. (2007). *Electrochim. Acta* **52**, 2869.
25. Szunerits, S., Knorr, N., Calemczuk, R., and Livache, T. (2004). *Langmuir*, **20**, 9236–9241.
26. Shinohara, Y., Kim, F., Shimizu, M., Goto, M., Tosu, M., and Hasegawa, Y. (1994). *Eur. J. Biochem.*, **223**, 189–194.

27. MacKenzie, C. R., Hirama, T., Deng, S.-J., Bundle, D. R., Narang, S. A., and Young, N. M. (1996). *J. Biol. Chem.*, **19**, 1527.
28. Liu, S., Vareiro, M. M. L. M., Fraser, S., and Jenkins, A. T. A. (2005). *Langmuir*, **21**, 8572.
29. Kambhampati, D. K. M., Robertson, J. W., Cai, M., Pemberton, J. E., and Knoll, W. (2001). *Langmuir*, **17**, 1169–1175.
30. Phillips, K. S., Han, J., Martinez, M., Wang, Z., Carter, D., and Cheng, Q. (2006). *Anal. Chem.*, **78**, 596.
31. Satriano, C., Edvardsson, M., Ohlsson, G., Wang, G., Svedhem, S., and Kasemo, B. (2010). *Langmuir*, **26**, 5715–5725.
32. Satriano, C., Marletta, G., and Kasemo, B. (2008). *Surf. Interface Anal.*, **40**, 649.
33. Szunerits, S., and Boukherroub, R. (2006). *Langmuir*, **22**, 1660.
34. Szunerits, S., and Boukherroub, R. (2006). *Electrochem. Commun.*, **8**, 439–444.
35. Szunerits, S., Coffinier, Y., Janel, S., and Boukherroub, R. (2006). *Langmuir*, **22**, 10716–10722.
36. Szunerits, S., Nunes-Kirchner, C., Wittstock, G., Boukherroub, R., and Chantal, G. (2008). *Electrochim. Acta*, **53**, 7805–7914.
37. Zawica, I., Wittstock, G., Boukherroub, R., Szunerits, S. (2008). *Langmuir*, **24**, 3922–3929.
38. Zawica, I., Wittstock, G., Boukherroub, R., Szunerits, S. (2007). *Langmuir*, **23**, 9303–9309.
39. Manesse, M., Stambouli, V., Boukherroub, R., and Szunerits, S. (2008). *Analyst*, **133**, 1097–1103.
40. Szunerits, S., Niedziółka-Jönsson, J., Boukherroub, R., Woisel, P., Baumann, J.-S., Siriwardena, A. (2010). *Anal. Chem.*, **82**, 8203–8210.
41. Sun, W.-L., Stabler, C. L., Cazalis, C. S., and Chaikof, E. L. (2006). *Bioconjugate Chem.*, **17**, 52.
42. Maalouli, N., Barras, A., Siriwardena, A., Bouazaoui, M., Boukherroub, R., and Szunerits, S. (2013). *Analyst*, **138**, 805.
43. Liu, L.-H., and Yan, M. (2010). *Acc. Chem. Res.*, **43**, 1434.
44. Lockett, M. R., Weibel, S. C., Philips, M. F., Shortreed, M. R., Sun, B., Corn, R. M., Hamers, R. J., Cerrina, F., and Smith, L. M. (2008). *J. Am. Chem. Soc.*, **130**, 8611.
45. Touahir, L., Niedziółka-Jönsson, J., Galopin, E., Boukherroub, R., Gouget-Laemmel, A. C., Solomon, I., Petukhov, M., Chazalviel, J.-N., Ozanam, F., and Szunerits, S. (2010). *Langmuir*, **26**, 6058.

46. Zynio, S. A., Samoylov, A. V., Surovtseva, E. R., Mirsky, V. M., and Shirshov, Y. M. (2002). *Sensors*, **2**, 62.
47. Wang, L., Sun, Y., Wang, J., Zhu, X., Jia, F., Cao, Y., Wang, X., Zhang, H., and Song, D. (2009). *Talanta*, **78**, 265.
48. Castel, X., Boukherroub, R., and Szunerits, S. (2008). *J. Phys. Chem. B*, **112**, 15813.
49. Castel, X., Boukherroub, R., and Szunerits, S. (2008). *J. Phys. Chem. C*, **112**, 10883.
50. Manesse, M., Sanjines, R., Stambouli, V., Boukherroub, R., and Szunerits, S. (2008). *Electrochem. Commun.*, **10**, 1041.
51. Manesse, M., Sanjines, R., Stambouli, V., Boukherroub, R., and Szunerits, S. (2009). *Langmuir*, **25**, 8036.
52. Touahir, L., Jenkins, A. T. A., Boukherroub, R., Gouget-Laemmel, A. C., Chazalviel, J.-N., Peretti, J., Ozanam, F., and Szunerits. S. (2010) *J. Phys. Chem. C*, **114**, 22582.
53. Szunerits, S., Maalouli, N., Wijaya, E., Vilcot, J.-P., and Boukherroub, R. (2013). *Anal. Bioanal. Chem.*, **405**, 1435–1443.
54. Subramanian, P., Lesniewski, A., Kaminska, I., Vlandas, A., Vasilescu, A., Niedziolka-Jonsson, J., Pichonat, E., Happy, H., Boukherroub, R., and Szunerits, S. (2013). *Biosens. Bioelectron.*, **50**, 239–243.
55. Subramanian, P., Barka-Bouaifel, F., Bouckaert, J., Yamakawa, N., Boukherroub, R., and Szunerits, S. (2014). *ACS Appl. Mater. Interfaces*, **6**, 5422–5431.
56. Penezic, A., Deokar, G., Vinaud, D., Pichonat, E., Subramanian, P., Gašparovi, B., Boukherroub, R., and Szunerits, S. (2014). *Plasmonics*, **9**, 677–683.
57. Choi, S. H., Kim, Y. L., and Byun, K. M. (2010). *Opt. Expr.*, **19**, 458.
59. Tawill, N., Sacher, E., Mandeville, R., and Meunier, M. (2012). *Biosens. Bioelectron.*, **37**, 24.

Chapter 3

Biophysics of DNA: DNA Melting Curve Analysis with Surface Plasmon Resonance Imaging

Arnaud Buhot, Julia Pingel, Jean-Bernard Fiche, Roberto Calemczuk, and Thierry Livache

SPrAM (UMR 5819, CNRS, UJF, CEA), INAC, CEA Grenoble
17 rue des Martyrs, 38054 Grenoble cedex 9, France
arnaud.buhot@cea.fr

3.1 Introduction

In the past 30 years, surface plasmon resonance (SPR) has emerged as a powerful technology for biomolecular interaction studies. This is principally due to its main advantages: the access to real-time kinetics of the interactions, which gives valuable information on the reaction rates and the sensitivity to a small change of optical properties near the surface, which allows the detection of unlabelled targets. Those advantages are crucial compared to other technologies limited to final point detection or implying target labeling like fluorescence. More recently, a third advantage was brought by SPR imaging (SPRi) performances, which permit to follow multiple interactions in parallel. Nonetheless, one

---

Introduction to Plasmonics: Advances and Applications
Edited by Sabine Szunerits and Rabah Boukherroub
Copyright © 2015 Pan Stanford Publishing Pte. Ltd.
ISBN 978-981-4613-12-5 (Hardcover), 978-981-4613-13-2 (eBook)
www.panstanford.com

of the advantages of the technique, the sensitivity to optical changes, is also one of its main drawbacks for low detection limit performances. Indeed, SPR is affected not only by the targets binding to the probes but also by all the other molecules adsorbing at the surface, leading to a non-specific signal. Furthermore, this high sensibility implies strong dependence to external parameters such as temperature, pH, and salt concentration.

On the application side, the emergence of biosensors and DNA microarrays is essential for personalized medicine and point-of-care testing. For example, single nucleotide point (SNP) mutations either genetic or somatic present known implications on predisposition, diagnosis and prognosis of numerous cancers for which chemo- or radiotherapies should be adapted to each patient. Thus, their detection by DNA microarrays would be an important step forward for personalized medicine. For the moment, such detection is limited by isothermal hybridization protocols requiring strong probe design optimization. In solution, the melting curves analysis has proved particularly efficient for SNP detection following polymer chain reaction (PCR) amplification, [1] but is limited in the number of SNP detected due to the need for labeling. Combining DNA microarray format with accurate melting curves analysis would open the door for the simultaneous detection of numerous SNPs. In the last ten years, we have developed such a technology using surface plasmon resonance imaging (SPRi) [2, 3] with possible implications for other applications [4–6]. Recently, alternative technologies have also emerged based on fiber optics SPR [7], fluorescence [8], second harmonic generation on silica surfaces [9], or electrochemistry [10].

The chapter is organized as follows. In Section 1.2, we present the SPRi apparatus temperature regulation and discuss the constraints brought by the high temperatures on signal detection and grafting chemistries. Then, various examples of DNA melting curves analysis are presented to illustrate the interest of temperature-controlled scans coupled to SPRi detection. In Section 1.3, we particularly lay emphasis on two physico-chemical studies of DNA hybridization on solid support like microarrays: the effects of formamide denaturant on DNA melting temperatures and the screening by the salt concentration of the electrostatic penalty. In Section 1.4, we present preliminary results in view of the multiple detection of SNP in biological samples. Finally, Section 1.5 presents a discussion on biotechnological opportunities of temperature-controlled SPRi detection.

3.2 Temperature Regulation of SPRi for DNA Melting Curves Analysis

In this section, we describe the modifications involved for temperature regulation of a SPRi apparatus from GenOptics commercialized by Horiba Jobin Yvon (see Fig. 3.1 for the scheme of the device).

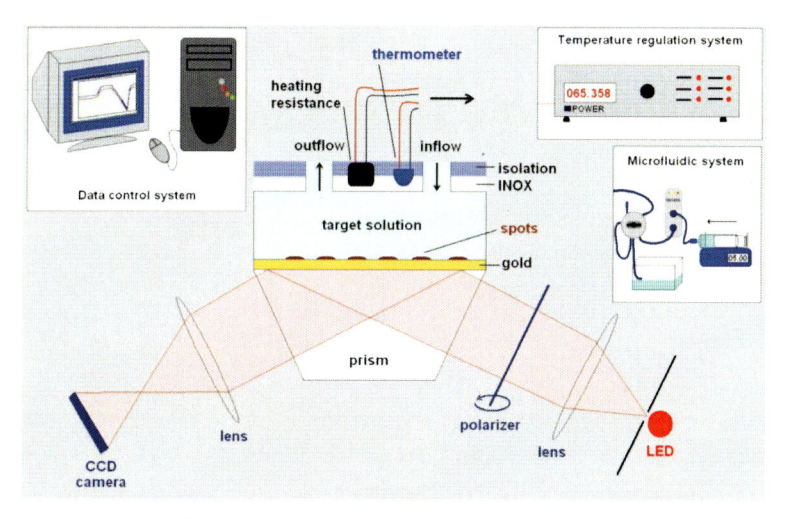

Figure 3.1 Surface plasmon resonance imaging apparatus with temperature regulation.

3.2.1 SPRi Apparatus with Temperature Regulation

The SPRi apparatus exploits surface plasmon excitation in the Kretschmann configuration. A high index gold-coated glass prism is mounted in the optical detection system. The incident light wave at 638.2 nm is delivered by a light-emitting diode (LED) placed behind a pinhole. A first optical lens parallelizes the light beam, which is polarized before falling on the prism under total internal reflection conditions. The reflected light is focused by a second lens and imaged by a CCD camera. The prism is inserted in a holder and pressed against a temperature-controlled flow cell of about 1 cm in diameter and 5 µL in volume where the height is fixed by an aluminum ring. The flow cell is made of stainless steel in which the inlet and the outlet of the flow system are screwed. On the metal, a heating resistance with maximum power of 12 W and a negative temperature coefficient (NTC) thermo-resistance are fixed. Both are inserted in isolating material to prevent heat loss. The whole SPRi apparatus is placed in a temperature-controlled incubator and the fluidic system is connected from the exterior to the flow cell. The flow is regulated by a syringe pump transporting the running buffer first through a degasser. Then, a six-way valve permits the injection of samples in the milliliter range into the system depending on the length of the loop connected.

Data acquisition and image processing are done by an SPRi software provided by GenOptics (SPRi-ViewIt). It permits the regulation of the acquisition time, the area of interest of the image and the definition of masks in order to obtain the reflectivity of the probe spots. The system automatically normalizes the reflectivity to the luminosity observed in the transverse electric mode and subtracts background noise from a reference image acquired while covering the light source. In this way, we obtain reflectivity curves for each spot inside the defined mask. The acquisition frequency is around 0.5 Hz with averaging over five images so that one data point is taken every 2 sec. The differential image permits to follow reflectivity changes on the prism by subtracting a reference image to the currently acquired one.

The integrated temperature regulation device permits rapid heating and cooling as well as direct measurement of the temperature on the prism. The temperature regulation is performed by a potential-integral-derivative (PID) controller system with high stability (below 0.05°C). A homemade LabView interface ensures a temperature regulation either at constant value or through linear scans at different rates up to 85°C. The thermometer is calibrated with good accuracy (around 0.05°C). The solution entering the flow cell on the prism is heated during its passage through narrow channels in the steel heating cell, which itself is surrounded by the incubator temperature. Therefore, the temperature on the prism slightly depends on the flow rate applied to the solution and, especially during controlled cooling, on the surrounding temperature. Measurements by SPR at different temperatures up to 80°C show an increasing difference of the temperature sensed on the prism for a flow rate change from 0 to 5 mL/h. For the highest temperature applied, the temperature on the prism may differ by about 1°C from the measured temperature, limiting the precision of the temperature measurements. Since this effect is flow rate dependent, and thus dependent on the conditions of each experiment, no systematic correction of the temperature has been applied. However, as soon as the rate of the temperature scans and the flow rate are identical between two experiments, the systematic error ΔT due to the flow rate is similar for each scan and the reproducibility of the observed temperatures is excellent. Finally, depending on the scan rates considered, the incubator temperature is set 5° to 10° below the minimal temperature of the flow cell to ensure rapid cooling.

3.2.2 Equilibrium versus Out-of-Equilibrium Melting Curves

In order to perform thermodynamical analysis of biomolecular interactions, temperature scans on a large range is beneficial. However, the SPR signal is highly sensible to any optical change and especially to the change of water or buffer optical index with temperature.

As can be seen on Fig. 3.2, the change of water optical index is responsible of a large part of the SPR signal shift detected during a temperature scan from 25 to 70°C. This amounts to 15% of reflectivity while the target hybridization signal (blue curve on Fig. 3.2) represents only few percents (around 1–3% depending on the length of the targets). Two strategies may be considered to remove the water or buffer dependence: either subtracting a negative control spot without hybridization on the same microarray and temperature scan or subtracting the SPR signal observed for two temperature scans (one with DNA target hybridization and the other without) on the same spot. We have found that the subtraction of the SPR signal on two different spots always leads to an incomplete removal of the temperature effect of the buffer mainly due to inhomogeneities between the grafting procedure from spot to spot affecting the plasmon curve and SPR response. The second strategy leads to better results at the expense of two consecutive temperature scans. It is essential to stress at this stage the importance of the reproducibility of the temperature scans in order to obtain a complete removal of the temperature effect of the buffer.

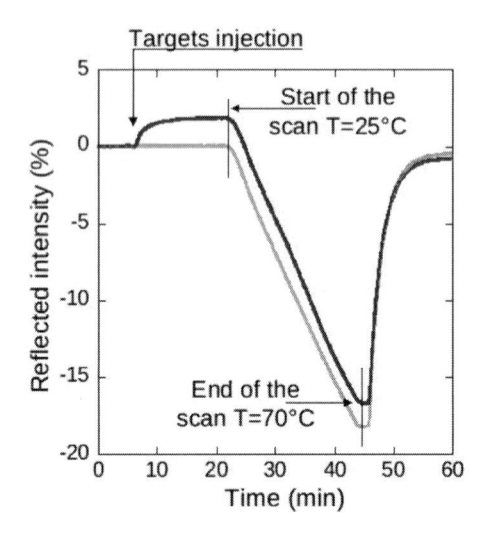

Figure 3.2 Effect of the temperature scan on the shift in reflectivity observed from SPRi on two different spots: one without target hybridization (gray) and one with target hybridization performed before the temperature scan (blue).

Furthermore, two possible target injection protocols exist leading to different results for the DNA melting curves. For physicochemical studies, the equilibrium thermodynamics of the DNA hybridization on a solid support is expected and in order to obtain such equilibrium, the presence of the injected targets on the flow cell during one of the temperature scans is essential (see Section 3.3.2 for details). For stability discrimination like in SNP detection, out-of-equilibrium melting curves analysis is possible. In this case, a first reference temperature scan is performed. Then, the targets are injected on the microarray at room temperature for hybridization. Once the hybridization of the targets is revealed by a corresponding shift in SPR signal, the flow cell is filled with the buffer and the second temperature scan is performed without targets in the flow cell. While with the first protocol we determine equilibrium melting temperatures T_m of the DNA duplex, the second protocol leads to rate dependent out-of-equilibrium dissociation temperatures T_d.

The last important point, for SNP detection applications, the complete denaturation of DNA targets requires high temperatures. Thus, the microarray is heated above 70°C repeatedly and stays at elevated temperatures for long periods increasing the risk of air bubble formation, which may disturb the SPR data acquisition. All buffers, including the one prepared for DNA target injection, have thus been carefully degassed before injection in the flow cell.

3.2.3 Stability of Grafting Chemistries at High Temperatures

The use of a large range of temperatures not only implies constraints on the detection technique as discussed previously, but also on the grafting chemistry of the probes. Two different chemistries have been considered to test the stability of the probe grafting at high temperatures (above 70°C).

3.2.3.1 Electro-copolymerization of poly-pyrrole

The electro-polymerization of pyrrole molecules and the formation of insoluble comb co-polymers presenting DNA probes is performed using a potentiostat device. An electrical impulsion of 2 V for 100 ms is applied to the cell containing mixtures of pyrrole and pyrrole modified DNA probes. Concentrations are from 1 to 10 µM for DNA modified pyrrole and 20 mM for pyrrole. The resulting polymers deposited onto the gold surface create stable spots of probes whose size depends on the reaction cell [11]. The charge necessary for the polymer synthesis serves as an indicator of the thickness of the deposited poly-pyrrole film.

Two different spotters are present in our laboratory. A homemade apparatus able to move a micropipette tip in the three directions to perform probe spots on the prism in a user defined matrix. Usually, spots are spaced by 800 µm and the spotting area is chosen to fit into the flow cell of the SPRi apparatus. Electrochemistry is done using a two-electrode system with the reference electrode connected to the counter electrode. A platinum wire of 125 µm diameter introduced inside the pipette tip serves as counter electrode whereas the gold surface is the working electrode. The spots have a diameter of about 400 µm and a thickness of 5 nm. The grafting density is estimated to be around 1 to 5 pmol/cm^2. The second spotter OmniGrid (GenOptics) is fully operated by a LabView interface. Instead of the pipette, spotting solutions are taken up in needles by capillarity. Spots have a diameter of 260 µm, which is the inner diameter of the Teflon covered needle. The thickness of the spots is smaller (about 1 nm) due to the reduced reaction cell.

Both deposition methods have advantages and drawbacks. In general, it is preferable to deposit a very thin poly-pyrrole layer since the plasmon curve is broadened when the layer increases. This increase leads to a reduced dynamic range of the biosensor. On the other hand, the deposition of a thicker layer results in the immobilization of more DNA probes that are able to hybridize and thus yields a higher mass and refractive index change for SPR signal. Notwithstanding, the deposit of DNA in a polymer layer may introduce an effective probe length dispersion since parts of the DNA strands may be hidden by the matrix and rendered inaccessible [2]. Thus, the spotting with the OmniGrid spotter robot leading to thicker layers may be a good approach once the DNA spotting conditions are optimized.

All microarrays are designed to present internal controls. A positive control DNA sequence differing from the target sequence is spotted which should not react with the actual target. However, when its complementary strand is injected, these spots give a specific signal. Additionally, every prism has some spots of poly-pyrrole without DNA. They are used as negative control since they should not respond to any DNA sequence injected. For all experiments, every sequence is deposited in multiplets to control the reproducibility (3–6 spots per sequence). The microarrays may be used for several experiments over a period of 2–3 months when stocked thoroughly desiccated under argon atmosphere at 4°C. For a storage period exceeding a few months, a signal loss around 50% may be observed. Interestingly, several temperature scans up to 85°C does not affect the stability of the probe grafting and the hybridization signal observed.

3.2.3.2 Thiol self-assembling monolayer

The second grafting chemistry considered is the self-assembling monolayer (SAM) of thiolated DNA, which takes place spontaneously on clean gold surfaces. The thiols form a monolayer whose concrete structure and thickness depend on the length of the chains, the interaction time, the head group, etc. The thiol-gold interaction is not covalent, but relies upon chemisorption of the thiol moiety to the gold surface and its stability depends on the experimental conditions like pH, temperature, incubation time, etc.

Gold-coated prisms are previously cleaned by plasma. The prism is then covered with 1 µM solution of thiolated DNA for 3–5 h in K_2HPO_4 at 1 M concentration and in a humid environment to avoid drying. After incubation, the prism is rinsed with deionized water and dried under argon or nitrogen stream. The prism is then immersed for 90 min in a 1 mM solution of 6-mercapto-1-hexanol (MCH), which serves as blocking thiols, replacing DNA that is not adsorbed by its thiol moiety and filling in imperfections of the layer [12]. After washing and drying, the chip is stored at 4°C and rapidly employed for SPRi studies. When only parts of the surface are functionalized by DNA, MCH also serves to block the remaining gold surface thereby rendering it insensitive to non-specific DNA adsorption.

The stability of thiol SAMs is lower than the electro-polymerization of pyrrole. The prisms may not be used on a long period (less than one month) and present signal loss after several temperature scan up to 85°C. In order to limit thiol desorption, only lower temperatures up to 75°C are considered, which may be detrimental in case of high melting temperatures for long probes.

3.3 Physico-Chemistry of DNA Melting at a Surface

Thermodynamical properties of DNA hybridization have been extensively studied in solution. As a consequence, the two-state phenomenological model with nearest-neighbor interactions became predictive [13]. When one strand, the probe, is grafted to a surface, potential interactions are present that prevent the use of solution-phase hybridization models. Theoretical models have been recently developed to take into account the interactions with the surface, between the probes and/or the targets [14–18]. However, an experimental validation of those models is still missing [19]. Dedicated physico-chemical experiments on the thermodynamical properties of surface-phase DNA hybridization and its dependence on controlled parameters are necessary. In this section, two such studies are presented to analyze first the effect of the presence of denaturant molecules on the dissociation temperature [20] and second to test the effect of salt concentration and the validity of an electrostatic model [21].

3.3.1 Effects of Denaturant Molecules

Numerous solvents are known to act as denaturing agents of DNA. Typical examples are formamide (FA), glycerol, dimethyl sulfoxide, and urea, among others. Although some denaturing compounds may interact with DNA bases by hydrophobic interactions, most denaturing agents interact by competition for hydrogen bonds of the double helix. On this respect, formamide is the most common and probably strongest DNA denaturing agent [22]. Those denaturants are particularly useful to adapt the melting temperatures of probe-target duplex in microarray applications.

It has been shown that formamide lowers linearly the melting temperature of free DNA in solution between 0.6 to 0.72°C/% FA (volume to volume). However, a question remains: What about the denaturation power of formamide when one strand of DNA is grafted on a solid support and is this linear dependence conserved? To obtain generalizable results, we used both grafting chemistries available, i.e. SAMs of thiols and poly-pyrrole electropolymerization. For those experiments, the PBS running buffer (450 mM NaCl, 0.05% Tween20, and 1 mM EDTA) was complemented with various volume fractions of formamide from 0 to 20% in steps of 5%. Hybridization experiments were carried out with targets at 250 nM concentration suspended in the running buffer for 10 min at 25°C with a flow rate of 83 µL/min. After 3 min of washing with the running buffer alone, a temperature scan was performed at 2°C/min up to 70°C. Furthermore, various probes were grafted on the microarrays. Besides perfect match (PM) probes fully complementary to the targets, some probes presenting one or two mismatches (MM) were also considered to analyze the impact of mutations on the formamide denaturant power [20].

Since formamide destabilizes the DNA duplex, the hybridization efficiency decreases with increasing percentage of formamide in the buffer. This effect is strongest on spots forming duplexes with two mismatches where less Watson-Crick base pairs are formed. At 20% formamide some probe/target duplexes with 2 MM have a signal to noise ratio below 10 and have been excluded. The dissociation temperature T_d defined as the temperature with 50% of the initial hybridization signal on the spots was considered to analyze the influence of formamide on the duplex stability. DNA melting curves were averaged over identical spots before normalization at low temperature. Figure 3.3a presents melting curves obtained for a perfect match duplex of 16 bases on thiol-grafted microarrays for various contents of formamide. The dissociation temperature is clearly decreasing with an increasing amount of formamide.

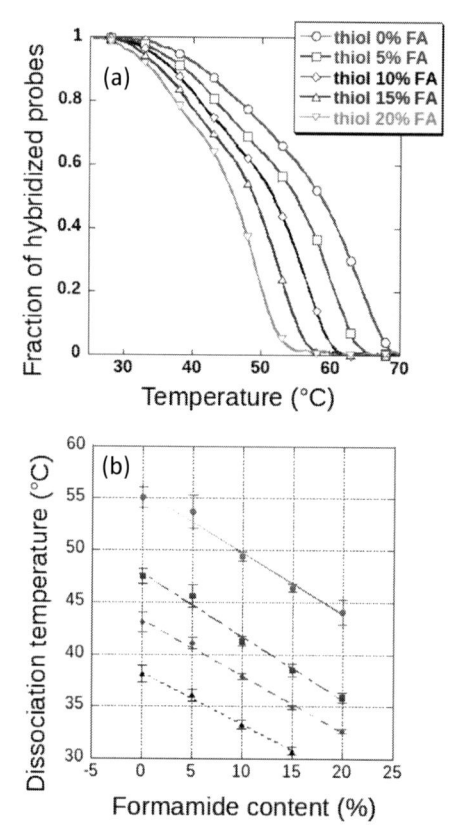

Figure 3.3 (a) Melting curves observed for a thiol-grafted PM probe and an increasing content of formamide denaturant from 0 to 20% by steps of 5%. (b) Dissociation temperatures observed for PM (red), 1 MM (blue and green) and 2 MM (black) as a function of the formamide content.

We observed higher dissociation temperatures with the thiol chemistry compared to the pyrrole grafting certainly due to differing hybridization environment and grafting density. However, the decrease on duplex stability due to an addition of formamide as observed by the reduced T_d is comparable for both grafting methods. Figure 3.3b shows measured dissociation temperatures and the standard errors on thiol functionalized surfaces for various probes (see [20] for further results and the sequences of probes and targets analyzed). The dissociation temperature decreases linearly with increasing volume fraction of formamide as formerly observed for DNA in solution and in polyacrylamide gel pads by Urakawa et al. [23]. Our results showed that formamide has no significant sequence-dependent destabilizing effect on the dissociation temperature and we obtained slopes from −0.49 to −0.65°C per percent of formamide. Urakawa et al. reported a higher T_d decrease for sequences where the mutation replaces a G/C base pair (−0.58°C/%FA) than for mutations replacing an A/T base pair (−0.52°C/%FA) but the position of the terminal mismatches plays a crucial role and also large experimental errors are biasing the appearing order. In our study, no systematic dependency on the mutation type, the number of H-bonds in the duplex or the duplex sequence can be determined. The averaged slopes of −0.57+/−0.05°C/%FA for poly-pyrrole surfaces and −0.59+/−0.05°C/%FA for thiol surfaces present no significant difference between the two surface chemistries. From Urakawa's data [23], an average formamide shift of −0.56°C/%FA is deduced for gel pads extending our result to another microarray technology. These values are lower than previously reported for DNA in solution; however, the results are obtained on immobilized oligonucleotides and for non-equilibrium thermal denaturation contrary to equilibrium melting of long DNA chains (>800 bp) in solution.

As a conclusion, the effect of formamide on dissociation temperatures depends little on the grafting chemistry and the probe sequence with an average decrease of −0.58°C/%FA on microarrays. Since formamide induces a linear variation, it is easy to shift the denaturation temperature to adapted values. For those reasons, the variation of the dissociation temperature found here may be applied to various biosensor and microarray platforms and thus serves in probe selection algorithms for the design of probe sequences.

3.3.2   Effects of Salt Concentration

The Langmuir model is the simplest approach to evaluate DNA hybridization on a surface, but is restricted to very limited conditions. In some cases, it may be useful to modify this model to take into account interaction effects, for example, the electrostatic interactions brought by the DNA charges. The influence of the electrostatic penalty caused by the charging of the surface upon hybridization of the targets has been studied by Halperin et al. [14] leading to a generalized Langmuir adsorption model. It depends on the charge at the surface through the probe length, the grafting density, and the salt concentration screening the electrostatic interactions.

In order to validate the model, we acquired equilibrium melting curves using the temperature regulated SPRi apparatus on both poly-pyrrole-grafted probes and thiol SAMs at various salt concentrations. Since we aimed at thermodynamical equilibrium, we performed heating and cooling cycles at controlled temperature scan rates to check for the presence or not of hysteresis in the melting curves. First experiments were carried out using static conditions, i.e. no flow applied during the temperature scan. Under these conditions, thermodynamic equilibrium was not reached (see Fig. 3.4). During the cooling process, the initial hybridization level was not recovered and we observed hysteresis due to an annealing process penalized by insufficient mass transport. In order to improve the convection of DNA targets without decreasing further the scan rate of 2°C/min, an agitation of the solution was applied during the temperature scans. The automated syringe pump pushes the liquid back and forth without any net flow. In this way, we were able to establish thermodynamic equilibrium over most of the temperature range considered. Furthermore, the melting curves are highly reproducible as can be seen from the two different temperature cycles performed at 157 mM NaCl salt concentration. The curves have been set to their hybridization fraction at the onset of the temperature scan with normalization applied to the whole heating and cooling cycle, so that any differences in heating and cooling would be conserved.

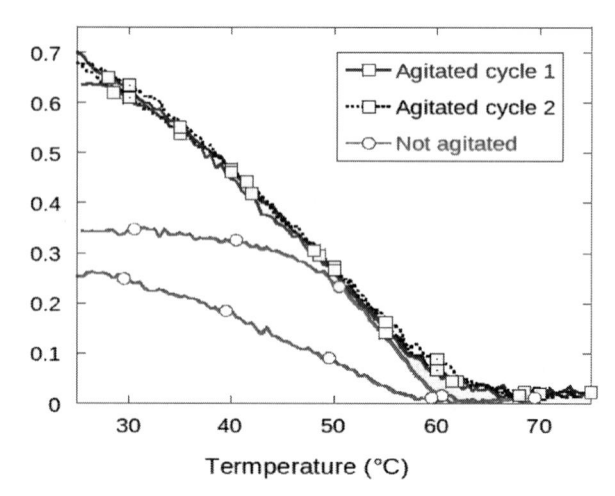

Figure 3.4 The melting curve for a temperature cycle with static conditions presents strong hysteresis (red), whereas two consecutive temperature cycles with back and forth injection of targets improving the mass transport convection lead to reproducible and equilibrium melting curves on microarrays.

The salt effects on the thermodynamical stability of DNA hybridization on microarrays is described by the following equation in the high salt regime [14]: $\theta/(1 - \theta) = c_t K(T) \exp[-\Gamma(1 + \alpha\theta)]$, where θ is the fraction of hybridized probes, $K(T)$ is the reaction constant, $\alpha = N_t/N_p$ takes into account the unequal length of targets N_t and probes N_p and $\Gamma = c_p/c_s$ characterizes the impact of electrostatic interactions with c_p the charge density of the unhybridized probe layer and c_s the salt concentration. In order to extract the thermodynamical parameters ΔH and ΔS, this expression may be written as follows:

$$\ln(\theta/(1 - \theta)) + c_p(1 + \alpha\theta)/c_s = -\Delta H/RT + \Delta S/R \qquad (3.1)$$

In the limit of low grafting density ($c_p = 0$), the expression reduces to the Langmuir model. However, for surface grafting densities of common microarrays in the range 1 to 10 pmol/cm^2 and for oligonucleotide probes corresponding to a charged layer thickness of 4 to 8 nm, the electrostatic interactions are relevant and c_p lies in the range 10 to 500 mM. Thus, larger salt concentrations ($c_s > c_p$) are necessary to screen the electrostatic penalty characterized by the coupling constant Γ.

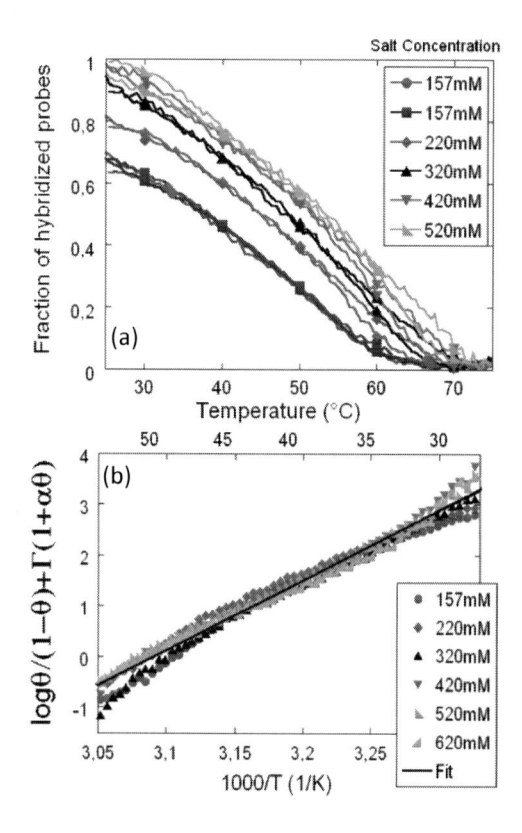

Figure 3.5 (a) Equilibrium melting curves observed on poly-pyrrole-grafted microarrays for various salt concentrations. (b) Rescaling of the equilibrium melting curves allowing the determination of the probe charge density c_p and the thermodynamic parameters ΔH and ΔS of the DNA hybridization on microarrays.

From our experimental data, we observe the salt concentration effects on equilibrium melting curves (see Fig. 3.5a). An increasing stability of the hybridization is observed upon shielding the electrostatic repulsion by an increase of the salt concentration. In order to validate the Langmuir model extended to incorporate the electrostatic interactions, we plot the left hand side of Eq. 3.1 versus the inverse of the temperature for each salt concentration. Such curves yield a linear dependency and all curves should collapse for a given value of c_p that is characteristic for our microarrays. Furthermore, from a simple linear regression, we determine the corresponding thermodynamic parameters ΔH and ΔS (see Fig. 3.5b) [21]. The optimal parameter for pyrrole electropolymerization microarrays is c_p = 13 ± 1 mM. This parameter was independent of the probe sequence as expected since it should only depend on the grafting density and the height of the charged layer. Thiol-grafted probes with 10 and 50 µM probe concentration in the SAM solution yielded larger c_p values, 39 and 53 mM, respectively, consistent with higher charge densities on thiol than on poly-pyrrole-grafted surfaces.

Gong and Levicky also studied the effects of salt concentration and probe density on the hybridization content at a fixed temperature using an electrochemical detection [24]. They established an empirical map of different hybridization regimes encountered at different salt concentrations and different probe densities on microarrays. Although this study spans only one order of magnitude in probe grafting density so that probes are always close enough to possibly interact, three different hybridization regimes have been found regarding the salt concentration and present a strong connection with the model of electrostatic interactions [14].

For completion, it should be mentioned that, although most experiments use monovalent cations from NaCl or KCl for charge screening, divalent cations are also relevant in DNA studies. Divalent cations shield the DNA backbone more efficiently than monovalent cations and already a small concentration (few millimolar) of $MgCl_2$ as commonly used for enzyme reactions sufficiently stabilizes DNA duplex to allow surface hybridization [25]. Further investigations of salt effects, either monovalent or divalent, on microarrays are still clearly needed.

3.4 Detection of Single Point Mutation from Melting Curve Analysis

As we have shown, physico-chemical characterizations of the DNA hybridization on microarray surfaces are an important step toward optimization for clinical applications [19]. In this section, we discuss the challenges one has to confront in order to unambiguously identify the injected target sequences or eventual target mixtures for SNP detection applications.

3.4.1 Detection with Oligonucleotides Targets

Melting curve analysis is a powerful tool to detect single point mutations. For a single target injected, the spots containing the complementary probes are expected to present the highest thermal stability. Figure 3.6a highlights this fact for the injection of M4c targets complementary to probes M4 (see sequences in [3]). Duplexes with one internal mismatch (probes N, M5, and M6) dissociate at lower temperatures while duplexes with two mutations (probes M1, M2, and M3) present even lower affinity. This SNP detection is even sensitive to the type of mismatch formed on the spot. The melting curves for spots with probes N, M5, and M6 are clearly separated. Surprisingly, the targets having two mutations present comparable melting temperatures. Those experimental results compare qualitatively with the predictions obtained using the phenomenological nearest-neighbor model of thermodynamical parameters for the solution phase hybridization [13]. However, we observe a broadening of the melting curves. The denaturation transition takes place over a wide range of temperature, which decreases the mismatch resolution and the differences in T_d for PM or MM targets.

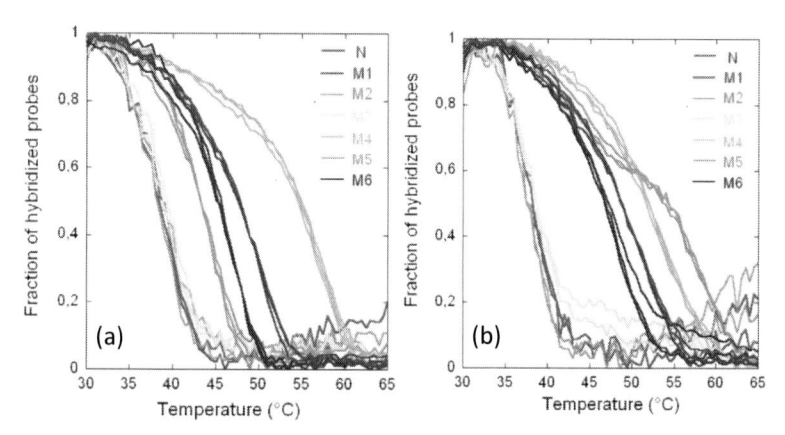

Figure 3.6 Melting curves for (a) homozygous and (b) heterozygous SNP detection with oligonucleotide targets. Colors are representative of the probes grafted to the spots (see [3] for sequences).

Being able to identify homozygous samples, the detection of heterozygous cases with 50%–50% target mixture of wild type and mutant is also crucial for clinical applications. In Fig. 3.6b, we present melting curves obtained after the injection of mixtures of an equal amount of M4c and M5c targets at a total concentration of 250 nM. A two-phase dissociation is clearly observed on spots corresponding to the probe M5. The first dissociation corresponds to the denaturation of the less stable duplex formed then the signal stagnates around $\theta = 0.5$ before a second dissociation phase reveals the departure of the more stable target (PM). The plateau is indicative of the proportions of the two targets injected. This interesting property is due to the fast hybridization of the targets on the spots with similar association rates. Reorganization is essentially based on the difference of the dissociation rate and would take a particularly long time at the room temperature hybridization temperature due to the high affinities of both targets towards the probes. As a consequence, we are able not only to detect the presence of a mutation in heterozygous cases but also to quantify the proportional abundance of mutations in the sample. This confirms the interest of melting curves analysis on microarrays for SNP genotyping. Furthermore, the method is universal in the sense that no particular optimization (probe length or sequence) is needed for the detection of different and multiple SNPs on the same microarray [3]. There is no need for a uniform dissociation temperature for all PM targets. The hybridization temperature could even be lowered for shorter probes and the temperature scan must just be sufficiently high to denature every duplex formed on the microarray.

3.4.2 Detection Limit of Somatic Mutations

Since we are able to detect mutations in target mixtures, it is interesting to determine the detection limit in fraction of mutation in the overall sample. The detection and, even better, the quantification of very low percentages of mutated versus non-mutated DNA would have important clinical applications for somatic mutations. Thus, the melting curves analysis on microarrays could prove useful for the analysis of biopsy samples. Biopsies are tissues extracted from specific sites at places in the body that are sometimes difficult to access, for example the stomach or the lung. The extracted sample contains a varying proportion of healthy and tumor cells with proportion of tumor cells as low as 1%. The extracted as well as amplified DNA will generally reflect the initial ratio of mutations. Detection methods should thus be able to find low abundant amounts of mutated DNA in a large background.

Experimentally, we obtained a detection limit of 5% mutations in samples containing a total target concentration of 250 nM [26]. While this detection limit is a low level, it is still insufficient for clinical applications. More sophisticated detection protocols are necessary. Since both detection and quantification are expected, a two-step approach may appear suitable. The first experiment would hybridize half of the sample on the microarray at room temperature before a temperature heating scan is performed and melting curves analyzed for the quantification of mutation ratio above 5%. The second experiment would use the second half of the sample and before the melting curve analysis; a more complex hybridization protocol is performed in order to amplify the mutation target hybridization on their complementary probes. Different strategies have been proposed and tested [26]. For example, successive temperature cycles accumulate the amount of mutation targets on their corresponding probes and may thus shift the plateau in the detectable range (above 5%). A second approach is the injection of the sample at high temperature followed by a slow cooling scan to room temperature. For each individual probe sequence, the complementary target will hybridize at higher temperature than mismatches. In this way, the mutation targets in low abundance hybridize on their PM probe before the competitor wild type targets. This protocol has the advantage of easy implementation to our system and assay time is optimized.

To complete this presentation of the detection of target mixtures with one allele in low abundance, it has to be said that simple isothermal hybridization at an optimum temperature of $T_m(MM) < T_{opt} < T_m(PM)$ will provide the most efficient accumulation of mutation target on its complementary probe. However, this approach is not flexible enough to be applicable to complex target mixtures sensing different SNP sites, unless substantial effort is done for optimization of probe length and sequences to obtain similar melting temperatures. Thus, the temperature cycle or cooling approaches have the advantage to be more universal and less dependent on the nature of the SNPs to be detected.

3.4.3 Homozygous and Heterozygous Detection of PCR Products

The first step for SNP genotyping and probably the main difficulty is the hybridization of the long PCR amplified DNA strands on microarrays. The crucial parameters influencing the hybridization conditions are clearly identified from the previous oligonucleotide experiments. Of course, target concentration is a main issue since diluted solutions will be limited by mass transport to the surface. For oligonucleotides, a detection limit around 10 nM was found. Due to lower diffusion constant, it seems preferable to work at high concentrations with PCR products in the limits of availability. The target conformation is also a crucial point since secondary structures blocking the hybridization part may be present. To improve surface hybridization, besides the use of helper sequences opening the secondary structures [27], the temperature of hybridization may play an important role. Furthermore, due to the length of the PCR product, long dangling tails may reduce the surface hybridization due to steric hindrance. To circumvent this effect, it might be useful to vary the grafting density to obtain higher spacing between the probes. When grafting density, target concentration and hybridization temperature are adapted to the hybridization configuration, surface hybridization might still be kinetically slow. So the last unknown parameter will be the hybridization time necessary to obtain detectable SPR signals. Due to the nearly 10-fold increased mass of the PCR product compared to oligonucleotides, we, however, expect a stronger SPR signal in spite of a lower surface coverage of the targets.

First of all, we have tested the interest of using a helper sequence to open the secondary structure of PCR products when the SNP site is impacted. The presence of the helper oligonucleotide in the sample solution increases significantly the hybridization rate and favors surface hybridization. Furthermore, we have tested the effect of hybridization temperature. A compromise was obtained around 45°C between the opening of the PCR secondary structures and the lower stability of the probe-target duplex stability. The hybridization of 1 µM purified PCR product in the presence of the helper oligonucleotide in PBS buffer for 3 h at 45°C with no-flow conditions leads to SPR signal intensities varying between 0.55% and 1.1% reflectivity shifts. Furthermore, hybridization signals are generally better the longer the probe, consistent with their higher stability. Contrarily to our expectations, we observed similar SPR signal for PCR products and oligonucleotides in spite of the 8-fold increased mass of the PCR products. Longer hybridization times might increase the surface capture; however, it is unlikely that the same target density as obtained for oligonucleotide targets may be reached. Since the hybridization signal seemed to indicate only minor increase after 1 h hybridization, the hybridization time was shortened accordingly for further experiments. We successfully hybridized targets corresponding to all three genotypes: homozygous wild-type C/C, heterozygous T/C and homozygous mutant T/T. The sequences of the PCR product considered in this study are 3'-CAC TTC AAG TAA AGG TTA GGC GGG AGG TAC CAC CGT CGC CCC TCG CAC CAC CGG CGT CAC GTT CCG GAC TTG GAC TCC TCG GGG TTG TTG AAG GAC AGG ATG ATG GCG GAG TGT GCG AAG GAG AGG TCT CAC TAG TTC ACA CTG GG**C/T** CA TTC ACT CCC ACT ACA GGG -biotin-5' corresponding to the allele **C/T**, respectively. Two different allele-specific oligonucleotide (ASO) probe lengths have been considered. The ASO-**G/A**24 or **G/A**24 comprises 24 bases, C̲A̲A GTG TGA CCC **G/A**GT AAG TGA G̲G̲G̲, while the two underlined bases on each side have been removed for the 20 base long probes. As already discussed, the largest problem for genotyping detection using melting curve analysis on microarray resides in the obtention of a detectable surface hybridization signal. Once achieved, temperature scans may be used to detect the genotype of the target without further optimization.

Figure 3.7 presents melting curves obtained for homozygous targets C/C (left) and T/T (right). The result of each individual spot on the microarray is shown and demonstrates very good signal reproducibility among different spots with the same probe (multiple curves of the same color in Fig. 3.7). It is interesting to notice the sharp transition observed for long targets compared to the oligonucleotide melting curves, suggesting a cooperative effect between targets. From the order of dissociation temperatures, the genotype is readily determined from 20 and 24 base long probes. In each case, the perfectly matched spot presents higher dissociation temperatures, i.e. G20 and G24 probes corresponding to PM show dissociation at higher temperatures for the injection of C/C target sample than A20 and A24 probes. Symmetrically, for T/T mutation targets, A20 and A24 probes present more stable hybridization than G20 and G24 probes specific for the wild type allele. Differences between PM and MM are more pronounced on shorter probes. The inset of each figure presents the derivative of the melting curves, which might be used for automated genotyping through the extraction of the minimum of each derivative melting curve. The corresponding temperatures determine which spot is perfectly matched to the target and which one forms a mismatch.

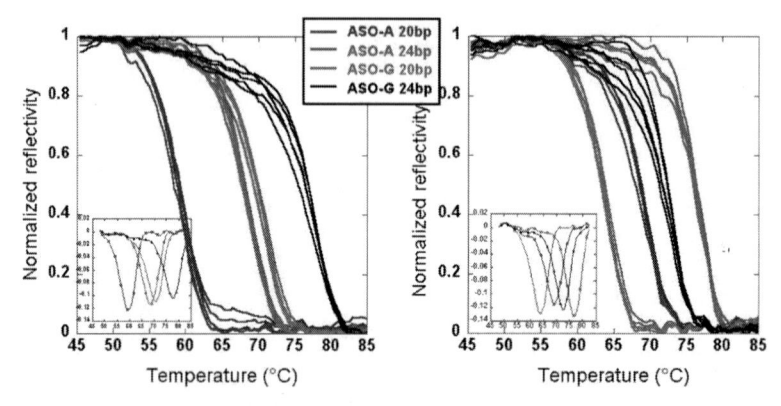

Figure 3.7 Melting curves of homozygous PCR products (C/C left) and (T/T right) on four different spots corresponding to the ASO G/A probes. Inset: Derivatives of the melting curves.

The next challenge is the detection of the heterozygous genotype C/T. Figure 3.8a displays dissociation curves for a PCR product with a target mixture C/T. As for the homozygous case, the dissociation is cooperative for target mixtures. As for oligonucleotide mixtures, we observe the two-step dissociation: at low temperatures, the target with one mismatch dissociates to the probe while at higher temperatures, the perfectly complementary target is denatured and washed away. Thus, a plateau in the melting curve is characteristic of heterozygous samples. Contrarily to observations on oligonucleotide melting curves, the plateau is not found at exactly the same proportion for each probe. 20 base long probes forming the less stable duplexes show an increased proportion of PM targets when compared to 24 base long probes. This indicates that dissociation at the hybridization temperature is not negligible and some reorganization takes place, reducing the plateau below 50%. Furthermore, although the plateau is not clearly identified in the melting curves, the two peak derivatives shown in Fig. 3.8b easily reveal its presence. As a conclusion, we are able to correctly identify the genotype of the injected PCR product. For correct identification, only two spots, complementary to each allele would be necessary in this approach.

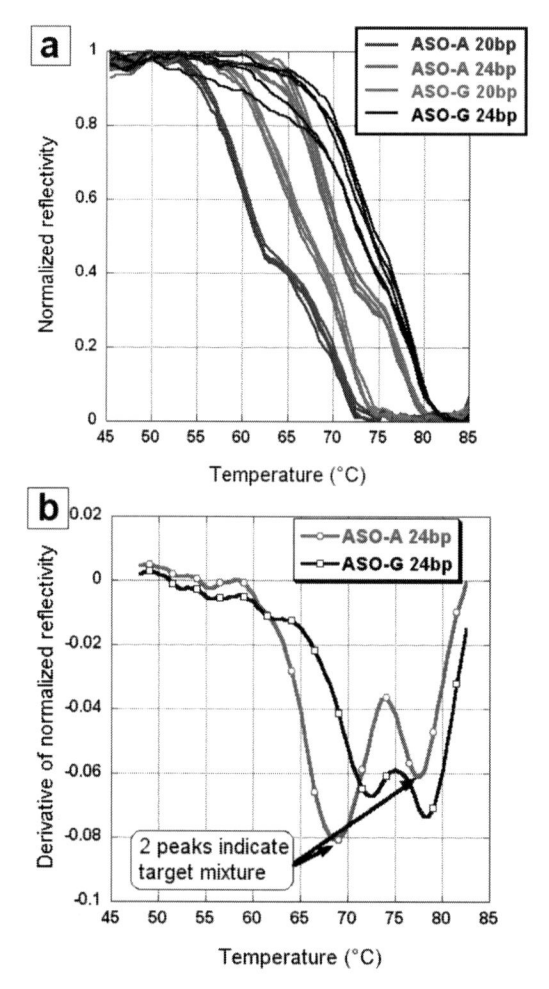

Figure 3.8 (a) Melting curves for the heterozygous case C/T. (b) Derivative of the melting curves for the 24 base long probes.

In fact, we observed an interesting feature of thermal denaturation for PCR amplified targets. The dissociation curves are not as flat as with oligonucleotide targets, but present a narrow transition over a short temperature range (lower than 10°C). This difference in the shape of the melting curves clearly indicates that the physico-chemical environment of the PCR product is distinguishably different from oligonucleotide targets. The cooperative melting transition observed with PCR product improves the possible selectivity of single point mutation detection since dissociation curves between matched and mismatched duplex are more easily distinguished. In biotechnology applications, such sharp melting curves have been observed on DNA-modified gold nanoparticles hybridized to oligonucleotide microarrays [28]. There, full width half maximum (FWHM) of the derivative melting curves as low as 3°C were observed, revealing even sharper transition than for the PCR product melting (FWHM of 6.5–8°C). Possible explanations are based on the formation of aggregates or networks that interconnect DNA targets and stabilize the hybridization until T_d is reached. Another model explained melting cooperativity by ion clouds shared between neighboring strands and confined zones that stabilize the hybrids. DNA cooperative melting on comb polymer-DNA has been analyzed [29] and seems to confirm the neighboring duplex cooperative model based on ion clouds. However, cooperativity is exclusively found when hybridization takes place in between structures that link DNA. Then, the dissociation of an individual free DNA strand is found coupled to the dissociation of surrounding DNA, leading to a switch-like dissociation. Here, hybridized targets are not physically linked to other targets. We, however, cannot exclude that the proximity between targets leads to entanglement and partial hybridization between dangling ends of neighboring targets. In this case, an interacting network might occur that, similar to the polymer combs exhibits cooperative melting. With the present data, we cannot, however, conclude the exact interpretation of our observations. However, the phenomenon seems to be independent of the presence of the helper and does not depend on the amount of surface hybridized targets.

In summary, we demonstrated here that point mutation detection using a temperature scan–based SPRi detection method can be applied to biological samples of high complexity. Problems related to secondary structures, hybridization with surface proximal tails, and low sample availability have been overcome. This work reveals that even our temperature scan–based method needs optimization for the hybridization protocol, once hybridization signal exceeds a signal-to-noise ratio of 10, genotyping of the targets is possible. Furthermore, the melting curves obtained on long targets show a sharp, reproducible dissociation profile as opposed to oligonucleotide samples with a large transition. This renders the discrimination between different curves easier and thus favors point mutation detection. The reason for this cooperative behavior is not completely clear but might be attributed to interactions of dangling ends with their environment.

3.5 Conclusion

In this chapter, we have shown that temperature-controlled SPR imaging devices open the study of the dynamics of DNA hybridization process. We have demonstrated the interest of DNA melting curves analysis on microarrays for a fundamental understanding of DNA hybridization on surfaces [2, 21]. Those results prove that biosensors and microarrays are much more complicated than expected. Further physico-chemical studies will be necessary to assess the assumptions in existing models and it is still difficult to imagine an integral theory that would apply to all kinds of substrates, grafting chemistries and experimental conditions [19].

Furthermore, we presented an approach based on melting curves analysis by SPRi detection on microarrays for typical medical studies. Those studies include the detection of low quantity of mutated DNA present in a large amount of wild type DNA [26] useful for biopsy samples of cancerous cells and the presence of SNP on PCR amplified samples [3] useful for the screening of genetic predisposition of cancer.

Finally, the use of temperature-controlled SPRi detection is not limited to DNA hybridization studies, but opens the door to a large spectrum of applications ranging from the in situ analysis of DNA repair protein activity [4] to the use of DNA or RNA aptamers as sensor probes for the detection of small molecules, proteins or even cells [5, 6].

References

1. Azam, M. S., and Gibbs-Davis, J. M. (2013). Monitoring DNA hybridization and thermal dissociation at the silica/water interface using resonantly enhanced second harmonic generation spectroscopy. *Anal. Chem.*, **85**, 8031–8038.
2. Belozerova, I., and Levicky, R. (2012). Melting thermodynamics of reversible DNA/ligand complexes at interfaces. *J. Am. Chem. Soc.*, **134**, 18667–18676.
3. Blake, R. D., and Delcourt, S. G. (1996). Thermodynamic effects of formamide on DNA stability. *Nucleic Acids Res.*, **24**, 2095–2103.
4. Corne, C., et al. (2008). SPR imaging for label-free multiplexed analyses of DNA N-glycosylase interactions with damaged DNA duplexes. *Analyst*, **133**, 1036–1045.
5. Daniel, C., Mélaïne, F., Roupioz, Y., Livache, T., and Buhot, A. (2013). Real time monitoring of thrombin interactions with its aptamers: Insights into the sandwich complex formation. *Biosens. Bioelectron.*, **40**, 186–192.
6. Daniel, C., Roupioz, Y., Gasparutto, D., Livache, T., and Buhot, A. (2013). Solution-phase vs. surface-phase aptamer-protein affinity from a label-free kinetic biosensor. *PloS One*, **8**, e75419.
7. Delport, F., et al. (2012). Real-time monitoring of DNA hybridization and melting processes using a fiber optic sensor. *Nanotechnology*, **23**, 065503.
8. Fiche, J. B., Buhot, A., Calemczuk, R., and Livache, T. (2007). Temperature effects on DNA chip experiments from Surface Plasmon Resonance imaging: Isotherms and melting curves. *Biophys. J.*, **92**, 935–946.
9. Fiche, J. B., Fuchs, J., Buhot, A., Calemczuk, R., and Livache, T. (2008). Point mutation detection by Surface Plasmon Resonance Imaging coupled with a temperature scan method in a model system. *Anal. Chem.*, **80**, 1049–1057.
10. Fuchs, J., Dell'Atti, D., Buhot, A., Calemczuk, R., Mascini, M., and Livache, T. (2010). Effects of formamide on the thermal stability of DNA duplexes on biochips. *Anal. Biochem.*, **397**, 132–134.
11. Fuchs, J., Fiche, J. B., Buhot, A., Calemczuk, R., and Livache, T. (2010). Salt concentration effects on equilibrium melting curves from DNA microarrays. *Biophys. J.*, **99**, 1886–1895.
12. Gibbs-Davis, J. M., Schatz, G. C., and Nguyen, S. T. (2007). Sharp melting transitions in DNA hybrids without aggregate dissolution: Proof of neighboring-duplex cooperativity. *J. Am. Chem. Soc.*, **129**, 15535–15540.

13. Gong, P., and Levicky, R. (2008). DNA surface hybridization regimes. *Proc. Natl. Acad. Sci. U S A*, **105**, 5301–5306.
14. Guedon, P., et al. (2000). Characterization and optimization of a real-time, parallel, label-free, polypyrrole-based DNA sensor by surface plasmon resonance imaging. *Anal. Chem.*, **124**, 14601–14607.
15. Halperin, A., Buhot, A., and Zhulina, E. B. (2004). Sensitivity, specificity and the hybridization isotherms of DNA chips. *Biophys. J.*, **86**, 718–730.
16. Halperin, A., Buhot, A., and Zhulina, E. B. (2004). Hybridization isotherms of DNA microarrays and the quantification of mutation studies. *Clin. Chem.*, **50**, 2054–2062.
17. Halperin, A., Buhot, A., and Zhulina, E. B. (2005). Brush effects on DNA chips: Thermodynamics, kinetics and design guidelines. *Biophys. J.*, **89**, 796–811.
18. Halperin, A., Buhot, A., and Zhulina, E. B. (2006). On the hybirdization isotherms of DNA microarrays: The Langmuir model and its extensions. *J. Phys. Condens. Matter.*, **18**, S463–S490.
19. Halperin, A., Buhot, A., and Zhulina, E. B. (2006). Hybridization at a surface: The role of spacers in DNA microarrays. *Langmuir*, **22**, 11290–11304.
20. Harrison, A., et al. (2013). Physico-chemical foundations underpinning microarray and next-generation sequencing experiments. *Nucl. Acids Res.*, **41**, 2779–2796.
21. Levicky, R., Herne, T. M., Tarlov, M. J., and Satija, S. K. (1998). Using self-assembly to control the structure of DNA monolayers on gold: A neutron reflectivity study, *J. Am. Chem. Soc.*, **120**, 9787–9792.
22. Marcy, Y., et al. (2008). Innovative integrated system for real-time measurement of hybridization and melting on standard format microarrays. *BioTechniques*, **44**, 913–920.
23. O'Meara, D., Nilsson, N., Nygren, P., Uhlén, M., and Lundeberg, J. (1998). Capture of single-stranded DNA assisted by oligonucleotide modules. *Anal. Biochem.*, **255**, 195–203.
24. Owczarzy, R., et al. (2008). Predicting stability of DNA duplexes in solutions containing magnesium and monovalent cations. *Biochemistry*, **47**, 5336–5353.
25. Pingel, J., Buhot, A., Calemczuk, R., and Livache, T. (2012). Temperature scans/cycles for the detection of low abundant DNA point mutations on microarrays. *Biosens. Bioelectron.*, **31**, 554–557.

26. Ririe, K. M., Rasmussen, R. P., and Wittwer, C. T. (1997). Product differentiation by analysis of DNA melting curves during the polymerase chain reaction. *Anal. Biochem.*, **245**, 154–160.
27. SantaLucia, J., and Hicks, D. (2004). The Thermodynamics of DNA structural motifs. *Annu. Rev. Biophys. Struct.*, **33**, 415–440.
28. Taton, T. A., Mirkin, C. A., and Letsinger, R. L. (2000). Scanometric DNA array detection with nanoparticle probes. *Science*, **289**, 1757–1760.
29. Urakawa, H., et al. (2002). Single-base-pair discrimination of terminal mismatches by using oligonucleotide microarrays and neural network analyses. *Appl. Environ. Micro.*, **68**, 235–244.

Chapter 4

Plasmon Waveguide Resonance Spectroscopy: Principles and Applications in Studies of Molecular Interactions within Membranes

Isabel D. Alves

Chemistry and Biology of Membranes and Nanoobjects UMR 5248 CNRS, University of Bordeaux, bat. B14 allée Geoffroy St. Hilaire, 33600 Pessac, France

i.alves@cbmn.u-bordeaux.fr

4.1 Introduction

Many important biological events and molecular interactions take place at the cell membrane. Such molecular interactions include external molecules that are prone to interact with cell membranes such as membranotropic peptides (cell penetrating, antimicrobial, viral and amyloid peptides), but also molecules that are already embedded in the membrane such as G-protein coupled receptors (GPCRs) that respond to a large variety of outside stimuli ranging from light to odorants and chemical ligands. In view of the great complexity of cellular membranes, lipid model systems have largely been employed to mimic the composition and structural organization of cellular membranes. Besides the lipid model system used, a crucial point in these studies concerns the technique employed to monitor such interactions. Important aspects to consider in the choice of the technique concern (1) the sensitivity, in view of the very low natural abundance of molecules such as the case of GPCRs, (2) the possibility to use lipid models systems whose properties can be properly characterized and that correctly mimic cell membranes while being stable and flexible, and (3) the property to be evaluated, ideally the technique should directly probe the molecular interaction without need for labeling since the addition of probes (radioactive or fluorescent) or reporter groups often lead to important structural modifications of the molecule under study.

A technique that can fulfill many of the above points is a relatively novel biophysical method named plasmon-waveguide resonance (PWR) spectroscopy. In this chapter, the principles underlying this technique will be presented. Plasmon-waveguide resonance combined with techniques for forming solid-supported proteolipid membranes allows kinetic, thermodynamic, and structural characterization of anisotropic thin films such as the case of lipid model systems and molecules therein embedded or interacting with such films. The anisotropic properties of such oriented systems can be directly probed, as can thermodynamic and kinetic properties that result from these interactions, without the need for any chemical modification.

In this chapter, the uses and power of this method in studies involving the activation and signaling of GPCRs as example of membrane proteins will be discussed. Additionally, applications regarding the interaction of membrane active peptides with lipid membranes will be presented. Finally, an overview of the current and future developments of this technique will be briefly highlighted.

4.2 Plasmon Spectroscopy

4.2.1 Description of Surface Plasmons

The phenomenon of anomalous dispersion on diffraction gratings due to the excitation of surface plasma waves was first described in the beginning of the twentieth century. In the late 1960s, optical excitation of surface plasmons by the method of attenuated total reflection (ATR) was demonstrated by Kretschmann [1] and Otto [2]. The potential of surface plasmon resonance (SPR) for characterization of thin films and monitoring processes at metal interfaces was recognized in the late seventies [3]. In 1982, the use of SPR for biosensing was demonstrated by Nylander and Liedeberg [4]. Since then, SPR has been intensively studied [5, 6] and found to be in good agreement with theoretical concepts based on Maxwell's theory of electromagnetism (EM). In Maxwell's theory of electromagnetism, plasmons are treated as charge density oscillations of the free electrons of a metal. These electron density fluctuations generate a surface localized electromagnetic (EM) wave, which nonradiatively propagates along a metal/dielectric interface. The electric field is normal to this interface and vanishes exponentially with distance. The EM characteristics are the same as those describing optically generated evanescent waves in total internal reflection techniques. The EM fields do not stop at the boundary, but penetrate a distance into the emerging medium in the form of a surface wave. Surface plasmon excitation is a resonance phenomenon that occurs when energy and momentum conditions between incident light photons and surface plasmons are matched according to the following equation:

$$\kappa_{SP} = \kappa_{ph} = (\omega/c)_{\varepsilon_0}^{1/2} \sin\alpha_0, \quad (4.1)$$

where

$$\kappa_{SP} = (\omega/c)(\varepsilon_1 \varepsilon_2/\varepsilon_1 + \varepsilon_2)^{1/2} \quad (4.2)$$

κ_{SP} is the longitudinal component of the surface plasmon wave vector, κ_{ph} is the component of the exciting light wave vector parallel to the active (metal) medium surface, ω is the frequency of the surface plasmon excitation wavelength (λ), c is the velocity of light in vacuo, ε_0, ε_1, and ε_2 are the complex dielectric constants for the incident, surface active and dielectric (or emerging) media, respectively, and α_0 is the incident coupling (resonance) angle.

To satisfy this relationship in a conventional resonator, plasmon excitation must occur through an evanescent wave generated by p-polarized incident light (electric vector perpendicular to the surface). This plasmon excitation geometry is ideal for application as a biomedical sensor, because the excitation light is totally separated from the emergent medium (e.g., a biological material to be analyzed). This allows the assay of such biological fluids regardless of their optical transparency (including, for example, blood, urine). The resonance conditions mentioned previously can be fulfilled by changing the incident light angle α at a constant wavelength. At resonance, the surface plasmons are generated at the expense of the energy of the excitation light leading to significant alterations of the totally reflected light intensity. Thus, a plot of the reflected light intensity *versus* incident angle is a quantitative measure of the resonance and constitutes a surface plasmon resonance spectrum. Alterations in the optical properties of the metal/dielectric medium interface (such as the immobilization of material) will affect the surface plasmon wave vector and, therefore, change the resonance characteristics, leading to changes in both the position and shape of the resonance curve. As can be seen from Eq. (4.1), these optical properties are fully described by the complex dielectric constant ε, which contains the refractive index, n, and the extinction coefficient, k (i.e., $\varepsilon = n - ik$) as well as the thickness, t, of a layer of material deposited at the interface. The sensitivity S, of measurements of such alterations can be defined as the change in reflectance (dR), measured at a specific angle, α_1 within the range of the resonance curve, divided by the change in one of the three optical parameters (dn, dk, and dt), i.e.,

$$S_{\alpha 1} = [dR_{(\alpha)}/dn\ dk\ dt]_{\alpha 1}$$

In general, the magnitude of changes in the experimental value of R is controlled by two factors: the shift of the position and the shape of the resonance spectrum. Both of these factors are related to the sharpness of the spectrum (i.e., its half-width); this defines the optical resolution (i.e., the smallest changes that can still be resolved as two different readings). Therefore, the overall sensitivity of a plasmon resonator will depend on both the magnitude of the electromagnetic field at the resonator surface and the optical resolution.

4.2.2 Types of Surface Plasmon Resonances

4.2.2.1 Conventional surface plasmon resonance

In the most straightforward case, for which the hypotenuse of a right angle prism is coated with a single high-performance metal layer (usually Ag or Au, although gold is often used in biosensors because of its resistance to corrosion), one can generate surface plasmons on the outer surface of the metal with visible light (typically 500 to 700 nm), as indicated in Fig. 4.1 for either a 55 nm Ag or a 48 nm Au layer (which are the optimal thicknesses for these metals). The electromagnetic wave created as a result of surface plasmon excitation is characterized by several important properties, which are very relevant in biosensor applications. First, there is an enormous increase in the intensity of the electromagnetic field generated by surface plasmons compared with that at the incident surface. It is also important to note that the energy of the field is proportional to the square of the electromagnetic wave intensity. This property further increases the sensitivity of the measurement. Second, the magnitude of the increase in the electromagnetic field intensity depends on the optical properties of the metal layer used to generate plasmons. Thus, as the calculation presented in Fig. 4.1 demonstrates, silver produces approximately a twofold higher intensity, which results in a fourfold higher overall sensitivity than gold [7].

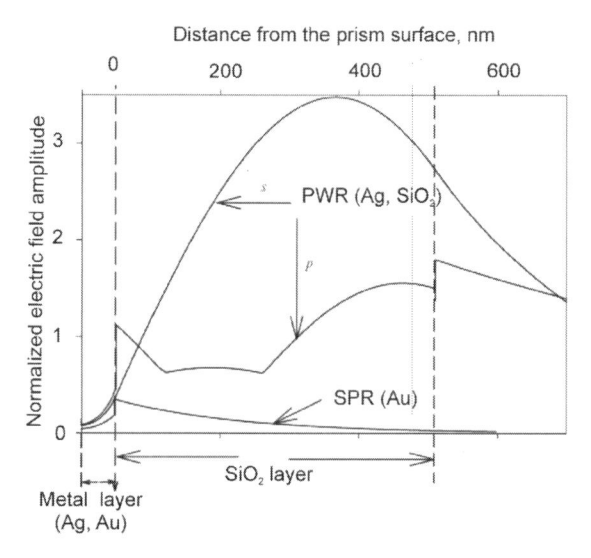

Figure 4.1 Electric field amplitudes for SPR and PWR resonators with 632.8 nm excitation wavelength. The sensitivity of the resonator is proportional to the amplitude of the field at the external surface layer (gold in the case of SPR and silica for PWR). Adapted from ref. 7.

4.2.2.2 Plasmon-waveguide resonance

In 1997, a variant of SPR that involves coupling of surface plasmon resonances in a thin metal film with guided waves in a dielectric overcoating, resulting in excitation of both plasmon and waveguide resonances was created [8–12]. The technique was called plasmon-waveguide resonance [7, 13, 14]. A plasmon-waveguide resonator contains a metallic layer (as in a conventional SPR assembly), which is deposited on either a prism or a grating and is overcoated with either a single dielectric layer or a system of dielectric layers. These layers are characterized by appropriate optical parameters so that the assembly is able to generate surface resonances on excitation by both p- and s-polarized light (Fig. 4.1). The addition of such a dielectric layer (or layers) to a conventional SPR assembly plays several important roles. First, it enhances the spectroscopic capabilities (because of excitation of resonances with both p- and s-polarized light components), which results in the ability to directly measure anisotropies in refractive index and optical absorption coefficient in a thin film immobilized onto the surface of the overcoating [15]. This allows information to be obtained about structural changes in the analyte, for materials that are uniaxially oriented at the interface. Second, it functions as an optical amplifier that significantly increases electromagnetic field intensities at the dielectric surface compared with conventional SPR (Fig. 4.1). This results in an increased sensitivity and spectral resolution (decreased resonance line widths) as shown in Fig. 4.2.

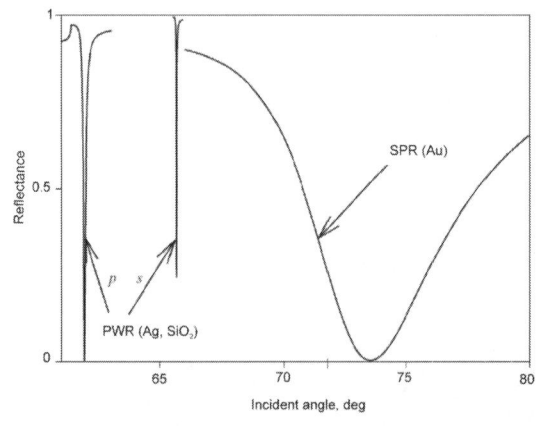

Figure 4.2 Calculated resonance curves for SPR and PWR resonators. Adapted from ref. 10.

The p- and s- resonances have different sensitivities, resulting from different evanescent electrical field intensity distributions (Fig. 4.1). In the simplest PWR sensor comprised of silver covered with silica, the s-polarization is significantly more sensitive (5-fold to 10-fold) than that of p-polarization. Both of these resonances are much more sensitive than conventional SPR that are usually based on a gold metal film that is approximately four- to five-fold less sensitive than a silver-based one. The simplest PWR sensor described previously further increases the sensitivity by a factor of 4 (for p-polarization) or 10 (for s-polarization), resulting in an overall 20- to 50-fold increase in sensitivity compared with a conventional gold-based SPR sensor. Plasmon-waveguide resonance sensitivity could be further amplified by modifying the simplest two-layered (silver and silica) sensor with more complex arrangements of dielectric layers. Such systems have been developed by Tollin's laboratory in order to distinguish between different microdomains in lipid bilayer membranes [16]. These systems are characterized by at least two orders of magnitude higher overall sensitivity that that obtained with conventional gold-based sensors. A third important characteristic of PWR is that the dielectric overcoating also serves as a mechanical and chemical shield for the thin metal layer. This allows reactive metals such as silver to be used in an aqueous environment, crucial for biosensor applications. The dielectric overcoat can also be used as a matrix where different chemistries can be performed to allow the immobilization of materials on the resonator surface. Figure 4.3 shows a general diagram of the PWR spectrometer. The sensor prism is pressed against a Teflon block that constitutes the PWR cell sample (volume of 500 μL).

The sensor prism surface is partially coated with thin layers of silver and SiO_2 on which the sample to be characterized is deposited. This entire unit is mounted on a rotating table that allows the incident angle of the laser beam to be continuously varied with 1 mdeg resolution so as to obtain a computer readout of the angular dependence of the reflected light intensity that constitutes a PWR spectra. To influence the PWR resonance spectra obtained with a sensor prism of a sample in contact with an aqueous solution, it is desirable for it to be immobilized within a distance from the surface corresponding to less than one wavelength of the monitoring light as the evanescent wave decreases exponentially when moving away from the surface. A variety of methods can be used for this purpose, including covalent attachment to the silica or physicochemical adsorption onto the surface. Here the focus will be on lipid membranes and molecules therein inserted or prone to interact with.

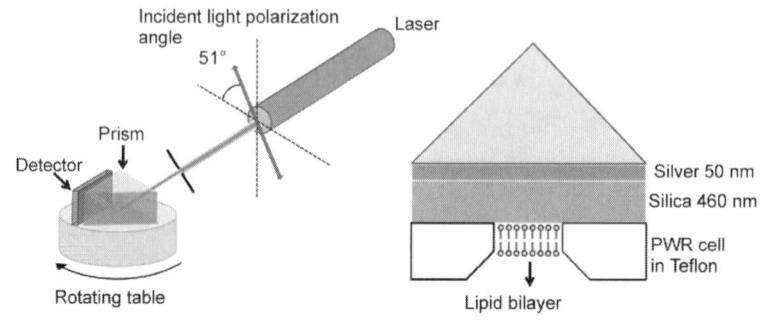

Figure 4.3 PWR setup. On the left the optical and mechanical components, the incident polarized light beam (a continuous wave laser He-Ne laser at 632.8 nm) is at 51° and the rotating table allows steps of 1 millidegree. On the right is a detailed view of the prism and the PWR cell sample with the lipid bilayer.

4.2.3 PWR Spectral Analysis

The optical properties of the system can be described by three optical parameters: refractive index (n), extinction coefficient (k) and thickness (t). Both n and k are tensors and they have different values along the measurement axes, t is a scalar quantity representing an average molecular length perpendicular to the plane of the film and is independent of polarization. The refractive index is a macroscopic quantity related to the properties of individual molecules through the molecular polarizability tensor, as well as to the environment in which these molecules are located (e.g., packing density and internal organization). Similarly, the extinction coefficient is related to the molecular optical transition tensor. Since common excitation wavelengths are far from the absorption bands of lipids and proteins, nonzero values of k are due only to scattering resulting from coating or imperfections in the deposited materials.

The distinction between thickness and the two other optical parameters is especially important when the molecules to be investigated are located in a 2D-matrix (such as a biomembrane, a lipid bilayer membrane or a thin film) that has a nonrandom organization and possesses long-range spatial molecular order. This molecular ordering creates optically anisotropic systems, having uniaxial optical axis resulting in two different principal refractive indices: n_e (also denoted as $n_{\parallel}$ or n_p) and n_0 (also referred to as $n_{\perp}$ or n_s), and two different extinction coefficients: k_p, and k_s [17]. The first of these indices is associated with a linearly polarized light wave in which the electric vector is polarized parallel to the optical axis (TM, transverse magnetic). The second one is observed with light in which the electric vector is perpendicular to the optical axis (TE, transverse electric). This is the fundamental basis on which measurement of the optical properties of anisotropic systems can lead to the evaluation of their structural parameters. In the simplified case in which a molecular shape can be approximated by a rodlike structure and the molecules are ordered such that their long axes are parallel (e.g., phospholipid molecules in a lipid bilayer membrane), one has an optically anisotropic system whose optical axis is perpendicular to the plane of the lipid bilayer [15, 17].

The optical anisotropy (A_n), can be characterized by values of the refractive index measured with two polarizations (i.e., parallel, n_p, and perpendicular n_s, to the optical axis) by the following equation:

$$A_n = (n_p^2 - n_s^2)/(n_{av}^2 + 2) \tag{4.3}$$

In this equation n_{av} is the average value of the refractive index and, for a uniaxial system is given by

$$n_{av}^2 = 1/3(n_p^2 + 2n_s^2) \tag{4.4}$$

The anisotropy in the refractive index reflects both the anisotropy in the molecular polarizability and the degree of long-range order of molecules in the system, and therefore can be used as a tool to analyze structural organization (molecular orientation).

From the Lorentz–Lorenz relation, for a pure substance, the mass density d, can be related to n [18, 19] by

$$d = ML = M/A((n_{av}^2 - 1)/(n_{av}^2 + 2)), \tag{4.5}$$

where M represents the molecular weight; L the number of moles per volume; and A the molar refractivity of the material. From the thickness and average refractive index value, one can calculate the surface mass density, i.e., mass per unit surface area [10]:

$$M = dt = 0.1M/A\{t\{(n_{av}^2 - 1)/(n_{av}^2 + 2)\}\}, \quad (4.6)$$

where the thickness is in nanometers and mass in µg/cm². Such a simple mass calculation becomes more complicated when the layer is formed from a mixture of substances, as is often the case in real measurements. This can still be dealt with depending upon the specific experimental conditions.

By the use of chromophore labeling of the molecules or part of the molecules under study and the use of appropriated incident light wavelength, the extinction coefficient anisotropy (A_k) can be calculated by

$$A_k = (k_p - k_s)/k_{av}. \quad (4.7)$$

In this equation, k_{av} is the average value of the extinction coefficient and, for a uniaxial system is given by

$$k_{av} = (k_p + 2k_s)/3 \quad (4.8)$$

k_{av} is also related to the surface concentration of chromophore:

$$C = (4\pi/\lambda)(k_{av}/\beta), \quad (4.9)$$

where C is the molar concentration of the chromophore and β the molar absorptivity.

From the refractive indices and extinction coefficients measured with two polarizations (n_p, n_s, k_p, and k_s), and the thickness of the membrane (t), one can calculate the following parameters describing the physical characteristics of the membrane: (1) the surface mass density (or molecular packing density), i.e., mass per unit surface area (or number of moles per unit surface area) [15, 17], which reflects the surface area occupied by a single molecule, (2) the optical anisotropy (A_n), which reflects the spatial mass distribution created by both the anisotropy in the molecular polarizability and the degree of long-range order of molecules within the system [15], (3) the surface chromophore density, and (4) their spatial distribution.

The experimental PWR spectra can be described by three parameters: spectral position, spectral width, and resonance depth.

In the case of lipid bilayers, these features depend on the surface mass and/or chromophore density, the spatial mass and/or chromophore distribution, and the membrane thickness. Thin-film electromagnetic theory based on Maxwell's equations provides an analytical relationship between the experimental spectral parameters and the optical properties [7, 13, 20]. This allows evaluation of the n, k, and t parameters from which the membrane physical properties can be assessed. The fact that there are three experimentally measured spectral parameters and three optical parameters allows, in principle, a unique determination of the n, k, and t values by fitting a theoretical resonance curve to the experimental one. Nonlinear least-square fitting of the experimental to theoretical spectra can be performed to obtain the optical parameters. In certain cases, other type of analysis is required; this arises from three principal reasons: (1) molecular interactions leading to a complex spectrum where it is no longer possible to obtain a unique determination of the optical parameters (e.g., they occur in a heterogeneous film), resulting in a final spectrum that represents a mixed population, (2) rapid conclusions about the interactions is required, which precludes a tedious and often difficult analysis, and (3) the resonance spectra are not good enough to be fitted by the theoretical curves (poorly resolved spectra usually are good enough to do some comparative experiments, but they cannot be used to quantify them by fitting procedures). In these cases, there are two other approaches available: (1) spectral simulation [16, 21], or (2) graphical analysis [22]. The spectral simulation procedure is based on the same principle as the fitting approach described previously, but is quicker and easier to apply.

Application of this method of data analysis allows important structural parameters of a lipid membrane to be obtained such as thickness, average surface area occupied by one lipid molecule (or molecular packing density), and degree of long-range molecular order. The graphical analysis procedure is based on the fact that PWR spectra are determined by two physical properties of a thin film such as a lipid membrane: (1) an average surface mass density, and (2) the spatial distribution of mass within the system that results from the structure of the deposited film. The separation of mass changes from those caused by structure is achieved by transforming the measured spectral changes (e.g., changes in the position of the spectra obtained either with p- or s-polarized exciting light) from an (s-p) orthogonal coordinate system into one reflecting (mass-structure). To perform such a transformation one must be able to place mass and structural axes within the orthogonal (s-p) coordinate system. This can be done if one knows the following two properties of the measurement system: (1) the mass sensitivity of the p- and s-axes in the (s-p) coordinate system (i.e., the sensor must be calibrated either theoretically or experimentally); (2) the optical symmetry of the measured system (i.e., whether the system is optically isotropic or anisotropic). For an anisotropic system, one must assume the direction of the optical axis (i.e., whether the optical axis is parallel to the p- or to the s-polarization direction). The axes of a new (mass/structure) coordinate system can then be scaled with the original (s/p) coordinates. Each point on the mass axis (Δ_m) can be expressed by changes of the original coordinates (Δ_s) and (Δ_p) as

$$(\Delta_m) = [(\Delta_s)_m^2 + (\Delta_p)_m^2]^{1/2} \qquad (4.10)$$

and on the structural axis:

$$(\Delta_{str}) = [(\Delta_s)_{str}^2 + (\Delta_p)_{str}^2]^{1/2} \qquad (4.11)$$

In this way, the contribution of structural changes and mass alterations are expressed in terms of angular shifts. Such spectral analysis has been performed in studies involving peptide interaction with membranes and ligand-activation of GPCRs [22–26].

4.3 PWR Applications

4.3.1 Lipid Bilayers

In terms of the lipid bilayers that have been used in SP experiments, one can find tethered lipid bilayers, polymer-supported bilayers, microarrayed layers and solid-supported bilayers among others. Herein, we will focus in solid-supported bilayers composed of commercially available lipids that have routinely been used in PWR studies and in a model more closely mimicking the cell membrane composed of cellular membrane fragments.

4.3.1.1 Solid-supported lipid bilayers

The principles behind the generation of a self-assembled solid-supported lipid membrane are the same that govern the formation of the so-called black lipid membrane prepared by the Mueller–Rudin procedure [27]. Briefly, a lipid bilayer forming solution is spread across a Teflon orifice that separates the silica waveguide surface from the aqueous phase in the PWR cell [10]. The hydrophilic SiO_2 surface is covered with a thin layer of water of condensation [28, 29] and attracts the polar groups of the lipid molecules with the hydrocarbon chains oriented towards the bulk lipid phase, which induces an initial orientation of the lipid molecules. The next step involves addition of aqueous buffer into the sample compartment of the PWR cell, which leads to the formation of the second monolayer, the thinning of the membrane and the formation of a plateau-Gibbs border that anchors the membrane to the Teflon spacer. Such an annulus of lipid solution not only anchors the bilayer, but also acts as a reservoir of lipid molecules. The formation of the lipid bilayer results in increased resonance angles for both polarizations, which indicate a mass increase in the vicinity of the sensor surface. A very important observation is that the spectral shifts are considerably larger for p- than for s-pol (as an example: 145 mdeg for p- and 95 mdeg for s-pol from ref. [30]), indicating that the lipid molecules are oriented with their longer axis along the p-pol and the shorter axis along the s-pol. The spectral shifts are very much dependent on the lipid composition of the membrane and vary even among different bilayers made from the same lipid composition reflecting the dynamic nature of the membrane [21]. The optical parameters of such membranes (n_p, n_s, k_p, k_s and t) have been reported [9, 21] and are in quite good agreement with values determined by other techniques. The process occurs in about 30 min and leads to the formation of a very stable, highly flexible bilayer. A great variety of lipid composition can be used to prepare such bilayer either laterally homogeneous or inhomogeneous (presence of microdomains), which allows one to probe the role of the lipid composition in membrane peptide/protein interaction and activity. Such lipid bilayers have been used by Tollin and collaborators for studies involving GPCR ligand activation and signaling and membrane active peptides and will be discussed below and the spectra for the formation of a lipid membrane can be visualized below in Fig. 4.4.

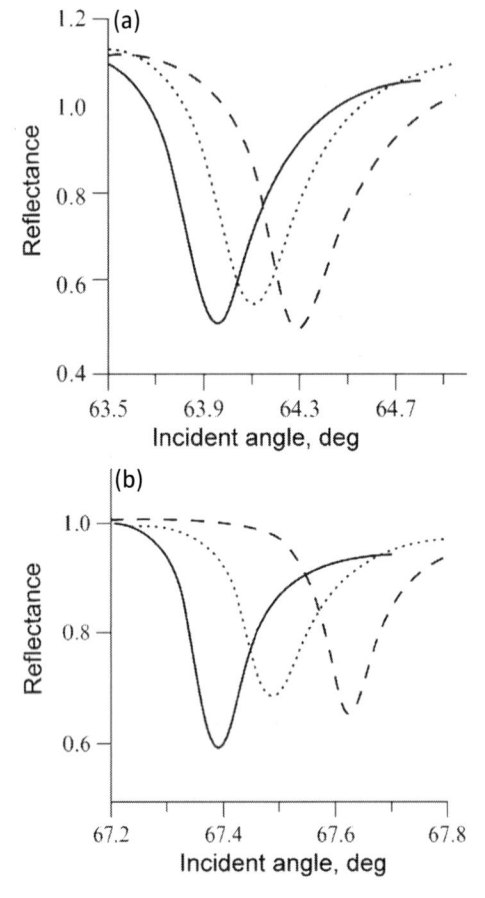

Figure 4.4 PWR spectra obtained for the buffer (10 mM Tris-HCl at pH 7.3) (solid line), an egg PC lipid bilayer (dotted line) and after addition of hDOR to the membrane (final cell concentration ~0.4 nM) with p- (a) and s-polarized (b) light. Adapted from ref. 30.

4.3.1.2 Membranes composed of cellular membrane fragments

A procedure has been reported for the formation of lipid membranes composed of cell membrane fragments derived from bacterial cells (chromathophores from *Rhidipseudomonas sphaeroides*) and mammalian cells (μ-opioid receptor and MC4 receptor transfected human embryonic kidney (HEK) cells and rat trigeminal ganglion cells) on the silica of a PWR spectrometer [31]. The use of membrane fragments (or even whole cells) deposited directly on the outer surface of the PWR sensor is based on previously reported studies showing that both lipid vesicles and whole cells adhere well and spread easily at the silica surfaces. The interaction of vesicles with surfaces (usually hydrophilic) was pioneered by McConnell's group [32, 33]. Several groups have been working since then in the understanding of the driving forces involved in the adsorption of the lipids onto the surfaces both from the theoretical and experimental point of view by the use of optical imaging techniques and optical biosensors. The general idea emerging from these studies is that cells initially interact with a solid surface by adsorption, followed by morphological changes (spreading) in which the cells increase the area in contact with the surface. A similar process occurs with the vesicles, following the spreading, vesicles undergo rupture, and the solid-supported bilayer is formed. The presence of calcium ions in the medium has shown to be important [34]. In the study by Salamon and collaborators, the binding of cytochrome C2 to membranes containing the photosynthetic reaction center and binding and activation of the μ-opioid and the melanocortin receptors by the respective agonist ligands was followed. Dissociation constants obtained are in very good agreement with those determined by these proteins purified and reconstituted in lipid model membranes as well as values obtained by radiolabeling assays, demonstrating the validity of the assay [31].

A solid supported lipid bilayer containing cell membrane fragments has also been developed by Alves and collaborators to investigate the interaction of cell-penetrating peptides (CPPs) with membranes [35]. The approach to prepare this membrane was the same as for a solid-supported lipid membrane described above, except that after spreading the lipid solution across the Teflon orifice contacting the prism surface, a buffer solution containing cell membrane fragments was injected in the cell. The presence of cell membrane fragments in these types of membranes was confirmed by the use of cell membrane fragments issued from cells expressing the neurokinin 1 (NK-1) receptor. Those membranes responded to the activation by substance P, an agonist to this receptor, in a very similar fashion to that of the reconstituted NK-1 receptor previously reported by PWR studies [36]. Cell membrane fragments issued from wild type CHO cells and CHO cells lacking glycosaminoglycans (GAGs) and sialic acid (SA) were used to investigate the role of these molecules in the binding affinity of penetratin, one of the first discovered CPPs.

4.3.2 GPCR Insertion into Membranes, Activation and Signaling

G-protein coupled receptors are the largest class of integral membrane proteins and the largest class of proteins in the human genome with more than 1000 different proteins now known. They are essential for most aspects of cellular communication in complex biological multicellular systems such as human beings and are the targets of most hormones and neurotransmitters that modulate behavior and metabolism. They are the target of approximately 50% of all current drugs. Despite their unquestionable interest, the study of their ligand-induced structural modifications following activation and signaling is challenging. The reasons rely on their low natural abundance in membranes, their amphipathic character, and difficulty in maintaining them in a functional state outside their natural environment and the lack of analytical methods that can analyze those proteins in an environment well mimicking the natural one. Plasmon-waveguide resonance together with the use of solid-supported bilayers appears as a very useful method for the reconstitution and study of these proteins. Indeed, these model systems are ideal for the study of GPCRs for the following reasons: (1) It provides a very fluid environment where lipids can move in and out of the lipid bilayer as needed due to the natural existence of the Plateau-Gibbs border; (2) the lipid membranes are easily prepared and stable and can be prepared from a variety of lipid compositions, expanding the use of this technique to lipid protein interaction studies; (3) the proteins to be incorporated into the bilayer do not need to be chemically modified as they are not attached to the plasmon surface; (4) since the bilayer is not covalently attached to the surface of the sensor it can be easily cleaned with detergent and reused many times; (5) the protein is incorporated into the lipid bilayer in a uniform manner, that is with the longer axis perpendicular and shorter axis parallel to the membrane plane. The molecules can be inserted into the lipid bilayer with either the N-terminus or the C-terminus facing the aqueous side of the membrane [37]. Whereas this last point may be interpreted by some as disadvantageous as a nonhomogeneous population of receptors is created, it can actually be advantageous as it allows one to bind either/both ligand and G-proteins (or other molecules that bind to GPCRs through the C-terminus face) to the proteolipid system, expanding the range of studies that are possible with this technique.

Following membrane formation, the insertion of the detergent-solubilized and purified GPCR into this membrane is done by detergent dilution below the critical micelle concentration (cmc). Care must be taken in the choice of the detergent used as to not perturb the lipid membrane; octylglucoside has been a good choice. Even though only a small amount (~5%) of the injected protein reaches and inserts into the membrane, this amount is sufficient for the studies due to the high sensitivity of the PWR. Following protein insertion and equilibration (PWR spectra is stable), unbound protein is removed by washing the cell sample. In Fig. 4.4 spectra are shown presenting the formation of a solid-supported lipid bilayer and the insertion of the human delta opioid receptor (hDOR) into this membrane. One can note that the insertion of this protein in the membrane leads to shifts in the resonance to higher angles for both polarizations (190 mdeg for p-pol and 130 mdeg for s-pol) and changes in the spectral depth, which are related with changes in the mass and the thickness of the bilayer.

Because the receptor protrudes from both sides of the lipid bilayer, one should expect the bilayer to become thicker upon receptor incorporation. These results are consistent with an increase in thickness upon incorporation of the hDOR into the membrane from 5.3 to 6.8 nm [10, 38] that correlates well with X-ray crystallography data for membranes containing rhodopsin [39]. Like in the case of lipid membrane formation, spectral shifts with *p*-polarization were larger than with *s*-polarization (indicating refractive index changes in the *p*-direction higher than for the *s*-direction), which is a consequence of the anisotropic structure (i.e., cylindrical shape) of the receptor molecules. This is also an evidence for the incorporation of the receptor into the bilayer with the expected orientation (i.e., the long axis oriented perpendicularly to the lipid bilayer), rather than just adsorbed to the surface of the bilayer, clearly reflecting a corresponding increase of the average long-range molecular order in the membrane resulting from receptor-lipid interactions.

Several aspects of the signal transduction pathways for the human delta opioid (hDOR) receptor have been investigated by our laboratory using PWR. The receptor has been shown to be active after incorporation into the lipid bilayer, in the PWR cell, based in its ability to recognize different classes of ligands with affinities very similar to the ones presented in the literature [30, 38] (Table 4.1). Furthermore conformational changes of the proteolipid system induced upon binding of different classes of ligands to the receptor have been monitored and showed that different classes of ligands produce distinctly different structures in the proteolipid system. Such studies are possible due to the potential of PWR to distinguish between changes in mass and changes in conformation occurring in anisotropic films, which arises from the existence of an additional layer of silica on the sensor surface that functions as a waveguide, allowing polarization to occur with both p- (perpendicular) and s- (parallel) polarized light. The different conformational states observed upon ligand binding to the receptor have been analyzed by fitting the PWR spectra to

Table 4.1 Affinity constants for binding of various ligands to the hDOR obtained from PWR and radiolabel methods reported in the literature

Ligands	K_D (nM) obtained from PWR	Ligand affinities (nM) from the literature*
DPDPE	16 ± 4	16
Deltorphin II	1 ± 0.2	0.7
SNC80	55 ± 10	56
NTI	0.024 ± 0.002	0.028
Naloxone	8 ± 2	10
TMT-L-Tic	2.9 ± 0.3	9
Morphine	520 ± 30	1101
Ethorphine	0.3 ± 0.1	0.2

*The references from which binding constants have been obtained can be found in ref. [31].

theoretical curves (obtained using Maxwell's equations) and the optical parameters extrapolated [38]. Such analysis have indicated that although both agonist binding and antagonist binding to the receptor cause increases in molecular ordering within the proteolipid membrane, only agonist binding induces an increase in thickness and molecular packing density of the membrane. This is interpreted as a consequence of mass movements perpendicular to the plane of the bilayer occurring within the lipid and receptor components. These results are consistent with models of receptor function that involve changes in the orientation of transmembrane helices and receptor elongation.

Using a similar setup, the interaction of G-proteins with the hDOR bound to different types of ligands was also investigated, followed by the interaction of GTPγS with the ligand-receptor-G-protein complex [37]. It should be pointed out that classical pharmacological methods only give indirect information about the interaction of GPCRs with G-proteins, because they are based on downstream responses. The GTPγS assay reflects a combination of both the affinity of the G-protein to the ligand-bound receptor, and the ability of the agonist-bound state to initiate GDP dissociation from the G-protein. Thus, such measurements do not directly probe interactions between the various signal transduction partners and suffer from the disadvantages of being very time consuming and dependent on the use of radiolabeled material. Another limitation of such experiments comes from the fact that cells usually express all the different G-protein subtypes making it very hard to understand the contribution of each individual subtype. Using PWR, the interaction of different G-protein subtypes (Giα1, Giα2, Gi iα3, and Goα) with hDOR, as well as the GTPγS interactions with the G-protein-agonist-bound hDOR complex, was directly monitored and dissociation constants obtained for each of the separate processes [40]. The results indicate a great degree of diversity and specificity in terms of the interactions of different ligand-activated receptor with different G-protein subtypes and there is no necessary correlation between the affinity of the G-protein to the ligand-receptor and its ability to undergo GDP/GTP exchange. There is evidence that G-proteins are differentially expressed in different tissues [41]. Furthermore, the α and $\beta\gamma$ subunits of the G-protein interact with different effectors leading to different signal transduction events in the cell [42]. The fact that different ligands from the same class can lead to preferential interaction with certain G-protein subtypes suggests that these preferences may lead to an additional source of drug side effects. Thus, certain ligands, by selecting specific G-protein subtypes over others may thereby be targeting specific signaling pathways that may lead to unwanted side effects. Alternatively, by designing ligands that activate only desired pathways, this might be avoided. Plasmon-waveguide resonance studies such as those presented here may therefore have a great impact on drug screening protocols and design, as well as providing new insights into the basis for differential physiological and/or pharmacological effects of drug activity.

4.3.3 Role of Lipids in GPCR Activation, Signaling, and Partition into Membrane Microdomains

Plasmon-waveguide resonance was applied to probe the effects of lipid composition on conformational changes of rhodopsin induced by light and due to binding and activation of transducin (Gt) [43]. In those experiments, octylglucoside-solubilized rhodopsin was incorporated by detergent dilution into solid-supported bilayers, composed of either egg phosphatidylcholine (PC) or various mixtures of a nonlamellar-forming lipid (dioleoylphosphatidylethanolamine; DOPE) together with a lamellar-forming lipid (dioleoylphosphatidylcholine; DOPC). Light-induced proteolipid conformational changes as a function of pH were measured and the pKa values for the MI-MII equilibrium were determined. Results obtained correlated well with previous flash photolysis studies, indicating that the PWR spectral shifts monitored metarhodopsin II formation [44]. The magnitude of these effects, and hence the extent of the conformational transition, was found to be proportional to the DOPE content. These results are consistent with previous suggestions that lipids having a negative spontaneous curvature favor elongation of rhodopsin during the activation process [44]. In addition, measurements of transducin (the G-protein that binds to rhodopsin following light activation) interaction with rhodopsin demonstrated that the affinity between both these proteins was influenced by the membrane lipid composition, being enhanced in the presence of DOPE in the membrane. The observation that the lipid composition of the bilayer can enhance the ability of a GPCR to become activated and to bind a G-protein has important implications for the regulation of receptor activity by sorting into membrane micro-domains [45].

We also have investigated the effects of membrane microdomains (lipid rafts) on GPCR function again using the hDOR as the example [16, 21]. In these experiments we first demonstrated that if we utilized a 1:1 mixture of dioleoylphosphatidylcholine (DOPC) and sphingomyelin (SM), we could unequivocally demonstrate using PWR spectroscopy that the lipid bilayer spontaneously segregated into two separate domains, the more fluid DOPC-rich domain (5.2 nm thickness), and the more rigid sphingomyelin-rich domain (5.9 nm thickness) [16]. The thickness and other optical parameters of the two microdomains were determined by spectral simulation involving initial calculations of single resonances for the pure components using Maxwell's equations applied to thin films, and then using sums of single curves with appropriate ratios to fit the experimental spectra (deconvolution) for the lipid mixtures. The details will not be discussed here [see ref. 16] except to say that in addition to the thickness of the separate domains already mentioned, the surface area per molecule and other optical parameters could be obtained. With these properties in hand, it then was possible to examine the segregation of the hDOR (either agonist DPDPE occupied receptor, or unliganded receptor) into the lipid domains and to examine the effects when DPDPE is added to the unoccupied receptor. The results obtained are very exciting. First, when the DPDPE-occupied hDOR was added to the bilayer system that contained a 1:1 POPC (palmitoyloleoyl PC) and SM mixture the receptor complex preferentially inserted into the SM-rich domain. On the other hand, when the unoccupied hDOR was added to the 1:1 POPC/SM bilayer, the receptor incorporated into both domains but prefers the POPC-rich domain (Fig. 4.5) [21]. Most interestingly, when an increasing concentration of DPDPE was added to the unoccupied hDOR, and the change in resonance position was monitored by PWR spectroscopy, the DPDPE affinity for the hDOR receptor in the SM-rich domain was about 30 times stronger than for the receptor in the PC-rich domain indicating a strong microenvironment effect on ligand affinity. Furthermore, we could monitor receptor trafficking that was occurring as a consequence of ligand binding. Thus, addition of the agonist DPDPE to the unoccupied receptor leads to receptor trafficking of the ligand–receptor complex to the SM-rich domain, which, as we have just noted, leads to a tighter binding of DPDPE to the receptor (Fig. 4.5). We have suggested that this trafficking may be accounted for by the occupied agonist receptor complex favoring the SM-rich domain because of hydrophobic matching between the more lipophilic sphingomyelin-rich lipid and the more lipophilic agonist occupied receptor. This is probably a consequence of the elongation that occurs in the receptor when DPDPE is bound, resulting in a better match to the longer fatty acyl chains in the SM molecules.

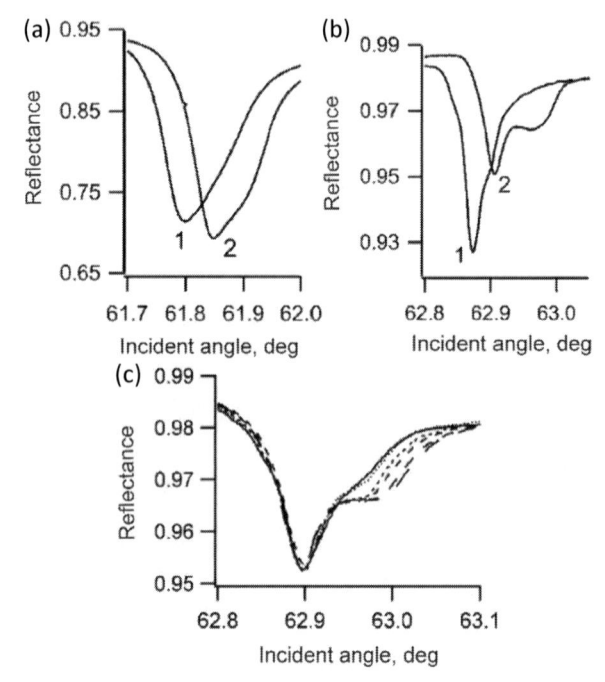

Figure 4.5 PWR spectra obtained for a solid-supported lipid bilayer containing a 1:1 mixture of POPC and SM (curve 1) and for the incorporation of the agonist (DPDPE)-bound hDOR (10 nM, final concentration in the sample cell) into the lipid bilayer (curve 2) with *p*- (a) and *s*-polarized (b) light. Panel c represents the time evolution of PWR spectra, obtained with *s*-polarized light, upon addition of an agonist (DPDPE) to the receptor when incorporated into a POPC/SM (1:1) lipid bilayer. With respect to the right shoulder, from left to right the spectra represent increasing time intervals: (s) 0, (solid line) 2, (- - -) 6, (---) 8, (- - -) 16, and (- - -) 20 min. Adapted from ref. 30.

4.3.4 Interaction of Membrane Active Peptides with Lipid Membranes

PWR has also been used to investigate the binding affinity of membrane active peptides like cell penetrating peptides, viral and amyloid peptides. The interaction of these peptides as well as their perturbation in the membrane organization was also investigated.

Studies on penetratin interaction with model membranes composed of zwitterionic and anionic lipids indicate a very distinct comportment with these two types of lipids. The affinity of penetratin for anionic lipids was found to be more than 10-fold increased relative to zwitterionic lipids [23, 35, 46]. Thus the electrostatic interaction between the positively charged residues in penetratin (+7) and the negatively charged lipid headgroups seem to be important in the peptide/lipid recognition. The response in terms of binding and reorganization in the lipid membrane observed was quite distinct for these two types of lipids, following a two-state binding event (with decreases in the angle resonance position at low concentration of peptide and increase at higher concentrations) in the case of zwitterionic lipids and one binding event in the case of anionic lipids. The responses were interpreted as follows: In the presence of anionic lipids, penetratin interacts with the lipid charges in the headgroups and leads to a decrease in the electrostatic repulsion between the lipid molecules and increase in the molecular ordering of the membrane. In the absence of anionic lipids and at low peptide concentration there is an initial contact with the lipids and a repulsion between the positive charges in the lipid and peptide that leads to a decrease in the packing density of the membrane (as noticed by the decrease in the resonance angle position). This leads to a deeper peptide insertion to reach the negative phosphate charges and to an overall increase in the membrane thickness [47]. Electrostatics does not seem to be the only important factor as the peptide also interacts strongly with membranes composed of phosphatidylethanolamine (PE). This lipid induces a negative curvature in the membrane and promotes the formation of hexagonal II phases in lipids (inverted micelles). The formation of inverted micelles has been proposed as one of the mechanisms to explain the passage of penetratin across the membrane [48, 49]. Finally in an attempt to shed light on the role of cell membrane carbohydrates in penetratin interaction with the membrane, PWR studies employing membranes containing cell membrane fragments derived from wild type cells (CHO WT) and CHO cells lacking glycosaminoglycans (GAG-deficient) as well as CHO cells lacking sialic acid (SA-deficient) were performed. The results indicate that the presence of GAG and SA in the membrane greatly enhances the peptide affinity to the membrane while did not affect the kinetics of the interaction [35].

Using PWR, the interaction of two peptides implicated in the early replication phase of HIV-1 infection, p6 and Vpr, with cell membranes was investigated. The molecular mechanism by which those proteins are transported to the plasma membrane after ribosomal synthesis is unknown. The studies indicate that p6 strongly interacts with membranes (with a K_D of ~40 nM), which may help explaining in part why Gag is targeted and assembles into membranes by coating itself with lipids. The presence of p6 in the membrane increases the affinity of Vpr to the membrane.

The interaction of amyloid peptides with lipid bilayers has also been investigated by PWR. Two amyloid peptides derived from the prion forming domain HET-s, named the nontoxic wild-type (WT) and a toxic mutant (M8) were investigated for their interaction with zwitterionic and anionic membranes [50]. The addition of WT or M8 did not cause significant changes in the resonance spectra of zwitterionic lipid bilayers, indicating no interaction with those membranes. The outcome was quite different in the case of anionic membranes, notably in the case of asolectin that contains about 30% anionic lipids in their composition. The addition of either WT or M8 peptides to this membrane lead to large increase in the PWR resonance angle indicating interaction with this membrane. Notably, M8 also lead to large changes in the spectra shape with a strong decrease in the spectral depth, an increase in the width and a split in the resonances with appearance of an additional resonance at higher resonance angles. Such spectral signature has been observed for lipid bilayers that are known to form lipid domains such as the case of lipid mixtures composed of a disordered and an ordered lipid such as an unsaturated PC (DOPC) and sphingomyelin (SM) and have been attributed to domains rich in PC and SM [16, 21].

4.4 PWR Ongoing Developments

Plasmon-waveguide resonance developments using the same general setup indicated above, namely the same incident light wavelength and detection system are under way in our laboratory to improve the signal/noise and to be able to obtain both polarization resonances within the same angular scan measurement. This latter is accomplished by placing the incident light polarized beam at 45°. To be able to monitor simultaneous (within the same angular scan) resonances with both polarizations is especially advantageous in the study of anisotropic systems such as lipid bilayers that are investigated in our laboratory.

One other development that is being set up in the laboratory concerns the coupling of PWR to impedance measurements. For that, on the outer prism surface (on top of the silver and silica surfaces), a thin layer of indium tin oxide (ITO), that functions as the working electrode, is placed. A reference and a counter electrode are placed on the buffer solution in the PWR cell sample (on the other side of the lipid membrane that is formed across the Teflon orifice) in a way to measure both cyclic voltammetry and impedance. Such studies are important to test and directly monitor the membrane integrity perturbation induced by membrane active peptides mentioned above that have been reported to act by pore formation (permanent or transient) in the membrane.

Yet another ongoing project in the laboratory consists in the practice of PWR in the IR region, the idea is to use a wavelength in the region of lipid absorbance so that additional information about the orientation of the lipid molecules and changes in the environment around the lipid IR-active groups such as the phosphates (water accessibility and others) could be monitored. For this, wavelength changes in the prism material, optical components, and detector are needed to use materials that are transparent in this spectral region.

References

1. Kretschmann, E., and Raether, H. Z. (1968). Radiative decay of non-radiative surface plasmons excited by light. *Naturforsch*, **23A**, 2135–2136.

2. Otto, A. (1968). Excitation of non-radiative surface plasma waves in silver by the method of frustrated total reflection. *Z. Physik.*, **216**, 398–410.
3. Gordon, J. G., and Ernst, S. (1986). Surface plasmons as a probe of the electrochemical interface. *Surface Sci.*, **8**, 456–459.
4. Liedeberg, B., Nylander, C., and Lundström, I. (1995). Biosensing with surface plasmon resonance: How it all started. *Biosens. Bioelectron.*, **10**, 1–9.
5. Raether, H. (1988). *Surface Plasmons on Smooth and Rough Surfaces and on Gratings* (Springer-Verlag, Berlin).
6. Boardman, A. D. (1982). *Electromagnetic Surface Modes.* John Wiley and Sons.
7. Salamon, Z., and Tollin, G. (1999). Surface plasmon resonance; Theory. In *Encyclopedia of Spectroscopy and Spectrometry.* (Lindon, J. C., Tranter, G. E., and Holmes, J. L., eds.), **3**, pp. 2311–2319. (Academic Press, New York).
8. Salamon, Z., and Tollin, G. (2001). Coupled plasmon-waveguide resonance spectroscopic device and method for measuring film properties in the ultraviolet and infrared spectral ranges. *US Patent # 6,330,387 B1.*
9. Salamon, Z., and Tollin, G. (2002). Coupled plasmon-waveguide resonance spectroscopic device and method for measuring film properties in the ultraviolet and infrared spectral ranges. *US Patent # 6,421,128 B1.*
10. Salamon, Z., Macleod, H. A., and Tollin, G. (1997). Coupled plasmon-waveguide resonators: A new spectroscopic tool for probing proteolipid film structure and properties. *Biophys. J.*, **73**, 2791–2797.
11. Salamon, Z., Macleod, H. A., and Tollin, G. (1997). Surface plasmon resonance spectroscopy as a tool for investigating the biochemical and biophysical properties of membrane protein systems. I. Theoretical principles. *Biochim. Biophys. Acta*, **1331**, 117–129.
12. Salamon, Z., Macleod, H. A., and Tollin, G. (1999). Coupled plasmon-waveguide resonance spectroscopic device and method for measuring film properties. *US Patent # 5,991,488.*
13. Salamon, Z., and Tollin, G. (1999). Surface plasmon resonance; Applications. In *Encyclopedia of Spectroscopy and Spectrometry.* (Lindon, J. C., Tranter, G. E., and Holmes, J. L. eds.), **3**, pp. 2294–3301. (Academic Press, New York).

14. Salamon, Z., Brown, M. F., and Tollin, G. (1999). Surface plasmon resonance spectroscopy; probing interactions within membranes. *Trends Biochem. Sci.*, **24**, 213–219.
15. Salamon, Z., and Tollin, G. (2001). Optical anisotropy in lipid bilayer membranes: coupled plasmon-waveguide resonance measurements of molecular orientation, polarizability and shape. I. Theoretical principles. *Biophys. J.*, **80**, 1557–1567.
16. Salamon, Z., Devanathna, S., Alves, I. D., and Tollin, G. (2005). Plasmon-waveguide resonance studies of lateral segregation of lipids and proteins into microdomains (rafts) in solid-supported bilayers. *J. Biol. Chem.*, **280**, 11175–11184.
17. Salamon, Z., and Tollin, G. (2001). Plasmon resonance spectroscopy: Probing molecular interactions at surfaces and interfaces. *Spectroscopy*, **15**, 161–175.
18. Born, M., and Wolf, E. (1965). *Principles of Optics*. (Pergamon Press, New York).
19. Cuypers, P. A., Corsel, J. W., Janssen, M. P., Kop, J. M., Hermens, W. T., and Hemker, H. C. (1983). The adsorption of prothrombin to phosphatidylserine multilayers quantitated by ellipsometry. *J. Biol. Chem.*, **258**, 2426–2431.
20. Macleod, H. A. (1986). *Thin Film Optical Filters*. Adam Hilger, Bristol.
21. Alves, I. D., Salamon, Z., Hruby, V. J., Tollin, G. (2005). Ligand modulation of lateral segregation of a G-protein coupled receptor into lipid microdomains in shingomyelin/phosphatidylcholine solid supported bilayers. *Biochemistry*, **44**, 9168–9178.
22. Salamon, Z., and Tollin, G. (2004). Graphical analysis of mass and anisotropy changes observed by plasmon-waveguide resonance spectroscopy can provide useful insights into membrane protein function. *Biophys. J.*, **86**, 2508–2516.
23. Alves, I. D., Correia, I., Jiao, C-Y., Sachon, E., Sagan, S., Lavielle, S., Tollin, G., and Chassaing, G. (2008). The interaction of cell-penetrating peptides with lipid model systems and subsequent lipid reorganization: thermodynamic and structural characterization. *J. Pept. Sci.*, **15**, 200–209.
24. Salgado, G. F., Vogel, A., Marquant, R., Feller, S. E., Bouaziz, S., and Alves, I. D. (2009). The role of Membranes in the organization of HIV-1 Gag p6 and Vpr: p6 shows high affinity for membrane bilayers which substantially increases the interaction between p6 and Vpr. *J. Med. Chem.*, **52**, 7157–7162.

25. Maniti, O., Alves, I., Trugnan, G., Ayala-Sanmartin, J. (2010). Distinct behaviour of the homeodomain derived cell penetrating peptide penetratin in interaction with different phospholipids. *PloS One*, 5(12):e15819.
26. Devanathan, S., Yao, Z., Salamon, Z., Kobilka, B., and Tollin, G. (2004). Plasmon-waveguide resonance studies of ligand binding to the human beta 2-adrenergic receptor. *Biochemistry*, **43**, 3280–3288.
27. Mueller, P., Rudin, D. O., Tien, H. T., and Wescott, W. C. (1962). Reconstitution of cell membrane structure in vitro and its transformation into an excitable system. *Nature*, **194**, 979–980.
28. Silberzan, P., Leger, L., Auserre, D., and Benattar, J. J. (1991). Silanation of silica surfaces. A new method of constructing pure or mixed monolayers. *Langmuir*, **7**, 1647–1651.
29. Gee, M. L., Healy, T. W., and White, L. R. (1990). Hydrophobicity effects in the condensation of water films on quartz. *J. Colloid Interface Sci.*, **83**, 6258–6262.
30. Alves, I. D., Cowell, S., Salamon, Z., Devanathan, S., Tollin, G., Hruby, V. J. (2004). Different Structural states of the proteolipid membrane are produced by ligand binding to the human delta-opioid receptor as shown by Plasmon Waveguide Resonance Spectroscopy. *Mol. Pharmacol.*, **65**, 1248–1257.
31. Salamon, Z., Fitch, J., Cai, M., Tumati, S., Navratilova, E., and Tollin, G. (2009). Plasmon-waveguide resonance studies of ligand binding to integral proteins in membrane fragments derived from bacterial and mammalian cells. *Anal. Biochem.*, **387**, 95–101.
32. Watts, T. A., Brian, A. A., Kappler, J. W., Marrack, P., and McConnell, H. M. (1984). Antigen presentation by supported planar membranes containing affinity-purified I. *Proc. Natl. Acad. Sci. U S A*, **81**, 7564–7568.
33. McConnell, H. M., Watts, T. H., Weis, R. M., and Brian, A. A. (1986). Supported planar membranes in studies of cell–cell recognition in the immune system. *Biochim. Biophys. Acta*, **864**, 95–106.
34. Richter, R., Mukhopadhyay, A., Brisson, A. (2003). Pathways of lipid vesicle deposition on solid surfaces: a combined QCM-D and AFM study, *Biophys. J.*, **85**, 3035–3047.
35. Alves, I. D., Bechara, C., Walrant, A., Zaltsman, Y., Jiao, C-Y., and Sagan, S. (2011). Relationships between membrane binding, affinity and cell internalization efficacy of a cell-penetrating peptide: penetratin as a case study. *PLoS One.*, **6**, e24096.

36. Alves, I. D., Delaroche, D., Mouillac, B., Salamon, Z., Tollin, G., Hruby, V. J., Lavielle S., and Sagan, S. (2006). The two NK-1 ligand binding sites correspond to distinct, independent and non-interconvertible receptor conformational states as confirmed by plasmon-waveguide resonance spectroscopy. *Biochemistry*, **45**, 5309–5318.
37. Alves, I. D., Salamon, Z., Varga, E., Yamamura, H. I., Tollin, G., and Hruby, V. J. (2003). Direct observation of G-protein binding to the human delta-opioid receptor using plasmon-waveguide resonance. *J. Biol. Chem.*, **278**, 48890–48897.
38. Salamon, Z., Cowell, S., Varga, E., Yamammura, H. I., Hruby, V. J., and Tollin, G (2000). Plasmon resonance studies of agonist/antagonist binding to the human δ-opioid receptor new structural insights into receptor-ligand interactions. *Biophys. J.*, **79**, 2463–2474.
39. Palczewski, K., Kumasaka, T., Hori, T., Behnke, C. A., Motoshima, H., Fox, B. A., Le Trong, I., Teller, D. C., Okada, T., Stenkamp, R. E., et al. (2000). Crystal structure of rhodopsin: A G-protein-coupled receptor. *Science*, **289**, 739–745.
40. Alves, I. D., Ciano, K., Boguslavski, V., Varga, E., Salamon, Z., Yamamura, H. I., Hruby, V. J., and Tollin, G. (2004). Selectivity, cooperativity and reciprocity in the interactions between the delta opioid receptor, its ligands and G-proteins. *J. Biol. Chem.*, **279**, 44673–44682.
41. Sondek, J., Bohm, A., Lambright, D. G., Hamm, H. E., and Sigler, P. B. (1996). Crystal structure of a G-protein beta gamma dimer at 2.1A resolution. *Nature*, **379**, 297–299.
42. Phillips, W. J., Wong, S. C., and Cerione, R. A. (1992). Rhodopsin/transducin interactions. II. Influence of the transducin-beta gamma subunit complex on the coupling of the transducin-alpha subunit to rhodopsin. *J. Biol. Chem.*, **267**, 17040–17046.
43. Alves, I. D., Salgado G. F. J., Salamon, Z., Brown, M. F., Tollin, G., and Hruby, V. J. (2005). Phosphatidylethanolamine enhances rhodopsin photoactivation and transducin binding in a solid supported lipid bilayer as determined using plasmon-waveguide resonance spectroscopy. *Biophys. J.*, **88**, 198–210.
44. Botelho, A. V., Gibson, N. J., Thurmond, R. L., Wang, Y., and Brown, M. F. (2002). Conformational energetics of rhodopsin modulated by nonlamellar-forming lipids. *Biochemistry*, **41**, 6354–6368.
45. Simons, K., and Toomre, D. (2000). Lipid rafts and signal transduction. *Nat. Rev. Mol. Cell Biol.*, **1**, 31–40.
46. Bechara, C., Pallerla, M., Zaltsman, Y., Burlina, F., Alves, I. D., Lequin, O., and Sagan, S. (2012). Tryptophan within basic peptide sequences triggers glycosaminoglycan-dependent endocytosis. *FASEB J.*, **27**, 738–749.

47. Salamon, Z., Lindblom, G., and Tollin, G. (2003). Plasmon-waveguide resonance and impedance spectroscopy studies of the interaction between penetratin and supported lipid bilayer membranes. *Biophys. J.*, **84**, 1796–1807.

48. Derossi, D., Joliot, A. H., Chassaing, G., and Prochiantz, A. (1996). The third helix of the Antennapedia homeodomain translocates through biological membranes, *J. Biol. Chem.*, **271**, 18188–18193.

49. Kawamoto, S., Takasu, M., Miyakawa, T., Morikawa, R., Oda, T., Futaki, S., and Nagao, H. (2011). Inverted micelle formation of cell-penetrating peptide studied by coarse-grained simulation: importance of attractive forces between cell-penetrating peptides and lipid headgroups. *J. Chem. Phys.*, **134**, 095103.

50. Ta, H. P., Berthelot, K., Coulary-Salin, B., Castano, S., Desbat, B., Bonnafous, P., Lambert, O., Alves, I., Cullin, C., and Lecomte, S. (2012). A yeast toxic mutant of HET-s amyloid disrupts membrane integrity. *Biochim. Biophys. Acta*, **1818**, 2325–2334.

Chapter 5

Surface-Wave Enhanced Biosensing

Wolfgang Knoll,[a,b] Amal Kasry,[a] Chun-Jen Huang,[c] Yi Wang,[b] and Jakub Dostalek[a,b]

[a]*AIT Austrian Institute of Technology, Muthgasse 11, Vienna 1190, Austria*
[b]*Center for Biomimetic Sensor Science, School of Materials Science and Engineering, Nanyang Technological University, 50 Nanyang Drive, Singapore 637553*
[c]*Graduate Institute of BioMedical Engineering, National Central University, Jhongda Rd. 300, Jhongli City, Taoyuan County 320, Taiwan 32001*

jakub.dostalek@ait.ac.at

5.1 Introduction

In 1983, a seminal paper by Bo Liedberg et al. introduced the first concept for the use of a particular type of surface waves, surface plasmon polaritons or surface plasmons for short, for biosensing [1]. Shortly thereafter, the first commercial implementation of surface plasmon resonance (SPR) spectroscopy for the detection of bioaffinity reactions appeared on the market [2]. Now, 30 years later, we look back to a remarkable success of this concept with numerous instruments being commercially available and a variety of modifications and extensions of the original principle being developed and described in the literature [3].

Our own contribution to the further extension of SPR in biosensing concerns mostly four different directions: (i) the introduction of the concept of surface plasmon modes as surface light that is bound to an interface between a (noble) metal and a dielectric medium but otherwise interacts with matter in a way totally equivalent to plane waves [4]. This allowed, e.g., for the introduction of the surface plasmon diffraction mode for biosensing [5–7], or for the development of surface plasmon microscopy [8–10]. (ii) The second extension was the combination of the field-enhancement associated with resonant excitation of surface plasmons and the principles of fluorescence detection [11, 12]. This allowed us to amplify fluorescence signal strength and extend the limit of detection (LOD) into the attomolar range [13], an analyte concentration regime that was and still is not accessible for a label-free detection mode. A few principles and examples of this surface plasmon fluorescence spectroscopy for the use in bioaffinity detection are summarized in Section 5.2. (iii) As introduced by D. Sarid in the early 1980s [14], if two surface plasmon modes travelling along (nearly) identical metal/dielectric interfaces interact by coupling their optical fields, e.g., through a metallic layer thinner than twice the decay length of the plasmon field inside the metal, i.e., in the range of a few 10 nm, two new Eigen modes appear: the long-range surface plasmon (LRSP) and the short-range surface plasmon (SRSP). We extended this concept for biosensor studies by coupling again the specific features of the optical field enhancements of LRSP with fluorescence detection principles demonstrating that this way a significant enhancement of the already attractive limit of detection can be achieved [15, 16]. This will be briefly summarized in Section 5.3. (iv) And finally, we extended the concept of evanescent wave sensor platforms to other guided optical modes, in particular, to the various waveguide modes that can be excited in the identical (Kretschmann) setup if a thin dielectric layer in the range of a few 100 nm to several microns is deposited onto the surface plasmon guiding metal surface [17]. In Section 5.4, we demonstrate this for the use of hydrogels as the wave-guiding layer [18]. These interfacial architectures can be designed such as to act as the guiding material, serve as the sensor matrix layer with the immobilized recognition sites, e.g., chemically attached antibodies [19], yet allow for the (nearly) unperturbed penetration of the analyte molecules by diffusion from the adjacent solution to the detection sites [20, 21].

5.2 Surface Plasmon Field-Enhanced Fluorescence Detection

One of the key principles for the design of surface sensor platforms is the correlation of the penetration length of the probing field, e.g., the decay of the shear field in the quartz crystal microbalance or the extend of the localized surface plasmon field for Au nanoparticles of different size with (the thickness of) the binding matrix of the sensor. For propagating surface plasmon modes (and in this short chapter we focus on these exclusively), the evanescent character of a surface plasmon mode manifests itself by an exponentially decaying optical field. The decay length into the metal is only a few or several 10 nm owing to the strong shielding effect of the nearly free electron gas in the metal; the decay length into the dielectric medium $L_p/2$ (where the bio-affinity reactions of interest happen) depends on (the complex) refractive index of the metal n_m, the employed laser wavelength, but also on the optical properties of the dielectric medium [4]. For the situation with a 50 nm thin Au layer in contact with water the (simulated) situation is shown in Fig. 5.1a: The evanescent field for a laser wavelength of λ = 632.8 nm reaches about $L_p/2$ = 90 nm into the aqueous phase (the decay length L_p is defined as a distance from a metal surface at which the field amplitude drops by a factor $1/e$). Correspondingly, the binding matrix used in the commercial chip CM5 from Biacore, a carboxymethylated dextran polymer brush, extends in the swollen state some 100–150 nm out into the buffer medium, thus matching the range that is probed by the optical surface plasmon field to the "slice" of the analyte solution with the bound species of interest (cf. Fig. 5.1b). This way, the number of binding events that can be probed by the optical sensor is considerably higher (by a factor of 3–5) than for a mere monomolecular arrangement of proteins bound directly to the transducer surface.

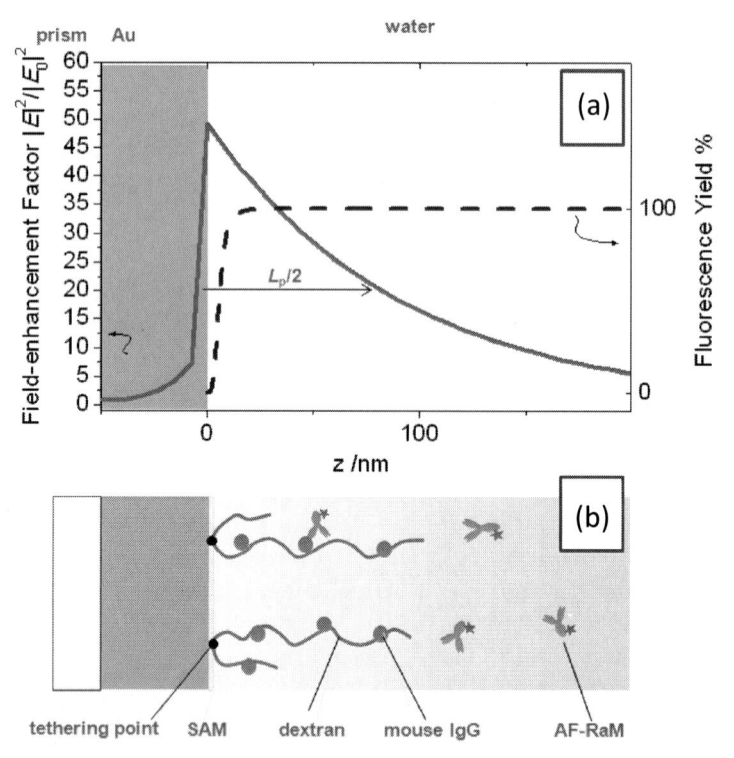

Figure 5.1 (a) Simulation of the spatial distribution normal to the interface of the optical field at resonant excitation of a surface plasmon mode between the thin Au layer as the metal and water as the dielectric medium; the peak intensity is located right at the interface, decaying quickly into the metal but also into the dielectric with a decay length of $L_p/2 = 90$ nm (for a laser wavelength of $\lambda = 632.8$ nm and Au refractive index $n = 0.125 + 3.56i$). For comparison, the dashed curve is a simulation of the fluorescence yield for a chromophore positioned at different distances away from the quenching metal layer. Only in the immediate proximity of the metal (up to a distance of about two Förster radii of 10–15 nm) a significant loss of fluorescence yield will lead to a reduced photon emission in surface plasmon fluorescence spectroscopy. (b) Schematics of a polymer brush (roughly to scale to the optical evanescent field in [a]) grafted to the sensor surface and functionalized by covalently attached antigens. To these binding sites chromophore-labeled antibodies can bind, still in the range of the resonantly enhanced optical field but outside the range of efficient fluorescence quenching.

Such a brush architecture as the binding matrix for the affinity reactions is also very beneficial for the fluorescence studies that we focus on in the following sections. As is also shown in Fig. 5.1a, the fluorescence yield of a chromophore near a metal surface that acts as a broadband acceptor system would be efficiently quenched for any bound (and labeled) analyte molecule that comes closer to the metal surface than about two Förster radii which amount to some 10–15 nm (cf. by the dashed curve in Fig. 5.1a) [22]. Any binding event that happens further away from that surface along the binding sites, e.g., chromophore-labeled antibodies binding to antigens that are covalently immobilized along the arms of the polymer brush (as sketched in Fig. 5.1b) results in a contribution to the fluorescence signal that is by no means weaker than from a free chromophore in solution. However, the evanescent character of the probing optical field with its finite decay length then limits the detected signal to the analyte molecules within the brush layer and, hence, is not overwhelmed by the abundance of analyte molecules in solution.

The experimental realization of a surface plasmon fluorescence spectrometer is schematically depicted in Fig. 5.2. Shown is an extension of a classical SPR spectrometer in the Kretschmann configuration with a coupling prism that matches the energy and

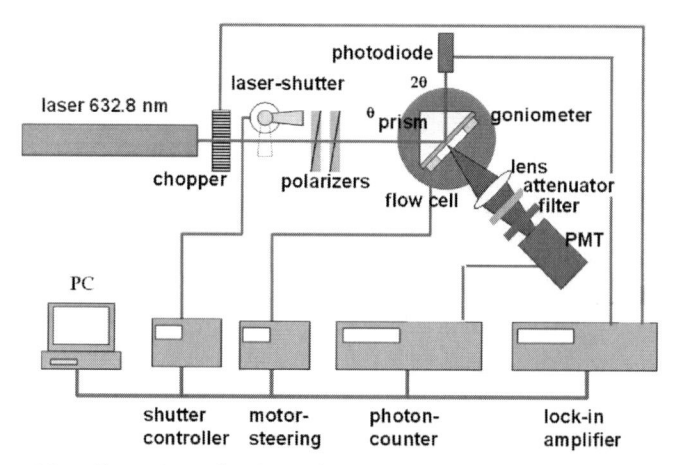

Figure 5.2 Extension of a Kretschmann surface plasmon spectrometer by a fluorescence detection unit consisting of a collection lens, an attenuator (if needed), a set of filters for the separation of scattered light, and a photomultiplier tube (PMT) or a (color) CCD camera for the microscopic mode of operation.

the momentum between the incoming laser photons and the evanescent surface plasmon modes. As one scans the angle of incidence in the normal angular $\theta/2\theta$ scan mode, an attached detection module rotates together with the prism collecting the emitted fluorescence photons through a spectral filter that differentiates the fluorescence from elastically scattered laser light. Typically, the detector is a photomultiplier or an avalanche photodiode but can be replaced in the imaging mode by a color CCD camera that allows then for multiplexed detection of fluorescence from an array of sensor elements in a parallel read-out mode [23].

An example for the sensitivity and LOD achievable with surface plasmon fluorescence spectroscopy (SPFS) is given in Fig. 5.3. The example describes the simplest case of an affinity reaction: The polymer brush was functionalized by the covalent attachment of mouse antibodies to which rabbit anti-mouse antibodies labeled with a fluorescent chromophore could bind. At the given concentrations this binding was purely diffusion controlled: After replacing the pure buffer in the flow-cell by an analyte solution of a given low concentration (in order to avoid depletion of the low analyte concentrations we always worked in a flow mode), the analyte molecules had to cross the unstirred layer from the bulk solution to the binding sites of the brush at the sensor surface by diffusion. According to Fick's law, this mass transfer limited process resulted in a linear increase of the fluorescence intensity with time with a slope which was directly proportional to the bulk concentration of the analyte solution running through the flow cell. This slope when plotted as a function of the bulk concentration of the employed analyte solution gave a calibration curve which could be recorded over almost six orders of magnitude in concentration (hence, in slope) as shown in Fig. 5.3. The intersection of this calibration curve with the 3σ baseline deviation level measured separately as the (fluorescence) stability limit of the set-up and background gives the limit-of-detection for this sensing platform: We obtain a LOD = 500 aM (5×10^{-16} M) [13].

The observed linear increase of the fluorescence intensity with time can be calibrated for the highest concentration against the simultaneously measured label-free signal from normal SPR spectroscopy in terms of how many protein molecules contribute to the observed fluorescence intensity. If extrapolated to the LOD concentration of c_0 = 500 aM one finds that the signal (i.e., the linear intensity increase with time) at this low concentration level originates from 10 antibody molecules that arrive in every minute at the brush matrix landing per every 1 mm² of sensor surface area. In other words, SPFS reaches a sensitivity regime close to the single molecule detection level.

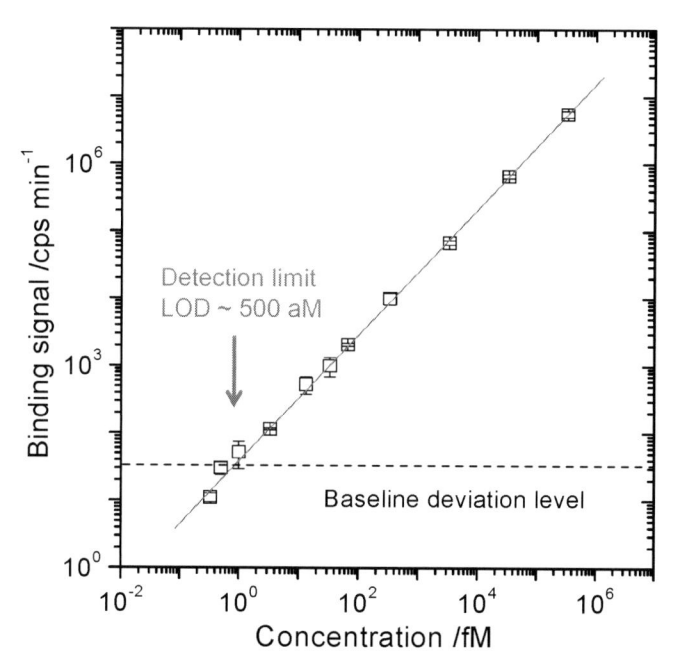

Figure 5.3 Extremely sensitive bio-sensors by SPFS: Shown is a plot of the slopes obtained from the binding kinetics as a function of the corresponding bulk concentration. The intersection of the fit to this calibration curve (red line) with the baseline (background fluorescence level) results in a LOD of 500 aM.

In most cases, the analyte is not directly fluorescently labeled; however, SPFS can still be applied even for the detection of unlabeled analyte molecules with a significant gain in sensitivity if, e.g., a sandwich assay is employed well-known from ELISA assays. The principles of this scheme and the results obtained for a brush modified with a capture antibody against prostate specific antigen (PSA) used in combination with a second fluorescence-labeled detection antibody against PSA are shown in Fig. 5.4 [24]. Irrespective of whether we used a one-step assay, i.e., the detection antibody was pre-incubated with the antigen PSA prior to the injection into the flow cell, or we employed a two-step assay, in which the PSA running through the flow cell was allowed to bind to the capture antibody for some time and then the fluorescently labeled detection antibody "decorated" the bound analyte, we obtained a LOD of better than 100 fM. This limit was not only 2–3 orders of magnitude lower than the clinically relevant analyte concentration; it was also significantly better than what could be achieved with the classical label-free detection, e.g., by SPR spectroscopy.

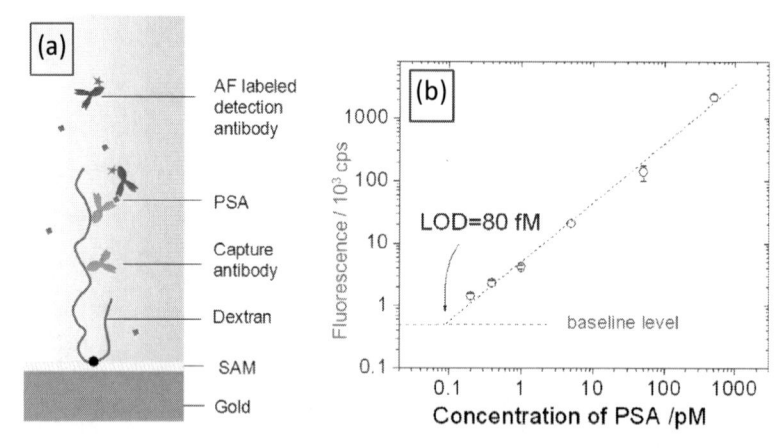

Figure 5.4 (a) Surface architecture of a dextran brush used in an antibody/antigen/antibody sandwich binding assay; (b) calibration curve of the SPFS biosensor for detection of prostate specific antigen.

5.3 Long-Range Surface Plasmon Fluorescence Spectroscopy

If one is dealing with a thin metal film that is sandwiched between two dielectric media of (nearly) identical refractive indices, $n_{b,d}$, plasmon modes excited at each of the two opposite interfaces will interact with each other provided the metal layer is sufficiently thin (thickness d_m of several 10 nm). Then, the optical fields within the metal overlap, which leads to an interaction that lifts off the dispersion degeneracy of the two identical evanescent waves and two new, coupled modes, appear—a symmetrical and an anti-symmetrical wave (referring to their transverse electric field distribution). The latter one, in particular, has attracted considerable interest because its electric field across the metal film that is responsible for the energy dissipation by the lossy metal is largely reduced and, thus, the propagation length of the mode is considerably increased. Hence, this mode is also called LRSP as opposed to the SRSP mode, which is subject to enhanced dissipation [14].

If we compare some spectroscopic features of regular SPs with those of LRSPs as it is done in Fig. 5.5, we see the following differences: (i) The angular scans of the two types of resonances show a significantly reduced line width that is associated with the excitation of LRSPs compared to that for regular SPs. This is a direct consequence of the reduced dissipation of the energy of the optical field in the metal layer, which is also the reason for the extended propagation length as mentioned before; (ii) if we calculate the field distribution normal to the metal/dielectric interfaces we see a further consequence of the reduced dissipation,

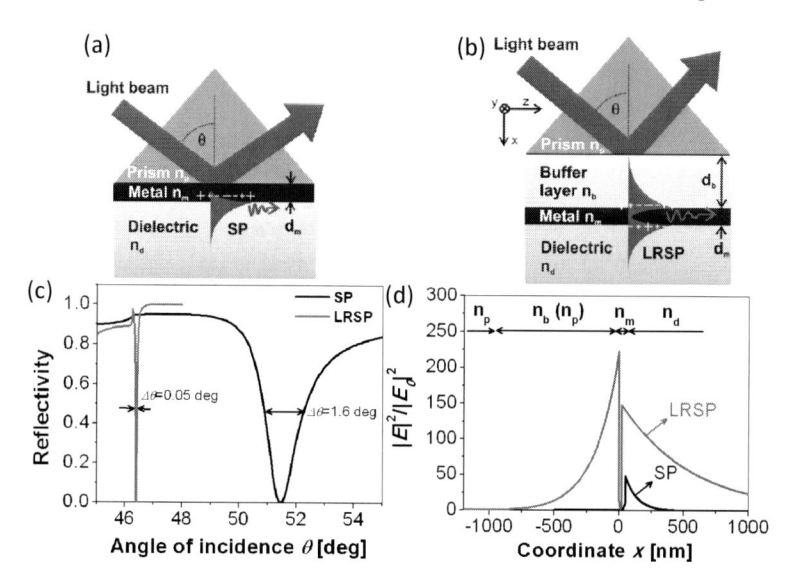

Figure 5.5 Schematic of the ATR coupled with Kretschmann geometry to regular SPs (a) and LRSPs (b). Example of simulated angular reflectivity spectra at the wavelength of λ = 633 nm and gold films supporting regular SPs (d_m = 50 nm) and LRSPs (d_m = 20 nm). The gold refractive index n_m = 0.125 + 3.56i was assumed and n_p = 1.845, n_b = 1.310, n_d = 1.333, and d_b = 850 nm.

i.e., a significantly enhanced optical field intensity at the interface that also extends into the dielectric media; (iii) and finally, the optical field reaches much farther out into the dielectric. The latter two features are displayed once more in greater detail in Fig. 5.6 where optical field simulations are summarized for LRSPs propagating along gold layers of different thicknesses d_m, ranging from 40 to 15 nm, as indicated. For comparison also the field distribution of normal SPs is shown. One can see that the thinner the metal layer is the higher are the field intensities at the interface metal/dielectric and the farther these fields extend into the dielectric medium.

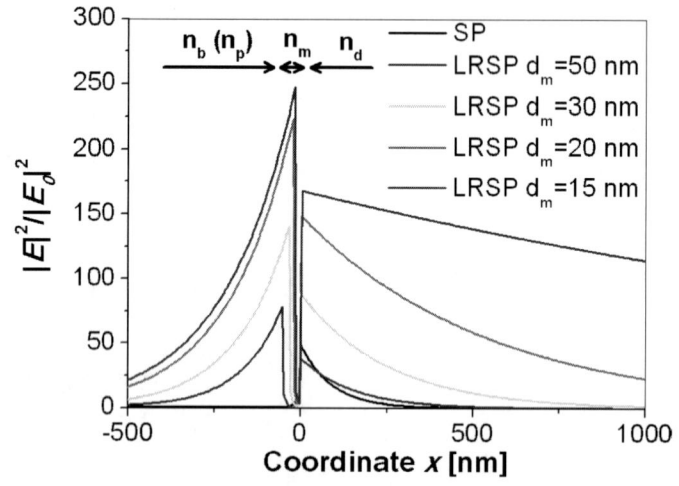

Figure 5.6 Simulated comparison of the field profile of LRSPs propagating along a gold film with varied thickness d_m = 15, 20, 30, and 50 nm. The parameters of layer structures are identical to those used in Fig. 5.5 and the thickness of the buffer layer d_b was adjusted to achieve full coupling to LRSPs.

It was argued [25] that the reduced width of the angular scans of LRSPs would result in a significantly enhanced sensitivity in optical biosensing with surface plasmons: Any thin layer, e.g., a bound protein layer, would result in a much stronger change of the reflected intensity at a given angle of observation because of the higher slope in the reflectivity scan. However, this argument does not take into account the larger extend of the optical field of LRSP modes that can reach out into the dielectric some few µm (cf. Fig. 5.6). In the overall refractive index architecture seen by the propagating LRSP wave an adsorbed protein layer changes the refractive index from $n = 1.33$ to about $n = 1.45$ of only a very thin slice of a few nm in thickness which has only a minor impact on the dispersion of LRSP modes. As a consequence, the shift of the resonance curve induced by this protein layer probed by LRSPs, hence, is largely reduced compared to regular SPR with the much stronger confinement of optical field of a normal surface plasmon mode nearer to the metal/dielectric interface. However, for fluorescence spectroscopy with LSRPs only the enhanced optical field is relevant and counts towards the obtainable sensitivity increase as we will demonstrate below [15]. As one can see in Fig. 5.6, this intensity enhancement is significant only for the very thinnest metal layers and, hence, the question arises as to whether one can actually prepare metal films in this thickness range that would allow for the translation of this field enhancement into a sensitivity increase in biosensing.

Furthermore, the experimental realization of the general concept of LRSP excitation requires an interfacial architecture with an ultrathin metal layer that is sandwiched between two dielectric media with nearly identical refractive index. In biosensing one of these dielectric media will be an aqueous sample or buffer with a refractive index close to $n_d = 1.33$. Hence, the material of the opposite (proximal) side of the metal layer needs to be a material that (i) has an equivalently low refractive index and that (ii) can be prepared in a way that it can be coated with the thin metal layer, typically by evaporation (cf. also the comparison of the two architectures for SPR and LRSP excitation given in Fig. 5.5). The materials of choice are two commercially available fluorine containing polymers, one called Teflon AF and the other Cytop; both have a refractive index near that of water ($n_b = 1.31$ for Teflon AF and $n_b = 1.34$ for Cytop at the wavelength of $\lambda = 633$ nm), both can be prepared in the required thickness range of typically up to 1 µm, and both are sufficiently smooth at their surface and are well suited for the evaporation of the ultrathin metal layer in the required thickness range. This is demonstrated in Fig. 5.7(b) where AFM images demonstrate the bare polymer support and the Au layers with nominal thicknesses of $d_{Au} = 30$ nm, 22.5 nm, and 15.8 nm.

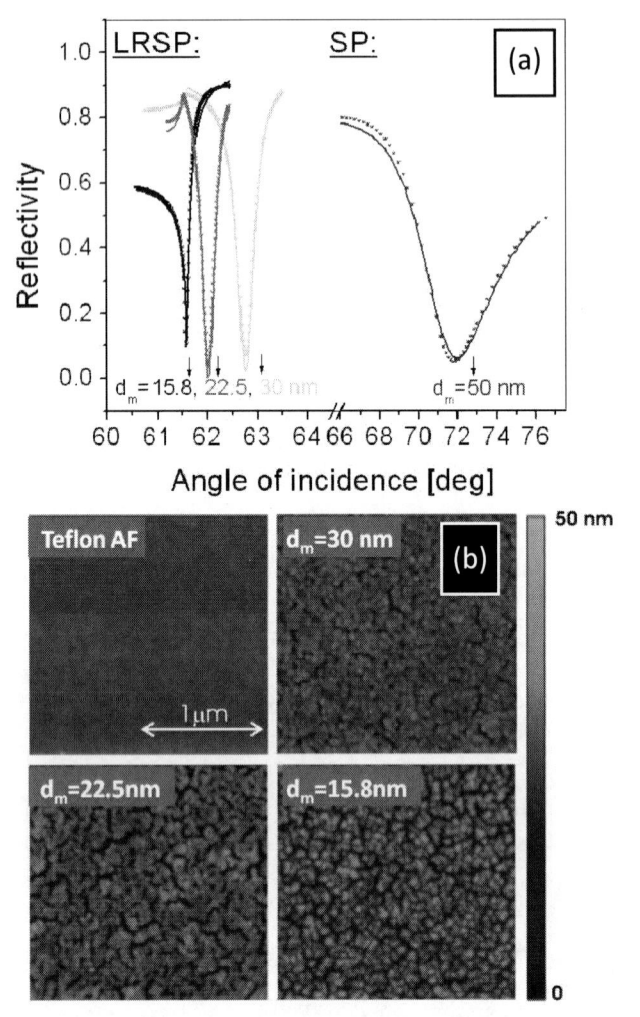

Figure 5.7 (a) Angular reflectivity and (b) morphology of gold films supporting LRSPs. The thickness of gold films d_m between Teflon AF and water was varied as indicated in the graphs.

Shown are further the angular reflectivity scans for the excitation of the LRSP modes in these three thin Au layers taken after having been brought into contact with water. For comparison, also the angular scan of a regular SPR experiment is displayed. The full curves given and fits to the experimental data with complex refractive index of the Au film ($n_m = n'_m + i \cdot n''_m$) revealed that n_m is almost identical for that of bulk Au for the thickness $d_m > 25$ nm [26]. However, below this thickness island morphology of gold film on fluoropolymers with low surface energy become pronounced which leads to the enhanced damping of LRSPs and decreased field intensity enhancement $|E|^2/|E_0|^2$. This problem can be overcome by chemical modification or plasma treatment of surfaces of fluoropolymer layers [27, 28].

A first demonstration of the obtainable fluorescence intensity increase and, hence, sensitivity enhancement is given in Fig. 5.8. Compared are the angular reflectivity scans for regular SPR in the Kretschmann prism-coupling mode and for a slightly modified layered sample architecture that allows for LRSP excitation. Furthermore, the simultaneously recorded angular scans of the fluorescence intensities are given for both, SPR and LRSP excitation, respectively. In order to demonstrate specifically the significant gain in fluorescence intensities measured for LRSP as a consequence of both, the higher optical field and the longer decay length into the dielectric, we chose a model sample architecture

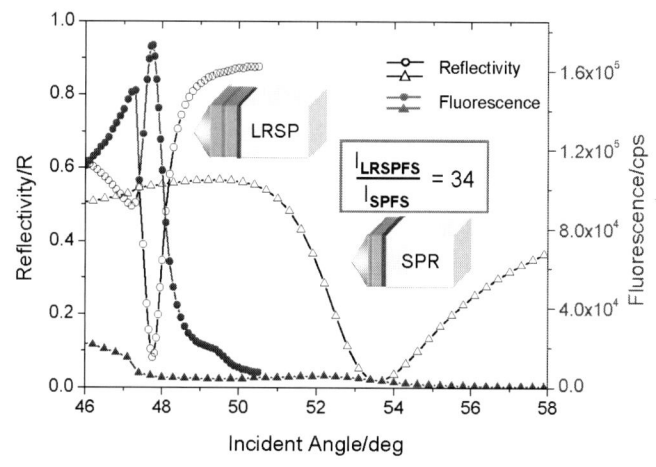

Figure 5.8 Comparison of the angular reflectivity scans recorded with normal SPR (open triangles) and in the LRSP configuration (open circles), together with the simultaneously measured angular fluorescence intensity curves (full triangles for SPFS and full circles for the LRSP fluorescence, respectively). The sample consisted of the prism, a 500 nm-thick cladding layer in the case of the LRSP excitation, an Au layer of 40 nm in each case, a 500 nm Teflon coating on top of the Au in both cases, and the chromophore labeled protein layer adsorbed from solution.

that allowed us to position the fluorescent chromophore layer some 500 nm away from the surface of the Au layer. For SPFS, with its exponential decay of the optical intensity at a decay length of about 90 nm only, this distance to the Au/dielectric interface results in a very weak excitation field and, hence, rather low level of fluorescence emitted by the chromophores (cf. Fig. 5.8, open black and blue full triangles). However, as expected (cf. the simulations in Fig. 5.6) the fluorescence intensity at resonant excitation of a LRSP wave was still significant, with a peak intensity ratio $I_{LRSPFS}/I_{SPFS} = 34$ [15]. This can be directly translated into a gain in sensitivity for biosensing applications [29–31].

5.4 Optical Waveguide Fluorescence Spectroscopy

Rather than going into more details of LRSPFS and giving more examples for its use in monitoring bioaffinity reactions we turn our attention to another version of surface wave enhanced biosensing, which is again a direct extension of the platforms discussed so far, i.e., guided optical waves, or (optical) waveguides for short [17]. The basic structural feature of an interfacial architecture that is able to guide light is schematically given in Fig. 5.9: On top of the metal layer that couples the laser photons in the total internal reflection mode of the Kretschmann prism setup and that is functionalized for biosensing applications by a hydrogel that carries the binding sites for the bioaffinity reactions with the analyte molecule from solution [19], new optical modes appear in the reflectivity scan. These modes are guided within the hydrogel layer provided this matrix layer has a refractive index, n_h, slightly higher than that of the adjacent buffer medium, n_d, and is sufficiently thick to ensure that the mode equation for guiding light in this slab waveguide format is fulfilled [32]:

$$\tan(\kappa d_h) = \frac{\gamma_d n_h^2/\kappa n_d^2 + \gamma_m n_h^2/\kappa n_m^2}{1-(\gamma_d n_h^2/\kappa n_d^2)(\gamma_m n_h^2/\kappa n_m^2)}. \tag{5.1}$$

This equation holds for transversal magnetic polarization (TM) and β is the propagation constant of guided modes, n_m states for the refractive index of the substrate, and d_h is a thickness of the hydrogel film. Terms $\kappa^2 = (k_0^2 n_h^2 - \beta^2)$, $\gamma_m^2 = \beta^2 - k_0^2 n_m^2$ and $\gamma_d^2 = \beta^2 - k_0^2 n_d^2$ are the transverse propagation constants in the polymer film, the metal, and the liquid, respectively. For the asymmetric configuration that is characteristic for this type of biosensor format with the metal as the substrate, a hydrogel layer with a thickness in the swollen state of a few µm, a segment density profile typical for soft polymer cushions at interfaces [33], and a refractive index just a little larger than that of the superstrate, i.e., the analyte solution, typically a few guided optical modes can be observed (very schematically shown and indexed in Fig. 5.9 as TM_1 and TM_2). Other than in the case of SP or LRSP modes with their optical field intensities being maximum at the metal surface and then decaying exponentially into the analyte matrix/solution here the optical architecture of the multilayer can be tuned in such a way that a maximum of the optical field is guided nearly completely inside the sensor matrix. This means that the density profile of the fluorescently labeled analytes (or of the fluorescent detection antibodies, cf. the sandwich assay in Fig. 5.4) bound to the capture sites within the hydrogel matrix have a maximum overlap with the probing optical field. Together with the high field enhancement factors that operate inside the waveguide layer this in turn leads to a further significant enhancement of the achievable sensitivity for biosensing applications.

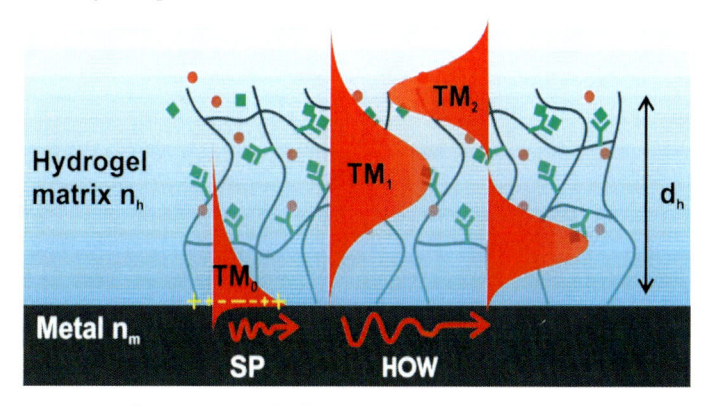

Figure 5.9 Schematics of a hydrogel film attached to a surface of a metal and supported surface plasmon (SP) and hydrogel optical waveguide (HOW) modes for transverse magnetic (TM) polarization.

An example of the two types of fluorescence spectroscopic modes, or more precisely, the two different optical field profiles normal to the waveguide layer are displayed in Fig. 5.10 for a LRSP-supporting layer structure and a hydrogel film with a thickness of d_h = 1.8 μm on the top. One can see that the optical intensity enhancement for the waveguide mode by far exceeds the one for the long-range surface plasmon mode. Furthermore, the field distribution can be tuned in order to probe molecular binding event at specific slice of the hydrogel structure. Both effects lead to the already mentioned higher sensitivity that can be achieved with hydrogel optical waveguide (HOW) spectroscopy as is shown below.

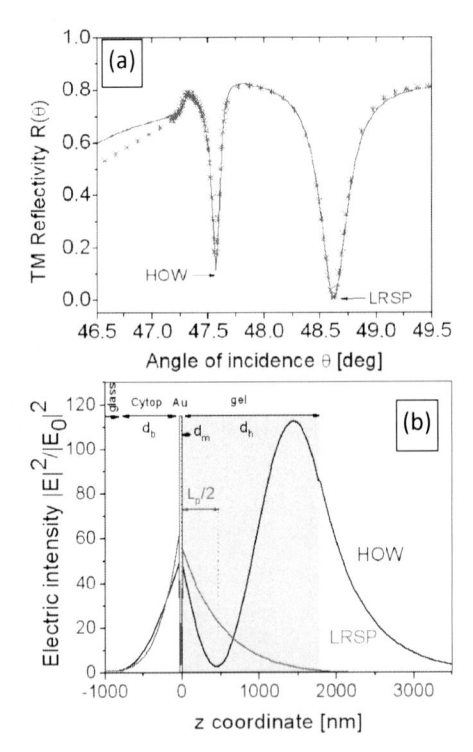

Figure 5.10 Example of (a) angular reflectivity with LRSP and HOW resonances and (b) respective profiles of electric field intensity upon the resonant coupling to these modes. The layer structure consists of a glass substrate with 715 nm-thick Cytop layer, 13.2 nm-thick gold film, and a NIPAAm-based hydrogel with a thickness of d_h = 1.8 μm that is swollen in phosphate buffered saline.

Figure 5.11 documents the implementation of LRSP and HOW modes for the amplified fluorescence immunoassay [21]. In this experiment, a mouse IgG was immobilized on a pNIPAAm-based hydrogel film that simultaneously served as a waveguide and a binding matrix. Afterwards, series of phosphate buffered saline samples spiked with Alexa Fluore 647-labeled anti-mouse IgG were successively flowed over the surface and a fluorescence signal associated with the affinity binding inside the matrix was measured. A comparison of obtained calibration curves for hydrogels films of different thicknesses (d_h = 0.06, 0.36, and 1.8 µm, as indicated) and fluorescence excitation via LRSP are presented. These results are compared with those obtained for the probing of the interface with the thickest hydrogel (d_h = 1.8 µm) by HOW mode.

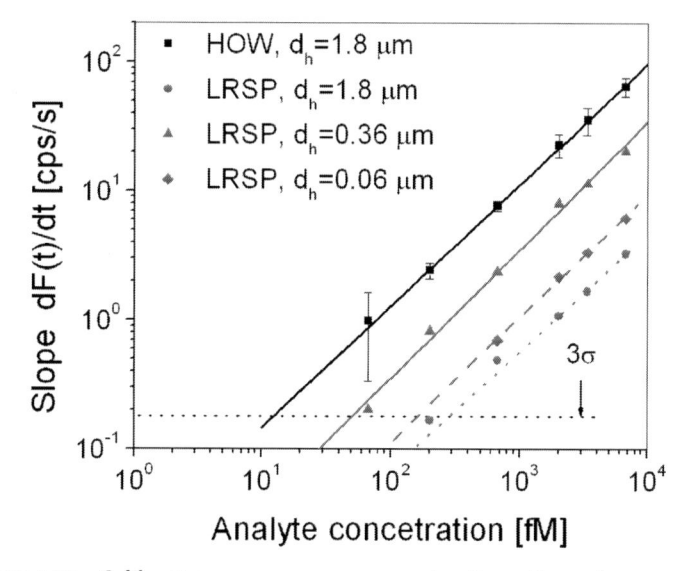

Figure 5.11 Calibration curves measured for the fluorescence immunoassay in a pNIPAAm hydrogel matrix with a thickness of d_h = 0.06, 0.36 and 1.8 µm. The assay utilized Alexa Fluor 647 labels that were excited at the wavelength 633 nm by resonantly excited LRSP and HOW modes as indicated in the graph.

The experimental data were all taken at analyte concentration regimes where mass-transfer limited diffusion leads to a linear increase with time of the fluorescence intensity F after injection of the analyte solution into the flow cell (cf. also Fig. 5.3). Again, plotting the slopes of these curves dF/dt as a function of the corresponding bulk concentration results in calibration curves that intersect with the background stability limit (3σ) giving the LOD of the corresponding experiment. As one can see in Fig. 5.11 that, indeed, the LOD for HOW is almost an order of magnitude lower than that for the LRSPFS results. However, the interpretation of those data needs a more detailed discussion in order to understand the thickness dependence seen in the data. Far from equilibrium, the affinity binding of target analyte occurs preferably in a top slice of the hydrogel matrix with a finite thickness d_p. This penetration depth d_p depends on a number of parameters. Firstly, it decreases with increasing density of catcher molecules that are attached to the hydrogel matrix as the time the analyte can diffuse before getting captured is shorten. Secondly, it increases when increasing the diffusion coefficient of the analyte in the polymer networks as it allows travel farther into the gel within the time required for the capture. Let us note that for hydrogel matrices typically used in our laboratory, d_p typically reaches several 100 nm. Therefore, for the probing of the hydrogel binding matrix with LRSP modes the highest sensitivity is achieved when $d_h \sim d_p$ (which occurred for d_h = 0.36 µm in the example shown in Fig. 5.11). For larger thicknesses d_h, the sensitivity decreases as the affinity binding events occur outside the LRSPs evanescent field (see Fig. 5.10b). However, almost 10-fold increased sensitivity can be achieved for the detection scheme that utilizes HOW mode supported by a thicker hydrogel layers with d_h = 1.8 µm. The reason is that the excitation of these modes is associated with higher field intensity enhancement that is strongly confined at the top slice of hydrogel matrix where the target molecules preferably bind.

5.5 Conclusions

The race for the most sensitive platform for biosensing applications is not decided yet (and certainly will not be for a long period of time): Optical concepts compete with electronic read-out ideas and vice versa. Among the various actively pursued principles in the optical regime surface plasmon excitations play a prominent and promising role, either as the currently very fashionable electromagnetic resonances localized in different types of (noble) metal nanostructures (particles, shells, other nano-objects with more complex shapes (triangles, cubes, stars) and strings and/or arrays thereof) or in various formats involving propagating modes, typically bound to a metal/dielectric interface.

In this short chapter, we focused on the latter case of surface-bound electromagnetic waves that can be also used as the light source for the excitation of chromophores thus allowing for the development of sensor concepts that combine the field enhancement mechanisms of surface plasmon excitation with the intrinsic sensitivity of fluorescence spectroscopy. We demonstrated that this way, unprecedented sensitivities could be reached, e.g., for the monitoring of affinity binding reactions of (fluorophore-labeled) analytes of interest with their surface-immobilized receptor structures (demonstrated for the case of an antigen-antibody interaction) which led to the (limit of) detection of 5×10^{-16} M, corresponding to the quantitative detection of only a few protein molecules per mm^2 reaching the sensor surface in every minute.

Extending this concept of combining the field enhancements achievable at resonant excitation of surface plasmon modes with fluorescence detection principles to the use of long-range surface plasmon waves with their strongly enhanced optical fields (due to their strongly reduced interaction with the lossy metal layer resulting in largely de-damped modes) resulted in even higher fluorescence signals of the bound analytes. This was not only a consequence of the mentioned field enhancement; this could be also achieved by making use of the much higher penetration length of the LRSP mode into the analyte solution: By coupling the binding sites to the polymeric backbone of a grafted hydrogel layer we could use both the higher field strength and the larger penetration depth to further enhance the sensor signal recorded for long-range surface plasmon–enhanced fluorescence spectroscopy compared to those obtained when using regular surface plasmons as the excitation light source.

From there, it was only a minor step to explore other modes of excitation that can be observed in surface grafted hydrogels of sufficient thickness: Optical modes guided within the hydrogel layer with an optical field (distribution) that corresponds not only to an exponentially decaying profile with the maximum intensity at the metal-sensor layer interface but rather can exhibit an intensity distribution that covers the whole thickness of the sensor layer slice (in the μm range). This way, an even higher fraction of the guided modes are actually propagating within the sensor matrix resulting in a further enhancement of the fluorescence signal of the analyte molecules.

Optical detection principles based on propagating surface plasmon waves (but not limited to those) continue to challenge the world of electrical/electronic/electrochemical techniques—the race goes on....

Acknowledgment

This work was only possible through the competent and enthusiastic collaboration of many colleagues. Thanks are due, in particular, to Patrick Beines, Ulrich Jonas, Torsten Liebermann, Bernhard Menges, Thomas Neumann, Hueyoug Park, Benno Rothenhäusler, Fang Yu, Danfeng Yao, and Manfred Zizlsperger. This work was partially supported by the Austrian NANO Initiative (FFG and BMVIT) through the NILPlasmonics project within the NILAustria cluster (www.NILAustria.at) and by the Austrian Science Fund (FWF) through the project ACTIPLAS (P 244920-N20).

References

1. Liedberg, B., Nylander, C., and Lundstrom, I. (1983). Surface plasmon resonance for gas detection and biosensing, *Sens. Actuator.*, **4**, 299–304.
2. www.biacore.com. [cited].
3. See the special issue of *Plasmonics* that summarizes contributions to a Symposium held in Singapore, November 5–7 2014, commemorating 30 years of SPR in biosensing.
4. Knoll, W. (1998). Interfaces and thin films as seen by bound electromagnetic waves, *Ann. Rev. Phys. Chem.*, **49**, 569–638.
5. Yu, F., and Knoll, W. (2004). Immunosensor with self-referencing based on surface plasmon diffraction, *Anal. Chem.*, **76**, 1971–1975.
6. Yu, F., Tian, S., Yao, D., and Knoll, W. (2004). Surface plasmon enhanced diffraction for label-free biosensing, *Anal. Chem.*, **76**, 3530–3535.
7. Yu, F., and Knoll, W. (2005). Surface plasmon diffraction biosensor, *J. Opt. Phys. Mater.*, **14**, 149–160.

8. Rothenhäusler, B., and Knoll, W. (1988). Surface-plasmon microscopy, *Nature*, **332**, 615–617.
9. Hickel, W., Kamp, D., and Knoll, W. (1989). Surface-plasmon microscopy, *Nature*, **339**, 186–190.
10. Zizlsperger, M., and Knoll, W. (1998). Multispot parallel on-line monitoring of interfacial binding reactions by surface plasmon microscopy, *Progr. Coll. Polym. Sci.*, **109**, 244–253.
11. Liebermann, T., and Knoll, W. (2000). Surface-plasmon field-enhanced fluorescence spectroscopy, *Coll. Surf. A*, **171**, 115–130.
12. Neumann, T., Johansson, M. L., Kambhampati, D., and Knoll, W. (2002). Surface-plasmon fluorescence spectroscopy, *Adv. Funct. Mater.*, **12**, 575–586.
13. Yu, F., Persson, B., Lofas, S., and Knoll, W. (2004). Attomolar sensitivity in bioassays based on surface plasmon fluorescence spectroscopy, *Am. Chem. Soc.*, **126**, 8902–8903.
14. Sarid, D. (1981). Long-range surface-plasma waves on very thin metal films, *Phys. Rev. Lett.*, **47**, 1927–1930.
15. Kasry, A., and Knoll, W. (2006). Long range surface plasmon fluorescence spectroscopy, *Appl. Phys. Lett.*, **89**, 101106.
16. Toma, K., Dostalek, J., and Knoll, W. (2011). Long range surface plasmon-coupled emission for biosensor applications, *Opt. Express*, **19**, 11090–11099.
17. Knoll, W. (1997). Guided wave optics for the characterization of polymeric thin films and interfaces, in *Handbook of Optical Properties Vol. II: Optics of Small Particles, Interfaces, and Surfaces* (Hummel, R. E., and Wißmann, P. ed) CRC Press.
18. Wang, Y., Huang, C. J., Jonas, U., Wei, T., Dostalek, J., and Knoll, W. (2010). Biosensor based on hydrogel optical waveguide spectroscopy, *Biosens. Bioelectron.*, **25**, 1663–1668.
19. Aulasevich, A., Roskamp, R. F., Jonas, U., Menges, B., Dostalek, J., and Knoll, W. (2009). Optical waveguide spectroscopy for the investigation of protein-funcionalized hydrogel films, *Macromol. Rapid Commun.*, **30**, 872–877.
20. Raccis, R., Roskamp, R., Hopp, I., Menges, B., Koynov, K., Jonas, U., Knoll, W., Butt, H., and Fytas, G. (2011). Probing mobility and structural inhomogeneities in grafted hydrogel films by fluorescence correlation spectroscopy, *Soft Matter*, **7**, 7042–7053.

21. Huang, C. J., Dostalek, J., and Knoll, W. (2010). Long range surface plasmon and hydrogel optical waveguide field-enhanced fluorescence biosensor with 3D hydrogel binding matrix: On the role of diffusion mass transfer, *Biosens. Bioelectron.*, **26**, 1425–1431.

22. Knoll, W., Kasry, A., Liu, J., Neumann, T., Niu, L., Park, H., Robelek, R., and Yu, F. (2008). Surface plasmon fluorescence techniques for bio-affinity studies, in *Handbook of Surface Plasmon Resonance* (Schasfoort, R. B. M., and Tudos, A. J., ed.), p. 275–312.

23. Liebermann, T., and Knoll, W. (2003). Parallel multispot detection of target hybridization to surface-bound probe oligonucleotides of different base mismatch by surface-plasmon field-enhanced fluorescence microscopy, *Langmuir*, **19**, 1567–1572.

24. Yu, F., Persson, B., Lofas, S., and Knoll, W. (2004). Surface plasmon fluorescence immunoassay of free prostate-specific antigen in human plasma at the femtomolar level, *Anal. Chem.*, **76**, 6765–6770.

25. Wark, A. W., Lee, H. J. and Corn, R. M. (2005). Long-range surface plasmon resonance imaging for bioaffinity sensors, *Anal. Chem.*, **77**, 3904–3907.

26. Dostalek, J., Kasry, A., and Knoll, W. (2007). Long range surface plasmons for observation of biomolecular binding events at metallic surfaces, *Plasmonics*, **2**, 97–106.

27. Huang, C. J., Dostalek, J., and Knoll, W. (2010). Optimization of layer structure supporting long range surface plasmons for surface plasmon-enhanced fluorescence spectroscopy biosensors, *J. Vac. Soc. Technol. B*, **28**, 66–72.

28. Mejard, R., Dostalek, J., Huang, C. J., Griesser, H., and Thierry, B. (2013). Tunable and robust long range surface plasmon resonance for biosensor applications, *Opt. Mater.*, **35**, 2507–2513.

29. Wang, Y., Dostalek, J., and Knoll, W. (2009). Biosensor for detection of aflatoxin M_1 in milk based on long range surface plasmon enhanced fluorescence spectroscopy, *Biosens. Bioelectron.*, **24**, 2264–2267.

30. Wang, Y., Brunsen, A., Jonas, U., Dostalek, J., and Knoll, W. (2009). Prostate specific antigen biosensor based on long range surface plasmon-enhanced fluorescence spectroscopy and dextran hydrogel binding matrix, *Anal. Chem.*, **81**, 9625–9632.

31. Huang, C. J., Sessitsch, A., Dostalek, J., and Knoll, W. (2011). Long range surface plasmon-enhanced fluorescence spectroscopy biosensor for ultrasensitive detection of *E. coli* O157:H7, *Anal. Chem.*, **83**, 674–677.

32. Dostalek, J., and Knoll, W. (2012). Plasmonics, in *Polymer Science: A Comprehensive Reference* (Matyjaszewski, K., and Möller, M., ed.), Elsevier: Amsterdam. p. 647–659.
33. Beines, P. W., Klosterkamp, I., Menges, B., Jonas, U., and Knoll, W. (2007). Responsive thin hydrogel layers from photocrosslinkable poly(*N*-isopropylacrylamide) terpolymers. *Langmuir*, **23**, 2231–2238.

Chapter 6

Infrared Surface Plasmon Resonance

Stefan Franzen,[a] Mark Losego,[b] Misun Kang,[a] Edward Sachet,[b] and Jon-Paul Maria[b]

[a]*Department of Chemistry, North Carolina State University, Raleigh, NC 27695, USA*
[b]*Department of Materials Science and Engineering, North Carolina State University, Raleigh, NC 27695, USA*

franzen@ncsu.edu.

6.1 Introduction

Surface plasmon resonance (SPR) in the visible region of the electromagnetic spectrum is a mature field of investigation based on the optical properties of the noble metals Ag and Au [1–5]. Other transition metals (except perhaps Cu) are not widely used since their plasma frequencies are in the hard ultraviolet and it has not proven practical to observe SPR in these metals. However, it must be born in mind that SPR can be observed far below the plasma frequency. The extension of SPR on Au to the near-infrared (near-IR) has been demonstrated [6–7]. This success has not extended to visible SPR on other transition metals. Therefore, Au and Ag still dominate, and for practical purposes Au is the conductor of choice in technological applications. If modeled as a

Drude conductor, Ag and Au have plasma frequencies of 2.18 and 2.183 × 10^{15} s^{-1}, respectively, [8]. However, it is important to note that the Drude model does not describe the optical properties of the noble metals accurately and the model can only be used to describe the lower frequency region of the dielectric functions where the real part becomes less than zero. The Drude model does not capture inter- and intraband transitions, which define the optical properties of those metals in the UV-Vis region of the electromagnetic spectrum. Clearly, the noble metals have plasma frequencies far above the infrared region. The vast majority of plasmonic applications have therefore been in the visible region of EM spectrum to take advantage of relatively large optical response due to Au when the index of refraction changes near an Au surface excited in an appropriate configuration [1–3].

Are Au and Ag unique among conductors in the sense that they are the only materials appropriate for practical applications of SPR? A relevant comparison of the relationship between the plasma frequency ω_p and the dispersion curve for a surface plasmon polariton (SPP) for Ag, Au, and indium tin oxide (ITO) is shown in Fig. 6.1 [9]. Figure 6.1 shows the real and imaginary dielectric functions, $\varepsilon_1(\omega)$ and $\varepsilon_2(\omega)$, respectively. While the Ag and Au responses shown in Fig. 6.1 are obtained from measured dielectric functions, the ITO dielectric function is obtained from a Drude model. Unlike Ag and Au, the Drude model works well for conducting metal oxides (CMOs) since there are no interband transitions that can mix with the plasmon band.

The real response represents dispersion and the blue curves can be plotted as ω vs. κ in order to show the dispersion of a given material. The imaginary part of the response represents the absorption of light due to the SPP. This absorption has been defined as the localized surface plasmon resonance (LSPR) [10–11]. The comparison in Fig. 6.1 reveals a striking similarity between ITO and Ag. In fact, the spectrum of Au has significant contribution from d-band to p-band transitions. Thus, the SPP of Au is not pure at all, but show strong mixing with other bands. Ag shows this to a smaller extent. However, ITO has essentially no mixing with band-to-band transitions since the ITO band gap is at ~3.4 eV, which is far above the plasma frequency for ITO [12]. One could say that ITO has a "pure" plasmon. In this sense, ITO is not unique, but it is one of many conducting metal oxides that have similar properties.

Figure 6.1 leads to the suggestion that one could drive SPPs in the near- and mid-IR.

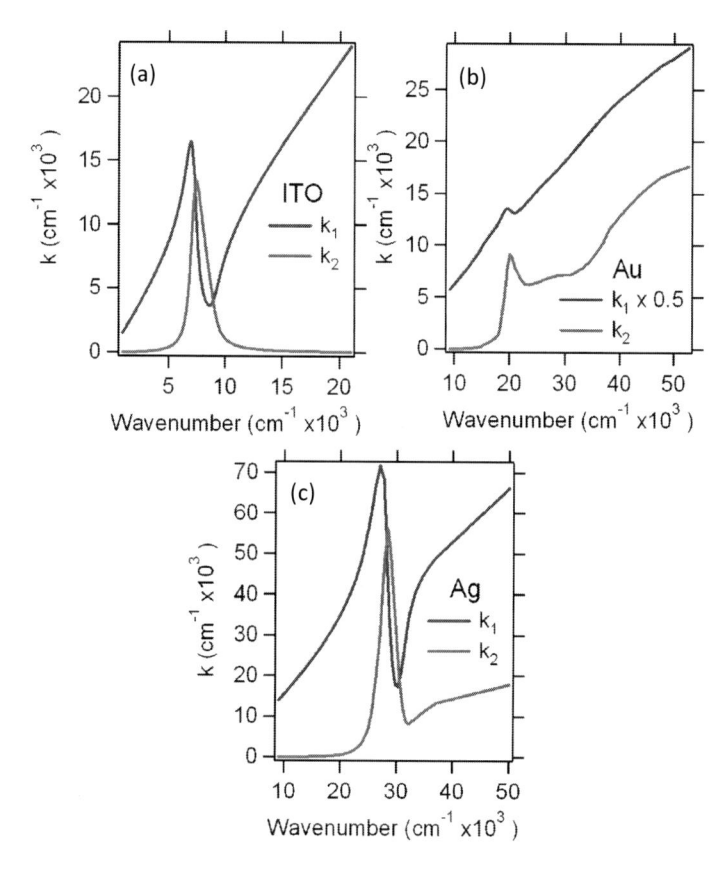

Figure 6.1 The dielectric function is plotted using the free electron optical constants (ITO) and experimental optical constants (gold and silver). The values are obtained from $\varepsilon_1(\omega) = n(\omega)^2 - \kappa(\omega)^2$ and $\varepsilon_2(\omega) = 2n(\omega)\kappa(\omega)$. The plots are shown for (a). Indium tin oxide, (b). Au and (c). Ag. Reproduced from reference 9 with permission from the American Chemical Society.

A pure plasmon means that plasmon that is described by the free electron or Drude model. The comparison of the optical properties of Ag and Au with ITO shown in Fig. 6.1 suggests that the optical properties of ITO are more similar to those of Ag than of Au, but the entire optical response is shifted to the near-IR. The plasmonic absorption (or localized surface plasmon resonance) should be observed at ~8,000 cm^{-1}, while that of Ag is ~33,000 cm^{-1} [9]. The dispersion of these materials shown in blue in Fig. 6.1 is quite similar to ITO and Ag. Actually, it is Au that is the most lossy optical material of the three compared in Fig. 6.1. Au is lossy because of the band-to-band transitions, but at the same time the mixing of these bands with the conduction electrons is thought to be the reason that one can observe a visible plasmon in the first place. The yellowish luster of Au is unique among metals and is related to the visible plasmon. These features have been considered in a range of applications for Au that have dominated the recent literature. It is important to recognize that other materials also have rich surface chemistries and possibilities for functionalization for applications [13–17]. In fact, there is an entire industry devoted coatings of conducting metal oxides for solar energy applications [18–19].

Even for Au and Ag, it is possible to shift both the SPP and LSPR to longer wavelengths. The SPP can be driven at larger angles and using higher index of refraction for an optical element known as the internal reflecting element (IRE) in order to shift the plasmonic response to the near-IR. On the other hand, the LSPR geometry is determined by the aspect ratio and aggregation state of particles. However, ITO has a SPP in the near-IR based on its material properties. One important distinction is that the optical properties of CMOs can be tuned based on their preparation since both charge carrier density and mobility can be altered in these materials.

In theory, SPPs can be driven by light at any frequency. Therefore, one might say that any conductor can have an SPP active in the mid-infrared region of the electromagnetic (EM) spectrum (100–4,000 cm^{-1}), assuming that the plasma wave number of the conductor is greater than about 5,000 cm^{-1} and that an appropriate angle can be accessed by the measuring instrument. On the other hand, there is a vast literature on conducting thin films, which have charge carrier densities 1–2 orders of magnitude lower than Au and therefore should have SPR signals observable in the near- and mid-IR. This chapter discusses research into thin-film materials and optical configurations and theoretical methods that can be used to observe SPR in novel thin-film materials [15–16, 20–24].

The search for new configurations that would provide perhaps greater sensitivity and also advantages in instrumentation led to research into the possibilities for optical exciting Au plasmon polaritons in the near-infrared region [6–7]. This configuration permits excitation of a SPP in the Kretschmann configuration, which is shown in Fig. 6.2a with an expansion in Fig. 6.2b [25]. The Kretschmann configuration is a totally internally reflecting geometry, which requires an IRE, which an dielectric function matched to that of the conducting layer, but greater than that of the sampled layer (Fig. 6.2a). This geometry results in an oscillating electric field parallel to the interface between the conductor and the sample layer, with evanescent waves extended perpendicular to the interface (Fig. 6.2b). Figure 6.2c shows a dispersion curve, which represents the angle-dependent frequency response of the material measured in the configuration shown in Figs. 6.2a and 6.2b. A SPP can be excited at the point where a light line intersects the dispersion curve. One of the technological innovations that initiated this field is a θ–2θ stage attached to a FTIR spectrometer [6, 26]. When attached to a near-IR FTIR spectrometer this device permits one to study plasmonic materials using near-IR light (4,500–11,000 cm^{-1}). This device opens up a wavelength range that permits the study of a number of other materials besides Au, and in fact first attempts to demonstrate SPP on CMOs were carried out using the FT-SPR attachment [15, 24]. The idea of using CMOs is based on the hypothesis that sufficiently high mobilities can be achieved to observe plasmonic extinction. Since the charge carrier densities are significantly lower in CMOs than in Au and Ag, the plasma frequency is also lower. However, clearly based on a decade of work in this field, there are many materials besides Ag and Au that can support SPR and other plasmonic phenomena [15–16, 21–24, 27–31].

Indium tin oxide was the first candidate for a plasmonic CMO. Based on the observation of the transition between a reflecting regime below ca. 6,000 cm^{-1} and a transparent region above this range, it was hypothesized that ITO would support a SPP if the proper excitation geometry were accessible [32]. One can visualize reflection in a conductor as a re-radiation of an electromagnetic field that is excited by an incident field. In order for reflection to occur, the electrons in the conductor must be capable of oscillating

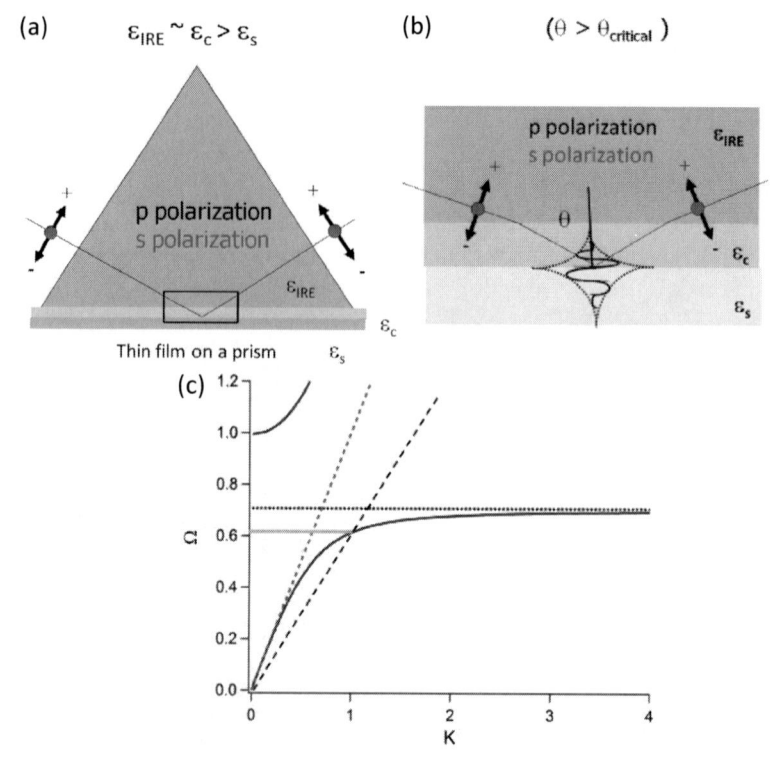

Figure 6.2 Geometry and conditions for SPR. (a) The concept of an internal reflecting element is shown with a conducting layer (c) and sampled layer (s). (b) The condition that the incident angle is greater than the critical angle is shown. The evanescent fields at the interface are represented by exponential functions. (c) The SPR dispersion curve is shown in reduced units for the frequency ($\Omega = \omega/\omega_p$) vs. wave vector ($K = \omega/c\omega_p$).

in phase with the exciting radiation. If the frequency of the exciting light is increased, one reaches a regime in which the electrons scatter with sufficient frequency that they can no longer oscillate in phase with the exciting light. Above this frequency, the material is an optical insulator and the light is transmitted (or scattered). At the frequency where the out-of-phase component of electron response appears due to scattering, there is absorption of radiation. This is the plasma frequency. Although the plasma frequency is well known in optics texts, we recapitulate the equations here to make the point that CMOs are conductors, which have a plasma frequency and therefore in theory can support SPPs [23].

6.2 The Hypothesis That Surface Plasmon Resonance Will Be Observed in Free Electron Conductors

In 2002, Brewer and Franzen made an observation that the reflectance of ITO decreased in the near-IR at a wave number that was proportional to the conductance of the thin film [32–33]. These relatively simple measurements of the resistivity using a four-point probe led to the hypothesis that the SPP could be observed in the mid-IR region in the Kretschmann geometry. The hypothesis was assumed that ITO was a free electron (Drude) conductor. Indium tin oxide is technically an n-type degenerate semiconductor. Depending on the Sn doping level and annealing conditions, it has charge carrier densities ranging from <10^{20} to 10^{21} cm^{-3}. Considering ITO as a representative material, we can construct a general model for the SPPs of free electron conductors.

SPR is a type of attenuated total reflection (ATR) spectroscopy. Any ATR spectroscopy requires an incident angle greater than the critical angle. However, it is not possible to couple into the conductor to excite a plasmon if the incident beam is in air or in any medium where the index of refraction is nearly 1.0. Instead, an IRE is required. An IRE is usually a prism that permits the light to reflect under the total internal reflection condition, which enables optical excitation of the SPP. To see how a free electron conductor can give rise to a SPP, we first consider the reduced Drude model. In reduced units, the free electron dielectric function is

$$\varepsilon(\Omega) = 1 - \frac{1}{\Omega^2 + \Delta^2} + i \frac{\Delta}{\Omega(\Omega^2 + \Delta^2)}, \quad (6.1)$$

where

$$\Delta = \frac{\Gamma}{\omega_p}, \quad \Omega = \frac{\omega}{\omega_p} \quad (6.2)$$

The plasma frequency is

$$\omega_p = \sqrt{\frac{ne^2}{m_e \varepsilon_0}} \qquad (6.3)$$

and the damping is

$$\Gamma = \frac{e}{m_e \mu}, \qquad (6.4)$$

where n is the charge carrier density, m_e is the effective mass, e is the electronic charge, μ is the mobility and ε_0 is the permittivity of vacuum. The solution to Maxwell's equations at the boundary of a conductor and an ambient gives a dispersion curve for the wave number of SPP as a function the wave vector ($k = c/\omega n \sin\theta$) of the incident light.

$$k(\Omega) = \Omega \frac{\omega_p}{c} \sqrt{\frac{\varepsilon_a \varepsilon(\Omega)}{\varepsilon_a + \varepsilon(\Omega)}} \qquad (6.5)$$

For example, a dispersion curve for an ambient that is vacuum ($\varepsilon_a = 1$) and a Drude free electron conductor can be calculated as follows:

$$k(\Omega) = \Omega \frac{\omega_p}{c} \sqrt{\frac{1 - \frac{1}{\Omega^2 + \Delta^2}}{2 - \frac{1}{\Omega^2 + \Delta^2}}} \qquad (6.6)$$

The reduced dispersion curve is plotted in Fig. 6.2A. Figure 6.2A shows clearly that an incident light beam cannot intersect the dispersion curve if it impinges on the conducting surface from vacuum ($n = 1$).

The minimum slope of the light is c/ω, which is only possible when the incident angle is 0°. Any other angle will give rise to larger slopes. This means that even the smallest slope in vacuum does not intersect the dispersion curve, which means physically that it cannot drive a SPP. However, if the light passes through an IRE as indicated in the sketch of Kretschmann configuration in Fig. 6.2b, then the light line can have a small slope due to the index of refraction of the IRE.

6.3 Confirmation of the Hypothesis That Conducting Metal Oxides Can Support Surface Plasmon Resonance

According to the hypothesis, one would expect that the SPP in ITO would be observed in the near infrared region if the appropriate angle could be accessed. It turned out that the angle range and frequencies available in the SPR attachment constructed by GWC, Inc. (Madison, WI) was in the correct range to observe the SPR effect. Although the GWC SPR attachment was designed without any intent to open up a new field of investigation for non-noble metals, [6–7] that was precisely the effect of the device. Once the effect was proven, we realized that better control over materials production was required. Thus, we began systematically producing ITO films with good process control. The first demonstration of a SPP on ITO (or any CMO) was published in the Journal of Applied Physics in 2006 [15]. Although the effect can be observed, in principle, for a number of different CMOs, it turns out the ITO has one of the highest charge carrier densities of any CMO. Thus, ITO was well suited to measurement within the limited wavenumber range available in the near-IR from 5000 cm^{-1}–9000 cm^{-1}. The first systematic application of the method appeared in 2008 when a series of measurements on ITO as a function of the film thickness was published [24]. Those observations shown in Fig. 6.3 reveal qualitative changes in the observed SPP depending on film thickness. Since SPR belongs to the class of ATR experiments, the observation of the signal depends on the thickness of the conductor.

If the conductor is thinner than the skin depth, then it cannot support a SPP since the charge carriers cannot oscillate along the surface. The response of such a thin film is instead the same as the of an LSPR. This was observed directly in ITO.

From the earliest measurements we have found that a model based on a combination of the Fresnel equations and the Drude model was able to predict the optical signal. The features shown in Fig. 6.3 are explained the theory in term of three regimes. The thinnest films, i.e., those less than the ~120 nm skin depth of ITO cannot support SPPs. Instead, we believe that the extinction in the thinnest films (Figs. 6.3A–6.3C) is a kind of LSPR, with a particular property that the polarization of the plasmon is perpendicular to the film. Between 71 nm and 121 nm the optical signal (Figs. 6.3D–6.3H) appears to be in transition between the perpendicular and parallel polarization. The normal region for SPR is observed between 120 and 200 nm (Figs. 6.3H–6.3J), where the film is thick enough to support SPPs polarized parallel to the surface, but still not so thick that there are significant losses. The intensity of the SPP decreases as the film thickness exceeds 200 nm as seen in Figs. 6.3H and 6.3I

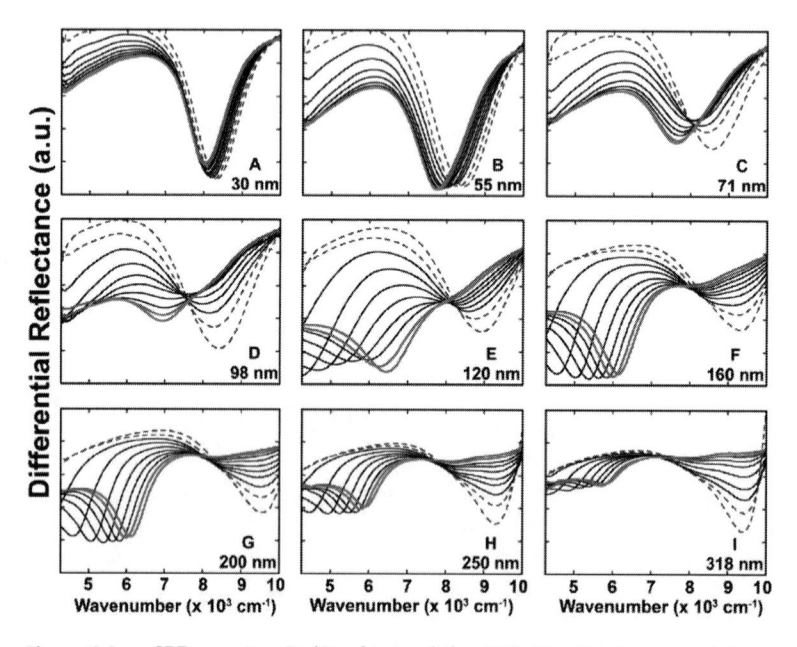

Figure 6.3 SPR spectra R_p/R_s obtained for ITO film thicknesses d from 30 (panel A) to 318 nm (panel M) and angles of incidence θ from 42 (red) to 54° (blue). Each line represents an angle increment of 0.83°. Reproduced from reference 23 with permission from the Optical Society of America.

The observation of SPR in a CMO is a general phenomenon and there have been studies using ellipsometry, microscopy and the Otto configuration for their characterization [27, 34–36]. These studies confirm the hypothesis that CMOs can be well characterized by the Drude model and therefore have relatively pure surface plasmons. The predicted properties of the LSPR have been observed in infrared absorbing nanoparticles as well [37–38].

6.4 The Effect of Carrier Concentration

The general applicability of the method for systematic study of materials properties was revealed in 2009 when the optical effect of varying thin-film fabrication methods was studied using FT-SPR [22]. There is a great deal of variability in commercial preparation of ITO. Systematic study of argon ion sputter gas pressure and annealing conditions has helped to systematize ITO thin-film production [14]. Once sputtered ITO films are usually annealed in forming gas (N_2/H_2) at 500°C for a period of 2–4 h. By controlling the pO_2 during annealing one reduce the charge carrier density as shown in Fig. 6.4a. Figure 6.4b shows that the electronic mobility is minimally affected by this difference in processing. Thus, annealing gives us a way to tune one parameter, charge carrier density, independently.

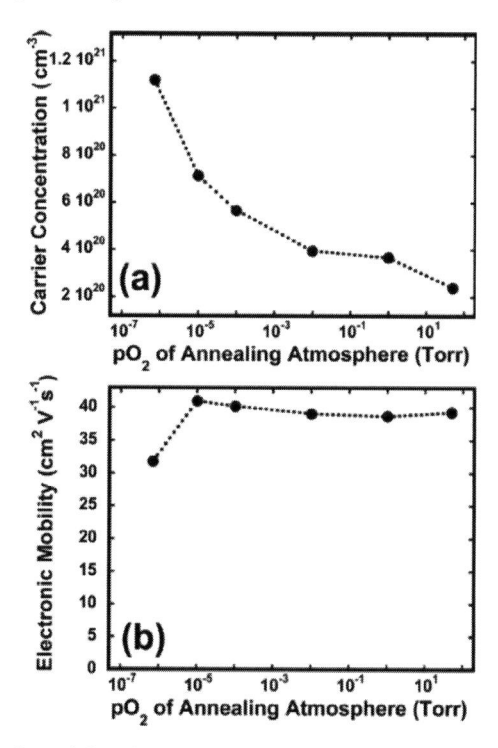

Figure 6.4 Plot of the charge carrier density and electronic mobility as a function of pO_2 during annealing. Reproduced from reference 21 with permission from the American Institute of Physics.

Figure 6.5 shows the experimental data for a series of ITO thin films prepared with systematic variation in their charge carrier density. The limiting angles are indicated by solid red and the dashed blue lines, which correspond to the highest incident angle measured, 52° and the lowest incident angle of 42°, respectively. Figure 6.5 shows that the polariton peaks move systematically from the near-IR towards the mid-IR, i.e., to lower energy for ω_p as n decreases, as predicted in Eq. 6.3. The exact position for the polariton in Fig. 6.5d is not known because it moves out of the range of the instrument and into the mid-IR. These graphs show that a slight change in the partial pressure of oxygen from 10^{-7} to 10^{-5}, graphs c and d, move the polariton peak to lower wave number by ca. 1000 cm^{-1}. In published studies, we have shown that theoretical graphs show a general agreement with the experimental data of the shape of the SPP curves and the location of the peaks for the carrier concentration series. The relatively small discrepancies in the modeling can be attributed to effective mass, since filling of vacancies by the oxygen will influence the effective mass of the thin film, therefore, should be slightly different from one to the next.

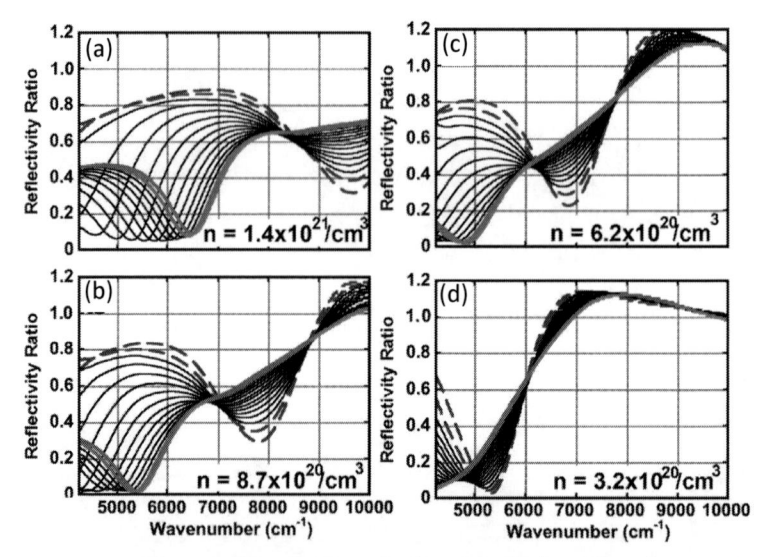

Figure 6.5 The experimental and theoretical data for the carrier concentration series. The partial pressure of oxygen for a is 0.01 mTorr, $b = 10^{-4}$ mTorr, $c = 10^{-5}$ mTorr, and $d = 10^{-7}$ mTorr. The corresponding charge carrier densities are given in each panel.

6.5 The Effect of Mobility

Electronic mobility in ITO films can be adjusted through the grain size of the material, which is controlled in turn by adjusting the sputter deposition pressure. Figure 6.6 shows the effect sputter gas pressure on both the charge carrier density and mobility. In this case, the charge carrier density is relatively constant, while the electronic mobility has a larger variation. There is also a maximum in the effect at around 9 mTorr Ar^+ pressure.

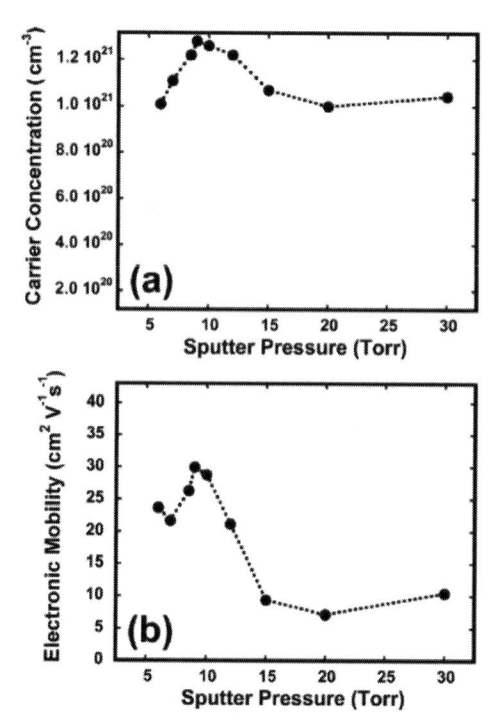

Figure 6.6 Plot of the charge carrier density and electronic mobility as a function of sputter pressure.

The systematic comparison shown in Fig. 6.7 focuses on the high pressure from 9 mTorr to 15 mTorr Ar^+. At a sputter pressure of 15 mTorr (Fig. 6.7c), the width of the peaks spans over 3000 cm^{-1} into the mid-IR while for 9 mTorr the peaks are not more than 2000 cm^{-1} across. As the pressure of the gas increases more collisions in the gas phase cause smaller grains in the film. The increase in grain boundaries results in electronic scattering, lowering the electronic mobility of the film. This lower electronic mobility mathematically relates to the optical damping constant, resulting in a widening of the polariton peak. The width of the SPP band is increasing as the mobility is decreasing. In Fig. 6.7, the experimental data show a broadening of the peaks as the pressure of Ar^+. Gas decreases below 9mTorr or increases above 9mTorr. The comparison of different series theoretical to experimental data in our published study shows how the general curve and the behavior of the SPP and other features are the same.

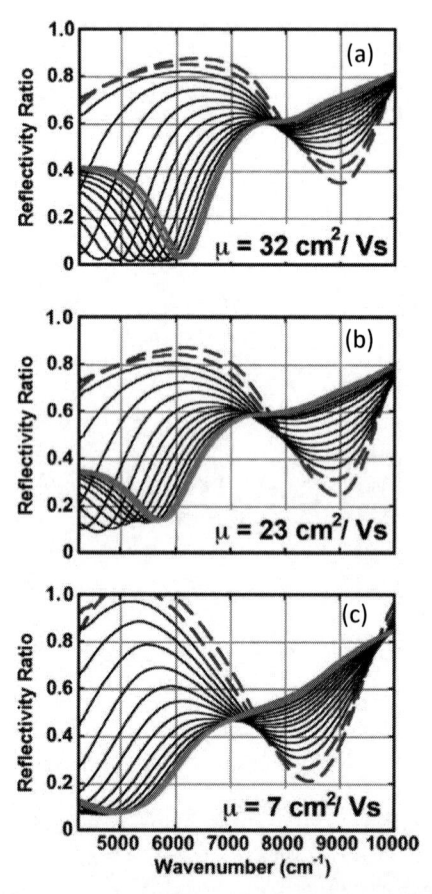

Figure 6.7 Varied pressure of the targeting gas changes the mobility cause a greater peak width, where the sputtering pressure of the target gas is 9 mTorr for (a), 12 mTorr for (b), and 15 mTorr for (c). The corresponding mobilities are given in each panel.

6.6 Hybrid Plasmons: Understanding the Relationship between Localized LSPR and SPR

While the use of the Fresnel equations and the Drude model provides good agreement with experimentally observed optical responses, the origin of differences in the response for the thickness series (Fig. 6.3) is not evident from the calculated response. The thinnest films have an extinction to higher energy than the SPR reponse. We proposed that the explanation for this phenomenon was linked to the relationship between the film thickness and the skin depth of the CMO. If the film is thinner than the skin depth then there should be no field response parallel to the surface because of the limitation on the electric field. One is forced to conclude that the optical response in such a thin film (e.g., 30 nm in ITO) is perpendicular to the film surface. The thin film is similar to a nanoparticle and we would predict that the extinction of a nanoparticle would occur in a similar region. According to the terminology adopted by van Duyne, the absorption of nanoparticles and other structures small than the skin depth of the conducting material are called localized surface plasmon resonance (LSPR) [10]. In ITO, the thinnest films have an absorption that is perpendicular to the thin-film surface [23]. This is the analog of the LSPR in a thin film. The hypothesis that the thin film supports a plasmonic response polarized perpendicular to the film surface is substantiated by the observation of hybrid plasmons composed of ITO/Au as shown in Fig. 6.8 [23].

The notion that plasmonic response can be tailored by altering the shape of a nanostructure is well established in studies of Au and Ag [10]. It is evident from the studies of ITO that similar optical features can be observed. The angle dip observed in Au and Ag in SPR experiments (see Fig. 6.8) can also be observed in ITO. However, ITO has a relatively low mobility (30 cm^2/Vs), which leads to a broad plasmonic response [24]. This may appear to be a disadvantage for near-IR plasmonic materials. However, research has shown that near- and mid-IR plasmonic materials can overcome this obstacle. The fact that there are literally hundreds of conducting metal oxides opens up the field for the investigation of new materials that have desirable properties. In order to understand these factors, we start with a thorough investigation of ITO.

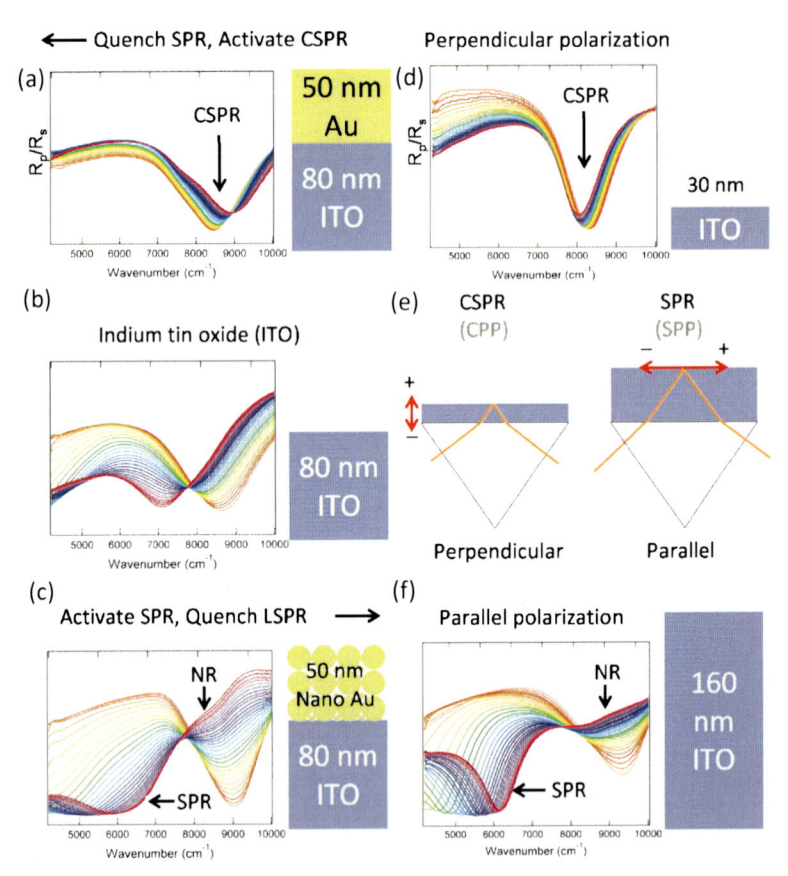

Figure 6.8 R_p/R_s data illustrating the effect of ITO thicknesses and Au overlayers on plasmon-polariton structures in reflectance spectra of ITO and hybrid ITO/Au films. Angles of incidence increase from 42° to 53° in steps of 0.35° as the colors advance from orange to violet. The charge separations of the capacitive (CP) and surface (SP) plasmons underlying the polariton structures of (a, d) and (c, f), respectively, are shown in (e). For the reference 80 nm ITO film (b) neither the SPP nor the CPP are fully activated. However, either the CPP (a, d) or the SPP (c, f) can be activated by controlling the properties of the Au overlayer (a, c) or the ITO thickness (d, f). The high energy feature in (d, f) is a non-resonant (NR) contribution. A metallic Au overlayer effectively reduces the ITO thickness, as seen by comparing (a) and (d). Alternatively, (c) and (f) show that the thickness of the ITO layer is effectively increased when the ITO is covered by a NP-Au overlayer fabricated so its effective carrier concentration matches that of the underlying film.

6.7 The Effect of Materials Properties on the Observed Surface Plasmon Polaritons

Once the SPP was observed in ITO, it was realized that ITO is an excellent testbed for the study of how SPPs are affected by charge carrier density, and mobility. The charge carrier density can be altered by changing the annealing conditions. The highest conductivity can be achieved by annealing ITO at 500°C in forming gas consisting of H_2/N_2 to remove O atoms from the lattice. Excess O tends to bind to octahedral Sn sites to expand the coordination sphere and pin the Sn electrons in a Sn-O bond, thus removing them from the available charge carriers. The effective O_2 partial pressure in the presence of forming gas is of the order 10^{-27} atm. If O_2 is added to the forming gas by bleeding in a small amount the charge carrier density can be lowered an controlled. This control of the charge carrier density permits ITO films to made with tuned ω_p. Such an effect is shown in Fig. 6.3. Indeed, as predicted by Eqn. 6.3, ω_p shifts to lower wave number proportional to $\sqrt{n}$.

By altering the pressure of the sputter gas, which consists of Ar ions, the grain size of the film can be altered. This alteration leads to a change in the mobility. The maximum mobility is ~30 cm^2/Vs but can be lowered as the grain size is decreased. The effect of lowered mobility is a broadening of the SPP. While the effect of altering the sputter gas pressure in ITO is to degrade the sharpness of the SPP, we learned from this experiment that the mobility is a key parameter in conducting metal oxides. This realization spurred a search for new CMOs with high mobility. However, we also realized from a study of a number of materials that the vast majority of CMOs have charge carrier densities lower than ITO. Thus, it was also necessary to push to the mid-IR for detection of SPPs in order to find the new materials.

6.8 Detection of Mid-Infrared Surface Plasmon Polaritons

While the observation of SPR in ITO presents a novel kind of effect, it is not obviously useful for new technologies. First, ITO has a relatively low mobility, μ, despite its high charge carrier density, n. This means that the observed SPPs are broad, which is undesirable for sensing applications. Moreover, the experiments conducted as function of the Ar⁺ sputter gas pressure revealed that ITO is already optimized. There is thus a limit to an improvement of ITO for sensing applications. Obviously, there is a large number of CMOs that have potential application for sensing. Indeed, some of these materials have higher mobility than ITO. However, the search for materials in the near-IR yielded relatively few materials with sufficiently high charge carrier density to be observed in the range of a near-IR FTSPR (4,500–11,000 cm^{-1}). Al-doped ZnO (AZO) is material that has a sufficient charge carrier density to support SPR in the mid-IR (1,000–4,000 cm^{-1}), but perhaps not the near-IR. A number of studies were conducted to produce AZO thin films with sufficiently high charge carrier density that they could be applied to near-IR FTSPR. The observed SPP was <5,000 cm^{-1} in all cases, which hampered the application of AZO. Recently published work on AZO for plasmonic applications is consistent with the need to develop more optical detection methods for plasmonic applications in the mid-IR [29]. Essentially all other interesting materials in this field have lower charge carrier densities than AZO. Thus, it became evident that the only way to advance the field was to develop mid-IR plasmonics.

By pushing the window of observation into the mid-IR one can potentially access a large number of materials. The issue is a relatively straightforward instrumentation problem. One solution is to adapt the θ–2θ stage already in use for near-IR FT-SPR experiments for mid-IR applications. The quartz optics must be replaced by CaF$_2$ optics for operation in the mid-IR. A second solution emerged, which is to adapt a variable-angle infrared reflection attachment for application to SPR, which is a relatively inexpensive modification of commercially available optics. For example, this has been demonstrated by the Booksh laboratory at the University of Delaware using a variable angle IR-SPR accessory, the Autoseagull (Harrick Scientific, Pleasantville, NY) and custom made variants of that device [39–40]. A third solution is to construct an appropriate sample holder for a variable angle infrared ellipsometer (IR-VASE). Of these approaches the IR-VASE provides the highest signal-to-noise ratio data and has the greatest angle range [21]. These innovations open up the field of near- and mid-IR FT-SPR for investigation.

6.9 The Search for High Mobility Conducting Metal Oxides

Once the instrumentation problem was solved, it became possible to examine a wide range of potential mid-IR plasmonic materials. The first material of choice was zinc oxide (ZnO), a well-understood material system. ZnO can be grown on a variety of IR transparent substrates, such as CaF_2, sapphire (Al_2O_3), MgO and Si. Using laser ablation techniques (pulsed laser deposition (PLD)), thin films of ZnO can be grown heteroepitaxially on the substrates mentioned above, allowing for high crystal quality and accurate control of the electrical properties of the deposited material [21].

The carrier concentration in ZnO can be adjusted by introducing Schottky defects according to the equation $0 = V_{Zn}^x + V_O^x$. The oxygen vacancies ionize further according to $V_O^x = V_O'' + 2e^-$ offering an avenue to adjust the free carrier concentration in ZnO crystals. Experimentally, this can be achieved by controlling the background oxygen pressure (pO_2) during deposition.

To demonstrate the feasibility of mid-IR SPR, a series of heteroepitaxial ZnO films on c-plane sapphire have been deposited. The epitaxial relationship can be summarized as (00.1)ZnO|| (00.1)Al_2O_3 and [2–1.0] ZnO||[1–1.0]. The films properties are summarized in Table 6.1.

Table 6.1 Thicknesses, charge carrier densities, and mobilities of AZO thin films

Sample	Thickness [nm]	Carrier concentration [cm^{-3}]	Mobility [cm^2/(V s)]	FWHM (00.2) [°]
A	400	8×10^{19}	20	0.5
B	600	4.5×10^{19}	18	0.6
C	800	6×10^{19}	18	0.6

Using the IR-VASE in a Kretschmann–Rather configuration, the mid-IR reflectivities for these films can be recorded as a function of incident angle and energy. The results are then plotted as reflectivity maps using a color scale to depict the reflectivity for a given angle/energy pair. In these maps, regions where coupling to the SPP can be observed are indicated by dark color shades (hence low reflectivities). Figure 6.9 summarizes the results for the three samples described in Table 6.1. The figure compares experimental data to simulated reflectivities. In all samples, a broad absorption between 2000 and 3000 cm^{-1} can be seen. The slight differences in the free carrier concentration can be seen as small shifts in the position of the SPR and the changing thickness changes the coupling efficiency thus changing the overall shape of the SPR in these maps. In general, good agreement between the simulated and experimental data is found.

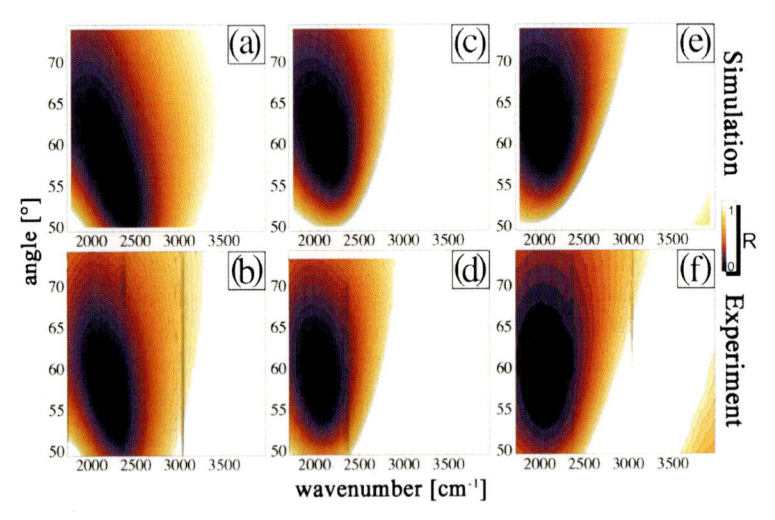

Figure 6.9 Reflectivity data for ZnO thin films and comparison to simulations. The data-pairs compare experiment to simulation for ZnO films with 400 nm (a,b), 600 nm (c,d) and 800 nm (e,f) film thickness. All data is depicted in the same parameter space.

These experiments successfully demonstrated the feasibility of mid-IR SPR; however, due to the poor electrical properties of ZnO at the high carrier concentration (hence high $[V_O'']$) needed for mid-IR interaction, the observed SPR is very broad and lossy. Mid-IR SPR applications based on this material would therefore not be competitive with the established materials such as the noble metals. Therefore, a CMO with significantly better transport properties has to be found to enable mid-IR SPR based technologies.

In general, as indicated by the Drude model, the damping term of the conduction electrons is mostly governed by carrier mobility, where higher mobility reduces the damping. Therefore, a material with the highest possible carrier mobility in the carrier range of interest ($8 \times 10^{19} - 5 \times 10^{20}$ cm^{-3}) would be the ideal candidate for future applications. Furthermore, since all the prospective materials are semiconductors, the free carrier concentration is tunable which allows shifting the resulting plasmonic phenomena over the mid-IR range. This tuning can be achieved either by doping with impurities, or by changing the defect equilibrium in a pure crystal such as in the case of ZnO.

6.10 Conclusion

The field of infrared plasmonics has grown from a hypothesis to a well-defined materials search for an optimized conducting metal oxide with a high mobility and tunable charge carrier density. The field required mid-infrared instrumentation since most of the materials of interest have charge carrier densities <10^{21} cm^3, which means that their SPR will be observed below 5,000 cm^{-1}. At present, examples of promising candidate materials are GaN, ScN, BaSnO$_3$, In$_2$O$_3$ and CdO. Further materials research is still needed to optimize the transport properties in these material systems; however, even with current materials mid-IR plasmonics promises exciting new possibilities. The use of plasmonic materials that compatible with semiconductors will permit utilization of the extensive experience in semiconductor processing and miniaturization. IR plasmonic materials offer the ability to monolithically integrate plasmonic elements into semiconductor stacks. This way, tightly integrated monolithic packages integrating solid state light sources with plasmonic crystals for, e.g., SPR sensing become feasible and offer an exciting outlook into the next generation of plasmonic technologies.

Acknowledgment

Financial support from the National Science Foundation (CHE-1112017) is gratefully acknowledged.

References

1. Huet, A. C., Fodey, T., Haughey, S. A., Weigel, S., Elliott, C., and Delahaut, P. (2010). Advances in biosensor-based analysis for antimicrobial residues in foods, *Trac-Trends Anal. Chem.*, **29**, 1281.
2. Karlsson, R. (2004). SPR for molecular interaction analysis: A review of emerging application areas, *J. Mol. Recognit.*, **17**, 151.
3. Malmborg, A. C., and Borrebaeck, C. A. K. (1995). Biacore as a tool in antibody engineering, *J. Immun. Method*, **183**, 7.
4. Brockman, J., Nelson, B., and Corn, R. (2000). Surface plasmon resonance imaging measurements of ultrathin organic films, *Ann. Rev. Phys. Chem.*, **51**, 41.
5. Jackman, J. A., Knoll, W., and Cho, N. J. (2012). Biotechnology applications of tethered lipid bilayer membranes, *Materials*, **5**, 2637.
6. Nelson, B. P., Frutos, A. G., Brockman, J. M., and Corn, R. M. (1999). Near-infrared surface plasmon resonance measurements of ultrathin films. 1. Angle shift and SPR imaging experiments, *Anal. Chem.*, **71**, 3928.
7. Frutos, A. G., Weibel, S. C., and Corn, R. M. (1999). Near-infrared surface plasmon resonance measurements of ultrathin films. 2. Fourier transform SPR spectroscopy, *Anal. Chem.*, **71**, 3935.
8. Murata, K., and Tanaka, H. (2010). Surface-wetting effects on the liquid-liquid transition of a single-component molecular liquid, *Nature Comm.*, **1**, 1–9.
9. Franzen, S. (2008). Surface plasmon polaritons and screened plasma absorption in indium tin oxide compared to silver and gold, *J. Phys. Chem. C*, **112**, 6027.
10. Haes, A. J., and Duyne, R. P. V. (2004). A unified view of propagating and localized surface plasmon resonance biosensors, *Anal. Bioanal. Chem.*, **379**, 920.
11. Willets, K. A., and Van Duyne, R. P. (2007). Localized surface plasmon resonance spectroscopy and sensing, *Ann. Rev. Phys. Chem.*, **58**, 267.
12. Odaka, H., Shigesato, Y., Murakami, T., and Iwata, S. (2001). Electronic structure analyses of Sn-doped In_2O_3, *Jpn. J. Appl. Phys. Part 1-Regul. Pap. Short Notes Rev. Pap.*, **40**, 3231.
13. Rhodes, C. L., Lappi, S., Fischer, D., Sambasivan, S., Genzer, J., and Franzen, S. (2008). Characterization of monolayer formation on aluminum-doped zinc oxide thin films, *Langmuir*, **24**, 433.

14. Cerruti, M., Rhodes, C., Losego, M., Efremenko, A., Maria, J.-P., Fischer, D., Franzen, S., and Genzer, J. (2007). Influence of indium–tin oxide surface structure on the ordering and coverage of carboxylic acid and thiol monolayers, *J. Phys. D: Appl. Phys.*, **40**, 4212.

15. Rhodes, C., Franzen, S., Maria, J.-P., Losego, M., Leonard, D. N., Laughlin, B., Duscher, G., and Weibel, S. (2006). Surface plasmon resonance in conducting metal oxides, *J. Appl. Phys.*, **100**, Art. no. 054905.

16. Losego, M. D., Guske, J. T., Efremenko, A., Maria, J. P., and Franzen, S. (2011). Characterizing the molecular order of phosphonic acid self-assembled monolayers on indium tin oxide surfaces, *Langmuir*, **27**, 11883.

17. Yan, C., Zharnikov, M., Golzhauser, A., and Grunze, M. (2000). Preparation and characterization of self-assembled monolayers on indium tin oxide, *Langmuir*, **16**, 6208.

18. Wu, Q. H. (2013). Progress in modification of indium-tin oxide/organic interfaces for organic light-emitting diodes, *Crit. Rev. Solid State Mater. Sci.*, **38**, 318.

19. Tada, A., Geng, Y. F., Nakamura, M., Wei, Q. S., Hashimoto, K., and Tajima, K. (2012). Interfacial modification of organic photovoltaic devices by molecular self-organization, *Phys. Chem. Chem. Phys.*, **14**, 3713.

20. Solieman, A., and Aegerter, M. A. (2006). Modeling of optical and electrical properties of In_2O_3: Sn coatings made by various techniques, *Thin Solid Films*, **502**, 205.

21. Sachet, E., Losego, M. D., Guske, J., Franzen, S., and Maria, J. P. (2013). Mid-infrared surface plasmon resonance in zinc oxide semiconductor thin films, *Appl. Phys. Lett.*, **102,** 051111.

22. Losego, M. D., Efremenko, A. Y., Rhodes, C. L., Cerruti, M. G., Franzen, S., and Maria, J. P. (2009). Conductive oxide thin films: Model systems for understanding and controlling surface plasmon resonance, *J. Appl. Phys.* **106**, Art. no. 024903.

23. Franzen, S., Rhodes, C., Cerruti, M., Gerber, R. W., Losego, M., Maria, J. P., and Aspnes, D. E. (2009). Plasmonic phenomena in indium tin oxide and ITO-Au hybrid films, *Opt. Lett.*, **34**, 2867.

24. Rhodes, C., Cerruti, M., Efremenko, A., Losego, M., Aspnes, D. E., Maria, J.-P., and Franzen, S. (2008). Dependence of plasmon polaritons on the thickness of indium tin oxide thin films, *J. Appl. Phys.*, **103**, Art. No. 093108.

25. Kretschmann, E., and Raether, H. (1968). Radiative decay of non radiative surface plasmons excited by light, *Zeit. Naturforsch. A*, **23**, 2135.
26. Frutos, A. G., and Corn, R. M. (1998). SPR of ultrathin organic films, *Anal. Chem.*, 70, 449A.
27. Kim, J., Naik, G. V., Emani, N. K., Guler, U., and Boltasseva, A. (2013). Plasmonic resonances in nanostructured transparent conducting oxide films, *IEEE J. Sel. Topic Quant. Electron.*, 19, Art. no. 4601907.
28. Naik, G. V., Schroeder, J. L., Ni, X. J., Kildishev, A. V., Sands, T. D., and Boltasseva, A. (2012). Titanium nitride as a plasmonic material for visible and near-infrared wavelengths, *Opt. Mat. Express*, **2**, 478.
29. Naik, G. V., Liu, J. J., Kildishev, A. V., Shalaev, V. M., and Boltasseva, A. (2012). Demonstration of Al:ZnO as a plasmonic component for near-infrared metamaterials, *Proc. Natl. Acad. Sci. U.S.A.*, **109**, 8834.
30. Emani, N. K., Chung, T. F., Ni, X. J., Kildishev, A. V., Chen, Y. P., and Boltasseva, A. (2012). Electrically tunable damping of plasmonic resonances with graphene, *Nano Lett.*, 12, 5202.
31. Naik, G. V., Shalaev, V. M., and Boltasseva, A. (2010). Semiconductor plasmonic metamaterials for near-infrared and telecommunication wavelength, *Metamater. Fundam. Appl. III*, 7754, doi:10.1117/12.863631.
32. Brewer, S. H., and Franzen, S. (2002). Optical properties of indium tin oxide and fluorine-doped tin oxide surfaces: Correlation of reflectivity, skin depth, and plasmon frequency with conductivity, *J. Alloys Comput.*, **338**, 73.
33. Brewer, S. H., and Franzen, S. (2002). Indium tin oxide plasma frequency dependence on sheet resistance and surface adlayers determined by reflectance FTIR spectroscopy, *J. Phys. Chem. B*, **106**, 12986.
34. Yasuhara, R., Murai, S., Fujita, K., and Tanaka, K. (2012). Atomically smooth and single crystalline indium tin oxide thin film with low optical loss, *Phys. Stat. Sol. C*, **9**, 2533.
35. Hinrichs, K., Furchner, A., Rappich, J., and Oates, T. W. H. (2013). Polarization-dependent and ellipsometric infrared microscopy for analysis of anisotropic thin films, *J. Phys. Chem. C*, **117**, 13557.
36. Dominici, L., Michelotti, F., Brown, T. M., Reale, A., and Carlo, A. (2009). DPlasmon polaritons in the near infrared on fluorine doped tin oxide films, *Opt. Express*, **17**, 10155.

37. Wang, T., and Radovanovic, P. V. (2011). Free electron concentration in colloidal indium tin oxide nanocrystals determined by their size and structure, *J. Phys. Chem. C*, **115**, 406.
38. Schelm, S., Smith, G. B., Garrett, P. D., and Fisher, W. K. (2005). Tuning the surface-plasmon resonance in nanoparticles for glazing applications, *J. Appl. Phys.*, **97**, Art. no. 124314.
39. Menegazzo, N., Kegel, L. L., Kim, Y. C., Allen, D. L., and Booksh, K. S. (2012). Adaptable infrared surface plasmon resonance spectroscopy accessory, *Rev. Sci. Instr.*, **83**, Art. no. 095113.
40. Menegazzo, N., Kegel, L. L., Kim, Y. C., and Booksh, K. S. (2010). Characterization of a variable angle reflection fourier transform infrared accessory modified for surface plasmon resonance spectroscopy, *Appl. Spec.*, **64**, 1181.

Chapter 7

The Unique Characteristics of Localized Surface Plasmon Resonance

Gaëtan Lévêque and Abdellatif Akjouj

Institut dElectronique, de Microélectronique et de Nanotechnologie,
Laboratoire Central, Cité Scientifique, Avenue Poincaré,
59652 Villeneuve d'Ascq Cedex, France

abdellatif.akjouj@univ-lille1.fr

The optical properties of metallic colloids smaller than the wavelength of the visible light have been empirically known since the antiquity and medieval ages. They were used to create colorful glasses by reduction of gold and silver oxides during the fabrication process. This phenomenon results from the absorption by the metal particles of a narrow portion of the visible spectrum of sun light. Indeed, at a specific wavelength depending on the shape, nature and environment of the nanoparticle takes place a resonant oscillation between the surface charge of the metal free electrons and the electromagnetic field scattered by the particle, the so-called localized surface plasmon resonance. The earliest publications about localized surface plasmons date back to 1857 with the experimental work of Michael Faraday (Faraday, 1857). On the theoretical side, Gustav Mie (Mie, 1908) and Richard Gans (Gans,

---

Introduction to Plasmonics: Advances and Applications
Edited by Sabine Szunerits and Rabah Boukherroub
Copyright © 2015 Pan Stanford Publishing Pte. Ltd.
ISBN 978-981-4613-12-5 (Hardcover), 978-981-4613-13-2 (eBook)
www.panstanford.com

1912) were the first to propose, at the end of the nineteenth century, a theoretical description of the phenomenon respectively for spheres or for spheroids in the quasi-static limit.

Appealing properties of localized surface plasmons are first the strong localization of light around the metal particle, which results from the large contribution of the evanescent waves to the mode structure. The optical field can be squeezed into very small volumes (few nanometers), associated to a large intensity enhancement called hotspots. They are usually located at sharp tips of metal particles (Martin, 2003), or inside gaps of a particle dimer (Novotny et van Hulst, 2011) or between particles and surfaces (Mubeen et al., 2012). Hotspots formation on rough metallic surfaces is widely used in material sciences, biology, and chemistry with surface-enhanced Raman spectroscopy, an improved version of Raman spectroscopy (Fleischmann et al., 1974) efficient enough for single molecule detection. Second, the dependency of the localized surface plasmon wavelength to the particle environment makes it interesting for sensing purpose in medical and biological applications, as a change of refractive index within the volume close the nanoparticle can be detected by a change in the transmission spectrum (Underwood and Mulvaney, 1994).

In the following, first we present the basics of localized surface plasmons for a single particle and then give two examples of systems of coupled metal nanoparticles, and finally present some properties of periodic plasmonic systems.

7.1 Localized Surface Plasmon Resonance of a Single Particle

In a typical experimental setup, plasmonic samples (obtained by either lithography or chemical synthesis) are characterized by measuring the transmission through the samples on a wide range of wavelengths. The profile of the different peaks gives access to the plasmon wavelength and lifetime, but much more information can be obtained on the mode nature through the shape of the peak and its shift under certain modification of the system (for example, the refractive index of the substrate, the size or aspect ratio of the particle, the distance between the particle and a substrate...). Experimentally, the extinction under a monochromatic excitation of pulsation ω corresponds to the decrease of the intensity of the incident light when it goes through the sample, through either absorption inside the sample or scattering of light out of the direction of the direction of propagation of the incident light. In the following, we will frequently refer to the extinction cross section, often calculated in theoretical investigations:

$$C_{ext}(\omega) = C_{abs}(\omega) + C_{diff}(\omega) = \frac{P_{abs}}{I_{inc}} + \frac{P_{diff}}{I_{inc}}$$

where P_{abs} and P_{diff} are respectively the power of the light absorbed inside the particle and scattered by the particle, where as I_{inc} is the intensity of the incident light.

Let us consider a metal nanoparticle of volume V, dielectric constant $\varepsilon(\omega) = \varepsilon'(\omega) + i\varepsilon''(\omega)$, placed in a transparent substrate of dielectric constant ε_B and refractive index $n = \sqrt{\varepsilon_B}$. The particle is illuminated by a monochromatic planewave $\mathbf{E}_0(\mathbf{r}, \omega)$, of amplitude $|\mathbf{E}_0|$. The absorption cross section is a function of the distribution of the electric field inside the particle through

$$C_{ext}(\omega) = \frac{\omega}{nc\varepsilon_0 |\mathbf{E}_0|^2} \Im\left[\int_V d\mathbf{r} \mathbf{P}(\mathbf{r}, \omega) \cdot \mathbf{E}_0^*(\mathbf{r}, \omega)\right]$$

with

$$\mathbf{P}(\mathbf{r}, \omega) = \varepsilon_0 [\varepsilon(\omega) - \varepsilon_B] \mathbf{E}(\mathbf{r}, \omega)$$

where ε_0 denotes the vacuum permittivity, $\Im$ the imaginary part, the star the complex conjugate, $\mathbf{P}$ the polarization density and $\mathbf{E}$ the total electric field inside the particle. Similarly, the absorption cross section is

$$C_{abs}(\omega) = \frac{\omega}{nc\varepsilon_0 |\mathbf{E}_0|^2} \Im\left[\int_V d\mathbf{r} \mathbf{P}(\mathbf{r}, \omega) \cdot \mathbf{E}^*(\mathbf{r}, \omega)\right]$$

The difference between the extinction and the absorption cross sections is the diffusion cross section:

$$C_{diff}f(\omega) = C_{ext}(\omega) - C_{abs}(\omega).$$

7.1.1 Single Particle in the Quasi-Static Approximation

The basic principles of localized surface plasmon resonances can be explained when the particle is much smaller than the incident light wavelength: the retardation due to the propagation of light inside the particle can be neglected and everything happens as if the electromagnetic field was time independent. This corresponds to the quasi-static approximation. In this limit, a certain class of nanoparticles has a very simple behavior: spheres and spheroids in general. Indeed, it can be shown that, in the quasi-static approximation, these particles present a lowest-order, dipolar mode in which the field inside the particle is homogeneous: the particle can be assimilated to a single scattering induced dipole, described in particular by its polarizability tensor $\bar{\bar{\alpha}}(\omega)$ at angular frequency ω. The relation between the induced dipole $\mathbf{p}$ and the incident electric field $\mathbf{E}_0$ at the location of the particle is then

$$\mathbf{p}(\omega) = V\mathbf{P}(\omega) = \varepsilon_0 \bar{\bar{\alpha}}(\omega)\mathbf{E}_0(\omega) \tag{7.1}$$

As we will always consider a monochromatic excitation in this chapter, the reference to ω will frequently be omitted in the following.

The reason why the polarizability is a 3 × 3 matrix is that in the most general situation where the particle is not isotropic, the induced dipole is not parallel to the incident field. However, in the particular case where the particle principal axes coincides with the reference frame axes (Ox), (Oy), and (Oz), the polarizability tensor is diagonal:

$$\bar{\bar{\alpha}} = \begin{bmatrix} \alpha_x & 0 & 0 \\ 0 & \alpha_y & 0 \\ 0 & 0 & \alpha_z \end{bmatrix}$$

Another useful expression is the relation between the particle polarization density $\mathbf{P}$ and the total electric field $\mathbf{E}$ inside the particle:

$$\mathbf{E} = \mathbf{E}_0 + \mathbf{E}_s = \mathbf{E}_0 - \frac{1}{\varepsilon_0 \varepsilon_B}\mathbf{L}.\mathbf{P}$$

The right member of this equation is the sum of the incident electric field $\mathbf{E}_0$ and the field $\mathbf{E}_s$ scattered inside the nanoparticle, proportional to the polarization density. In the most general case, the depolarization tensor **L** is, as for the polarizability, a 3 × 3 matrix (Yaghjian, 1980). The relation between the depolarization tensor and the polarizability tensor is then:

$$\overline{\overline{\alpha}} = V\Delta\varepsilon(\omega)\left[1 + \frac{\Delta\varepsilon(\omega)}{\varepsilon_B}\overline{\overline{L}}\right]^{-1}, \quad \text{with} \quad \Delta\varepsilon(\omega) = \varepsilon(\omega) - \varepsilon_B \quad (7.2)$$

Again, if the three principal axis of the particle are parallel to the (Ox), (Oy), and (Oz) directions, the tensor **L** is diagonal:

$$\mathbf{L} = \begin{bmatrix} L_x & 0 & 0 \\ 0 & L_y & 0 \\ 0 & 0 & L_z \end{bmatrix}, \quad \text{with} \quad L_x + L_y + L_z = 1$$

The values L_x, L_y, and L_z are obtained by computing, in the quasi-static limit, the scattered electric field $\mathbf{E}_s$ induced inside the spheroid by the surface charges $\sigma = \mathbf{P}.\mathbf{n}$, **n** being the unit vector normal to the particle surface, pointing outward.

In the quasi-static approximation, we find the expression of the extinction cross section using 7.1, when the incident field is parallel to the i principal direction of the spheroid:

$$C_{ext}^i = \frac{\omega}{nc}\Im\{\alpha_i\}$$

When averaged on the incident field orientation:

$$C_{ext} = \frac{\omega}{3nc}\Im\{\alpha_x + \alpha_y + \alpha_z\}$$

Using the preceding relations, it can be easily shown that the absorption cross section C_{abs} is equal to the extinction cross section C_{ext} so that the diffusion cross section is zero in the quasi-static approximation.

Finally, the localized surface plasmon wavelengths of the spheroidal particle are found by searching the maxima of the polarizability tensor components.

7.1.1.1 Case of the spherical particle

In the case of a spherical particle, both the polarization and the depolarization tensors are isotropic, and then can be reduced to scalars α and L. As for a sphere $L = 1/3$ (Yaghjian, 1980), we easily get the expression of the scalar polarizability:

$$\alpha(\omega) = 3V\varepsilon_B \frac{\varepsilon(\omega) - \varepsilon_B}{2\varepsilon_B + \varepsilon(\omega)}$$

If the particle is made of a transparent dielectric, the dielectric constant $\varepsilon(\omega)$ is positive and real. It follows that the polarizability is always finite and is a slow varying function of the wavelength of light in vacuum, $\lambda = 2\pi c/\omega$. However, as in the visible domain the dielectric constant of a metal particle is negative, the modulus of the polarizability can reach very large values when $\varepsilon'(\omega) \approx -2\varepsilon_B$. The wavelength λ_{LSP} for which the polarizability modulus is maximum corresponds to the excitation wavelength of the localized surface plasmon mode, which, in this quasi-static limit, is a dipole parallel to the direction of the incident electric field. If we take a Drude model for the dielectric constant of the metal nanoparticle, neglecting the absorption and with a plasma frequency $\omega_p = 2\pi c/\lambda_p$:

$$\varepsilon(\omega) = \varepsilon_\infty \left(1 - \frac{\omega_p^2}{\omega^2}\right)$$

or, in term of wavelength:

$$\varepsilon(\lambda) = \varepsilon_\infty \left(1 - \frac{\lambda^2}{\lambda_p^2}\right),$$

we find for the localized surface plasmon wavelength:

$$\lambda_{LSP} = \lambda_p \sqrt{1 + 2\varepsilon_B/\varepsilon_\infty} \qquad (7.3)$$

Obviously, when the losses are neglected the polarizability diverges at $\lambda = \lambda_{LSP}$, which is never the case in reality as absorption always takes place. In case of gold for example, localized surface plasmons of spherical particles embedded in a dielectric substrate with a moderate refractive index (like water or silica) occurs around λ = 520 nm, very close to the frequency domain where interband transitions, and then absorption, are important. In case of silver, the absorption is much lower and its dielectric constant is well approximated by a Drude model in the visible range. As an illustration, Fig. 7.1 shows the extinction spectra, normalized to maximum, for spherical gold and silver nanoparticles in the quasi-static approximation, in air ($\varepsilon_B = 1.00$), water ($\varepsilon_B = 1.77$) and silica ($\varepsilon_B = 2.25$) substrates. The dielectric constants for gold and silver were taken from Johnson and Christy (1972). As expected, the width of the plasmon resonance is much narrower for silver than for gold. However, gold is much preferred in experiments because silver easily oxidizes in a wet environment.

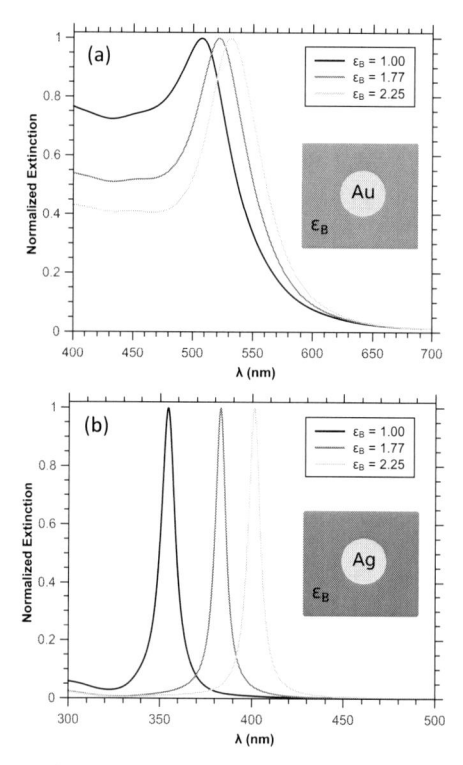

Figure 7.1 Normalized extinction spectra of a metal nanoparticle inside substrates of different dielectric constant: (a) gold; (b) silver.

The shift of the plasmon wavelength with the refractive index of the substrate is a very useful characteristic of a plasmonic nanoparticle, particularly for sensing applications. The so-called refractive-index sensitivity S corresponds to the shift in the resonance wavelength under a change of the refractive index of the substrate by 1. Equation 7.3 leads to

$$S = \frac{d\lambda_{LSP}}{dn} = \frac{2n\lambda_p}{\sqrt{1+2\varepsilon_B/\varepsilon_\infty}}$$

It appears clearly in Fig. 7.1 that the refractive-index sensitivity in water is much larger for a silver ($S \approx 109$ nm/RIU) than for a gold nanosphere ($S \approx 59$ nm/RIU) but is still orders of magnitude below sensitivities obtained in propagative-surface-plasmon-based sensing devices.

7.1.1.2 Case of the spheroidal particle

For a spheroidal particle in the quasi-static approximation, the field scattered inside the particle is generally not parallel to the induced dipole, and the depolarization tensor is a 3 × 3 matrix. We assume in the following that the three principal axis of the particle are parallel to the (Ox), (Oy), and (Oz) directions, so that both $\bar{\bar{\alpha}}$ and $\bar{\bar{L}}$ are diagonal, of elements α_i and L_i in each direction $i = x, y, z$. Then, to each i direction corresponds one specific localized plasmon mode excited at one specific wavelength. Each plasmon mode can be excited if the incident electric field has a non-zero projection onto the corresponding principal axis of the particle. Following Eq. 7.2, we have

$$\alpha_i = V\varepsilon_B \frac{\varepsilon - \varepsilon_B}{\varepsilon_B + L_i(\varepsilon - \varepsilon_B)}$$

and, using again a lossless Drude model:

$$\frac{\lambda_{LSP}^i}{\lambda_p} = \sqrt{1 + \frac{\varepsilon_B}{\varepsilon_\infty}\left(\frac{1}{L_i} - 1\right)}$$

The refractive-index sensitivity of each i plasmon mode is then

$$S_i = \frac{d\lambda_{LSP}^i}{dn} = \frac{n\lambda_p\left(\frac{1}{L_i} - 1\right)}{\varepsilon_\infty\sqrt{1 + \frac{\varepsilon_B}{\varepsilon_\infty}\left(\frac{1}{L_i} - 1\right)}} = \frac{1}{n}\left(\lambda_{LSP}^i - \frac{\lambda_p^2}{\lambda_{LSP}^i}\right)$$

This last equation shows that, in the quasi-static approximation, the refractive-index sensitivity of the metal nanoparticle is only a function of the plasmon resonance wavelength. The dependency on the shape, included here in the coefficients L_i, occurs only through the position of the plasmon wavelength (Saison et al., 2013).

Figure 7.2a shows the evolution of the two plasmon mode wavelengths as a function of the aspect ratio $r = b/a$ for a spheroid of semi-axes b along the z direction, a along the x and y directions. The expressions of the depolarization factors can be found in Bohren and Huffmann (1998). When r is close to zero, which means that the spheroid looks like a flat disk, $L_x = L_y \approx 0$, whereas $L \approx 1$. The wavelength of the plasmon mode parallel to the disk is then close to infinity, whereas the wavelength of the plasmon mode parallel to the thin edge of the disk is close to λ_p. However, when r is close to infinity, the spheroid looks like an infinite cylinder, for which $L = 0$, and then $L_x = L_y = 1/2$. In that case, the wavelength of the plasmon mode parallel to the cylinder axis is close to infinity, whereas the frequency of the plasmon mode parallel to the cylinder section is close to a limit lower than the bulk plasmon frequency by a factor of $\sqrt{1 + \varepsilon_B/\varepsilon_\infty} \approx 1.2$, when the nanoparticle is surrounded by water and $\varepsilon_\infty = 4.0$.

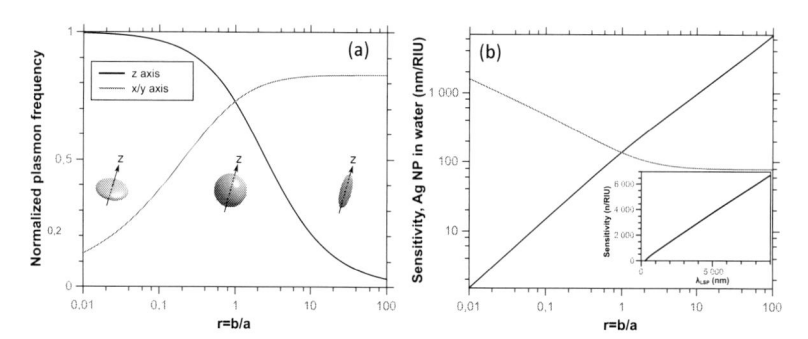

Figure 7.2 (a) Evolution of the two normalized plasmon frequencies $\lambda_p/\lambda_{LSP}^i$ with the spheroid aspect ratio $r = b/a$; (b) Evolution of the refractive-index sensitivities of a silver nanoparticle for the two plasmon modes as a function of r.

Figure 7.2b shows the evolution of the refractive-index sensitivity of the spheroid in water as a function of the aspect ratio r. The particle is supposed to be made of silver, described by a

Drude model with $\varepsilon_\infty = 4.0$ and $\lambda_p = 284.0$ nm. The inset shows the evolution of the sensitivity as a function of the particle surface plasmon wavelength, where it appears that it can reach very high values, either for very elongated or very flat particles, but at the expense of a large red-shift of the surface plasmon resonance wavelength.

Finally, if the investigation of the plasmonic properties of a spheroidal particle seems simple in the quasi-static approximation, the same problem becomes rapidly impossible to solve analytically for a larger particle, or with a different shape, or even simply coupled to a single dielectric interface. This is the reason why numerical simulations are frequently employed. A few methods are going to be briefly presented in the next section.

7.1.2 Beyond the Quasi-Static Approximation

When the size of the particle increases, retardation effects cannot be omitted any more. The distribution of the electric field of the dipolar mode inside the particle is not homogeneous, and the description of the particle response with a simple polarizability tensor is not valid. The first to propose a rigorous method to investigate the optical properties of a spherical metal particle in an homogeneous substrate was Gustave Mie (Mie, 1908).

Mie scattering method As a full description of the Mie method can be find in number of textbooks (Bohren and Huffmann, 1998), it will not be detailed here. Let us just remind to the reader that this method relies on the expansion of the electric field inside and outside the spherical nanoparticle on a set of vector spherical Bessel functions. In particular, the electric field outside the particle is written:

$$\mathbf{E}_s = \sum_{n=1}^{\infty} E_n (i a_n \mathbf{N}_n^{(3)} - b_n \mathbf{M}_n^{(3)})$$

where $\mathbf{N}_n^{(3)}$ and $\mathbf{M}_n^{(3)}$ are vectorial spherical harmonics and E_n an amplitude. The a_n and b_n coefficients are defined by the relations:

$$a_n = \frac{m^2 j_n(mx)[x j_n(x)]' - j_n(x)[mx j_n(mx)]'}{m^2 j_n(mx)[x h_n^{(1)}(x)]' - h_n^{(1)}(x)[mx j_n(mx)]'}$$

$$b_n = \frac{j_n(mx)[xj_n(x)]' - j_n(x)[mxj_n(mx)]'}{j_n(mx)[xh_n^{(1)}(x)]' - h_n^{(1)}(x)[mxj_n(mx)]'}$$

where j_n is the spherical Bessel function of order n, $h_n^{(1)}$ is the spherical Hankel function of first kind and order n, $m = \sqrt{\varepsilon(\omega)/\varepsilon_B}$ and the prime denotes the derivative with respect to the function argument (x or mx). The extinction, diffusion, and absorption cross sections are then expressed as follows:

$$C_{ext} = \frac{2\pi}{k^2} \sum_{n=0}^{\infty} (2n+1)\Re(a_n + b_n)$$

$$C_{sca} = \frac{2\pi}{k^2} \sum_{n=0}^{\infty} (2n+1)(|a_n|^2 + |b_n|^2)$$

$$C_{abs} = C_{ext} - C_{sca} \tag{7.6}$$

with $k = n\omega/c$.

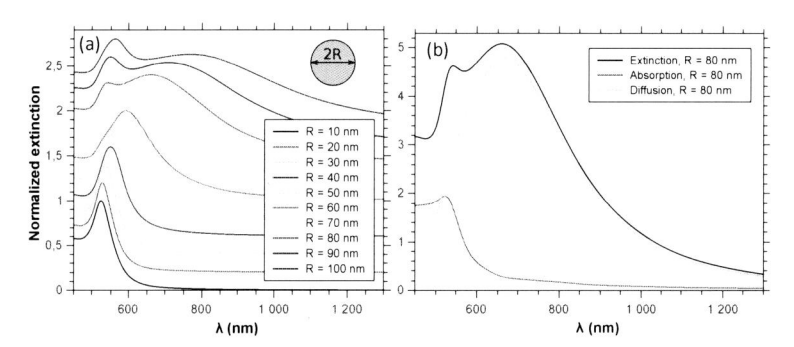

Figure 7.3 (a) Mie calculations of the extinction cross section, for a gold particle of increasing diameter in water; (b) Comparison between the extinction, absorption and diffusion cross sections for a particle of 80 nm radius.

As an example, Fig. 7.3a shows the extinction spectra of a spherical gold particle in water, with increasing diameter up to 200 nm. When the radius is small enough, only one mode appears around $\lambda = 520$ nm. This mode corresponds to the dipolar plasmon, at the wavelength found in the quasi-static approximation, which shifts to the red with increasing radius. However, a second mode at shorter wavelength appears as a shoulder in the main dipolar resonance for a radius larger than 50 nm, and is slightly red-shifted when the particle size increases. This mode is of quadrupolar nature, and appears when the coefficients a_2 and b_2 become non-zero in the expression 7.4. When comparing the absorption and scattering cross sections for R = 80 nm in Fig. 7.3b, we can easily notice that the long-wavelength mode is much more radiative than the short-wavelength mode, which is a signature of the quadrupolar nature of this mode. This can be confirmed by plotting the distribution of the surface charges (Fig. 7.4), computed with a Green's function method, briefly presented below. The distribution is represented at the time where the surface charge is maximal. Higher order modes exist as well; however, they are difficult to observe in the case of gold because of the large absorption in the blue.

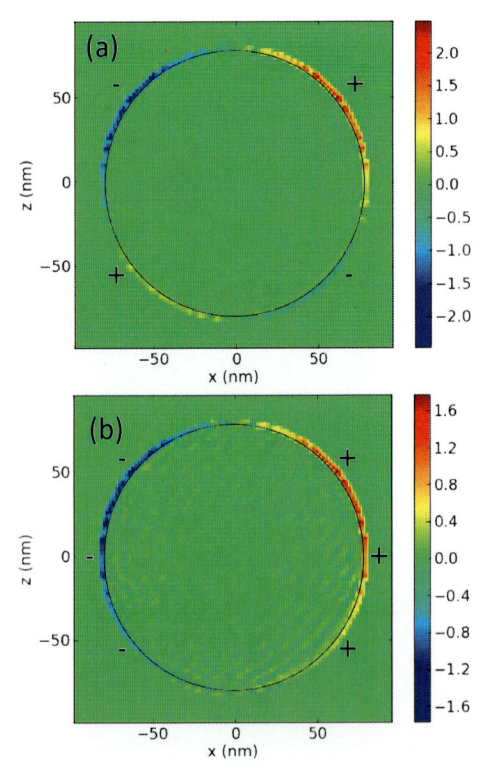

Figure 7.4 Distribution of the surface charges at maximum time, for the, (a), short wavelength and, (b), long wavelength modes.

When the shape of the particle is more complex, the Mie method is difficult to apply. However, expressions are available for shapes like cylinders and coated spheres, but not for spheroids, except under certain condition of illuminations. To go beyond these shapes, other numerical methods can be used, such as finite difference-time-domain (see Section 7.3), finite elements methods, or discrete dipole-approximation/Green's function formalism. We will briefly discuss in the following the case of a 3D parallelepiped particle, computed with the Green's tensor approach.

Green's tensor method. As for the Mie scattering method, the Green's tensor formalism is described in a number of articles (Martin and Piller, 1998; Paulus and Martin, 2001), and only a brief outline of the method is given here. The Green's tensor formalism allows to compute the electromagnetic field scattered by a small object embedded in a complex environment, under an arbitrary monochromatic illumination. If we consider a particle of volume V placed inside a multi-layered background, the total electric field in the whole space is given by the Lippmann–Schwinger equation:

$$\mathbf{E}(\mathbf{r}) = \mathbf{E}_0(\mathbf{r}) + \frac{\omega^2}{\varepsilon_0 c^2} \int_V d\mathbf{r}' \mathbf{G}_0(\mathbf{r}, \mathbf{r}') \mathbf{P}(\mathbf{r}') \tag{7.7}$$

The tensor $\mathbf{G}_0(\mathbf{r}, \mathbf{r}')$ is the Green's function of the system without the particle, and corresponds to the response of the environment to a localized dipolar excitation. It can be computed numerically for every couple $(\mathbf{r}, \mathbf{r}')$ in the whole space. As in the previous section, the polarization density is relied to the distribution of the total electric field inside the particle through:

$$\mathbf{P}(\mathbf{r}) = \varepsilon_0[\varepsilon(\mathbf{r}) - \varepsilon_B(\mathbf{r})]\mathbf{E}(\mathbf{r})$$

where ε_B, the dielectric constant of the background, can be a function of the position. Equation 7.7 then reads:

$$\mathbf{E}(\mathbf{r}) = \mathbf{E}_0(\mathbf{r}) + \frac{\omega^2}{c^2} \int_V d\mathbf{r}' \mathbf{G}_0(\mathbf{r}, \mathbf{r}')[\varepsilon(\mathbf{r}') - \varepsilon_B(\mathbf{r}')]\mathbf{E}(\mathbf{r}')$$

As the total electric field appears on both the left and the right side of the equality, the equation is self-consistent. It is solved numerically by discretizing the object in small cells.

Figure 7.5a shows the extinction, absorption, and scattering spectra of a parallelepiped gold particle of size $100 \times 100 \times 50$ nm^3, placed in water. The incident electric field is taken either parallel to the long edge (solid lines) or to the short edge (dashed line) of the particle. In the case of the long edge excitation, only one resonance can be noticed, at $\lambda = 700$ nm. This corresponds to the dipolar plasmon mode parallel to the long edge of the particle. Its position is largely red-shifted compared to the gold sphere because of the very flatten shape (see Fig. 7.1a). As for the 80 nm radius gold sphere, the scattering dominates the absorption in the contribution to extinction. When the incident electric field is parallel to the short edge, the long-edge dipolar plasmon cannot be excited because of its symmetry. However, two modes appear between 550 nm and 600 nm, of similar contribution to absorption and diffusion, which correspond to linear combination of short edge dipolar plasmon and quadrupolar mode (see Fig. 7.5b).

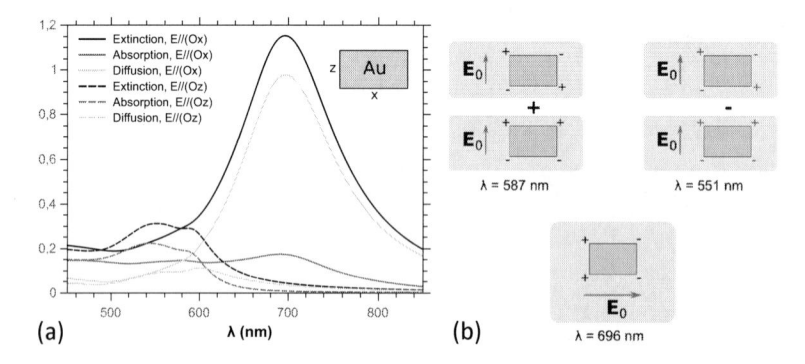

Figure 7.5 (a) Extinction, absorption, and diffusion spectra of a parallelepiped $100 \times 100 \times 50$ nm 3 gold particle in water. The x direction is parallel to one of the two long edges, whereas the z direction is parallel to the short edge. (b) Structures of the plasmons modes excited along the short edge or the long edge.

More generally, the number and structure of the surface plasmon modes of a localized plasmon mode can be large and complex, as shown in Zhang et al. (2011). These modes are combined together when several particles interact together and/or with a different environment, for example, a dielectric interface.

7.2 Examples of Coupled Plasmonic Systems

In this section, we give two examples of coupled plasmonic systems, consisting in a compact arrangement of a finite number of closely spaced metal nano-objects. When the separation between two particles is very small, typically few ten nanometers in the visible domain, the evanescent components of each individual localized plasmon modes overlap, resulting in a strong coupling between each particle. Their combination can result in delocalized collective modes of specific wavelength. Actually, we show next that the nature of the modes depends a lot on whether the localized plasmon mode of each individual particle is degenerated or not, i.e. in this case if the particles are identical or not.

7.2.1 Chain of Identical Particles

In the first example, we investigate the conduction of light along a chain of N identical gold particles deposited on a silica substrate (Lévêque and Quidant, 2008). The system is excited by a dipole located just in front of the first particle, its direction being parallel to the chain axis. The transmission of the chain is computed by evaluating the amplitude of the electric field just behind the last particle (see Fig. 7.6a). The excitation dipole couples to the collective surface plasmon modes arising from the electromagnetic coupling of the individual surface plasmons supported by each particle. The interaction of the dipole field with the chain was modeled using the Green's function method.

Figure 7.6b shows the transmission spectra for an increasing number N of particles. These systems are clearly multimodes, even if the strong absorption of gold makes the width of each mode too large to be able to resolve every single peak. The number of modes increases with N, they result from the recombination of the degenerated plasmon modes associated to each particle into delocalized collective modes. It appears clearly that the spectra are qualitatively different for chains with an even or an odd number of particles. In particular, the main peak around 760 nm exists only for an odd number of particles, and its position does not change with N. Actually, this mode is the optical analogue of the mechanical mode presented in Fig. 7.7 for a chain of oscillating masses, where one particle out of two does not move, making every other masses oscillating at the frequency of the single mass system. In the corresponding plasmonic mode, one particle out of two is "off", with no field inside. As a consequence, the neighboring particles are like isolated from all the others and the chain behaves as about $N/2$ independent gold particles. This type of system has been studied for their potential in light conduction with a very large lateral confinement; however, the propagation length is generally short, and does not exceed a few micrometers.

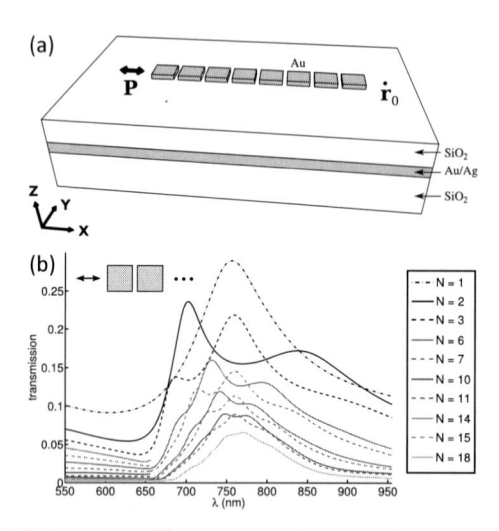

Figure 7.6 (a) System of N gold nanoparticles (dimensions $100 \times 100 \times 20$ nm^3) on a silica substrate, under dipolar excitation 100 nm away from the first particle. The period of the chain is 120 nm. (b) Amplitude of the electric field at a point $\mathbf{r}_0$ located 100 nm away from the last particle, as a function of the wavelength, and for an increasing number of particles.

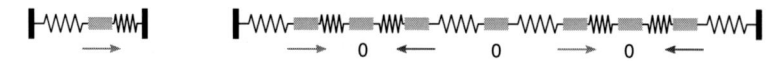

Figure 7.7 Mechanical equivalent of the plasmonic chain, as an oscillating masses system.

7.2.2 Chain of Different Particles

The second example presents a chain composed of six different gold particles (Lévêque and Martin, 2008), again in a near-field coupling regime (see Fig. 7.8a). The width d of every particle in the direction perpendicular to chain axis increases regularly along the chain. As a consequence, the individual plasmon modes parallel to that direction are non-degenerated. The entire system is illuminated from the silica substrate under total internal reflection by a plane wave with electric field parallel to the interface. It will excite the localized plasmon mode parallel to the longest side of each particle, whose resonance wavelength increases with d. A thin gold film, which role will be explained next, is placed 100 nm under the chain, inside the silica substrate.

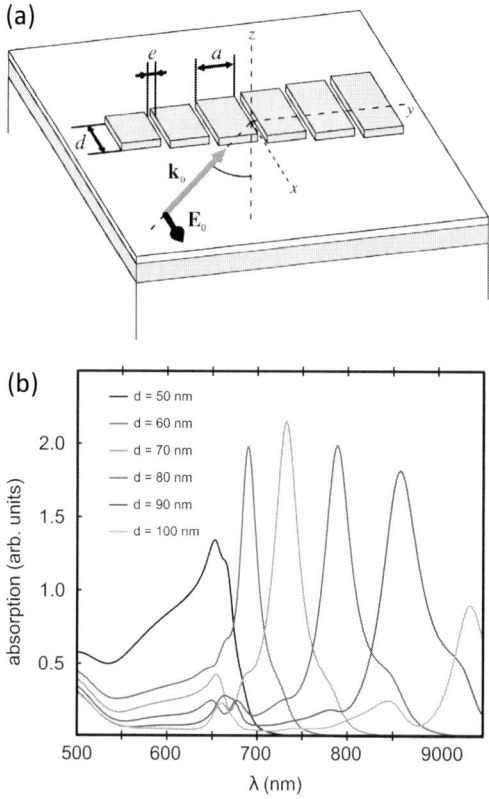

Figure 7.8 (a) Geometry of the system. The y side has a constant length $a = 50$ nm, while the x side of each particle increases from $d = 50$ nm to $d = 100$ nm with steps of 10 nm. The thickness is 20 nm, and the spacing between each particle is $e = 10$ nm. The incident wave propagates parallel to the yz plane, the incident electric field is parallel to the x axis. (b) Absorption spectrum computed for each particle in the chain.

Figure 7.8b shows the absorption spectra of each particle within the chain, which presents essentially one narrow absorption peak. This particularly remarkable feature comes from the fact that first the plasmon modes of every particle are not degenerated as in the previous case, but as well that the underneath gold film allows to concentrate the field in between the particle and the film, which decreases the interparticle coupling with respect to the film–particle coupling, and prevents the light from spreading on the neighbors when one particle mode is excited. Hence, the system exhibits six well-defined resonances. The associated modes are very well localized around each corresponding particle inside the chain, even with such a short distance between every object. The interest of such a multiresonant structure could be to create multiple addressable light spots whose excitation can be time-controlled on a femtosecond time scale using a custom designed light pulse.

7.3 Localized Surface Plasmon for a Periodic Nanostructure

In this last part of our chapter, we analyze the evolution of the surface plasmon modes in a 2D periodic nanostructure (NS), a layer of gold nanoparticles of diameter l, height h, and interparticle distance a are deposited on a transparent glass substrate, coated with dielectric films of diamond (NCD) and covered with a non-absorbing medium (air in our case). We focus our study on the variation in localized surface plasmon resonance (LSPR) structure as a function of dielectric coating thickness. The influence of the morphology and interparticle distance on the LSPR spectra of a glass/AuNS's interface without the NCD layer was investigated through the calculation of theoretical absorption spectra. The theoretical results were compared with experimental ones obtained on glass/AuNS's/ NCD surface.

7.3.1 Model and Simulation Method

Calculations are performed using finite difference time domain (FDTD) method which solves Maxwell's equations by discretizing both time and space and by replacing derivatives with finite differences (Yee, 1966; Tavlove, 1995). Our calculation is performed in a two-dimensional (2D) box (along x and y axes) with propagation along the y axis. Perfect Matching Layer (PML) conditions are applied at the boundaries y of the box, in order to avoid reflections of outgoing waves (Berenger, 1994). Along the x direction, the unit cell is repeated periodically and the structure is supposed to be infinite along the z direction. Space is discretized in both x and y directions using a mesh interval equal to $x = y = 1$ nm. The equations of motion are solved with a time integration step $t = x/4c$ and a number of time steps equals to 2^{20}, which is the necessary tested time for a good convergence of the numerical calculation.

The incoming pulse, having TM polarization, is generated at the bottom part of the unit cell, by a current source parallel to the x axis and having a planar profile along the x direction. The current is generated during a short period of time in such a way as to excite the electromagnetic waves in the frequency domain of interest. The transmitted and reflected signals are recorded as a function of time and finally Fourier transformed to obtain the transmission and reflection coefficients versus frequency. All the transmission (T) and reflection (R) spectra are normalized with respect to the one corresponding to a structure without the array of metallic nanoparticles. Finally, we calculate the absorption (A) by using the classical formula $A = 1 - (R + T)$. The absorption is reported in dB as a function of the wavelength.

Figure 7.9 illustrates the structure studied in this part of chapter. A layer of gold nanoparticles of diameter l, height h and interparticle distance a are deposited on a transparent glass substrate (refractive index $n_1 = 1.51$), coated with dielectric films of NCD with $n_2 = 2.4$ and covered with a non-absorbing medium such as air with $n_3 = 1$. The frequency-dependent complex permittivity of metal (gold) is described by the Lorentz–Drude model (Rakic et al., 1998).

$$\varepsilon(\omega) = \varepsilon_{r,\infty} + \sum_{m=0}^{M} \frac{f_m \omega_p^2}{\omega_m^2 - \omega^2 + j\omega\Gamma_m}.$$

where $\varepsilon_{r,\infty}$ is the relative permittivity at infinite frequency, ω_p the plasma frequency, and ω_m, f_m, and Γ_m are the resonance frequency, strength and damping frequency, respectively, of m-th oscillator. The Lorentz–Drude model uses M damped harmonic oscillators to describe the small resonances observed in the metal's frequency response. The values of the constants in Eq. (1) are taken from Rakic et al. (1998). This model allows us to fit experimental data for the frequency-dependent dielectric constant of metals such as gold (Rakic et al., 1998), including both the real and imaginary parts, with a good agreement in the visible wavelength range (400 nm $< \lambda <$ 1900 nm).

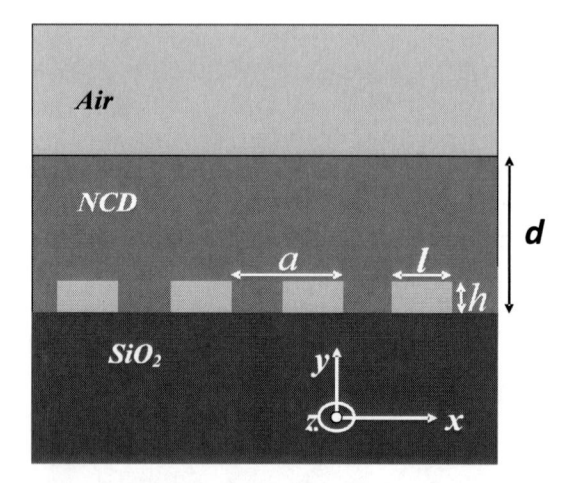

Figure 7.9 Representation of the structure studied in this part. The gold nanorod is characterized by the height h and the width l. The lattice parameter "a" is defined as the distance between two nearest neighboring gold nanorods. The input source is placed in the glass substrate and the detector in air.

7.3.2 Absorption Spectra for Au Nanostructures Array

The periodic structure studied in this paragraph is presented in Fig. 7.10a, where the gold NSs are deposited on the SiO_2 substrate and covered with air. We study the influence of changes in the height of the particles h (4–27 nm), the interparticle distance a (40–70 nm) and the particle diameter l (30–50 nm) on the LSPR. Figure 7.10b shows absorption spectra versus wavelength, calculated for glass/Au NSs interfaces of three values of the height h of the nanoparticles. The period of the Au NSs is fixed at a = 70 nm and the length of the Au NS at l = 25 nm. A decrease in the height of the Au NSs shifts λ_{max} to the red. The calculated maps of the electric field (Figs. 7.10c,d) at the monochromatic wavelengths

λ_{max} = 608 nm (the green peak in Fig. 7.10b) show a strong enhancement of the field located on the four corner of the Au nanoparticle.

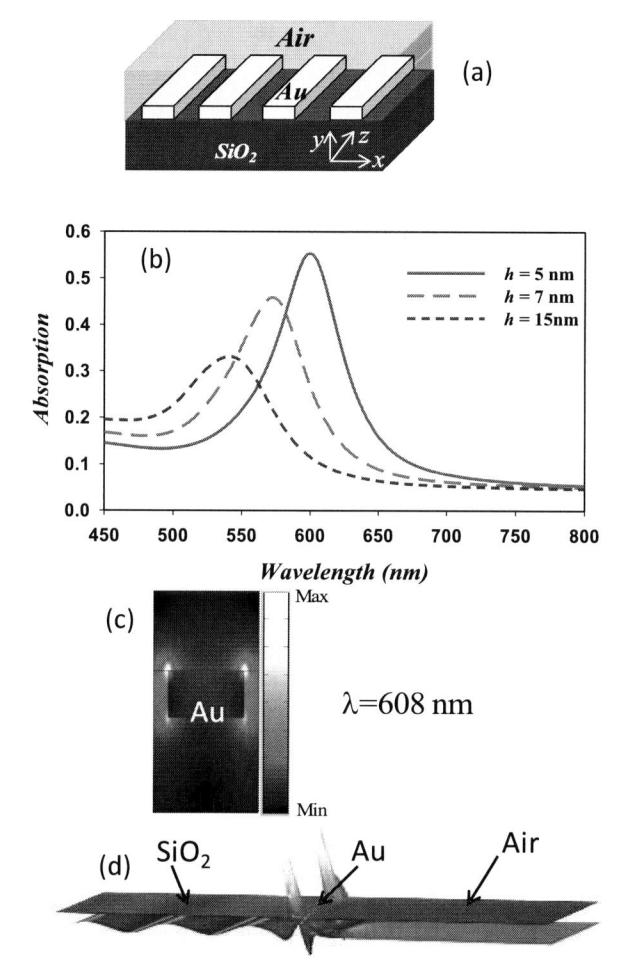

Figure 7.10 (a) The same as Fig. 7.9 but without dielectric films of diamond. (b) Change of absorption spectra versus wavelength of glass/Au NSs covered with a semi-infinite air medium. The transmission is calculated for Au NSs of different height (h = 5, 7 and 15 nm). The period of the Au NSs is fixed at a = 70 nm and the length of the Au NS at l = 25 nm. (c) and (d) Maps of the electric field for two monochromatic incident radiations at the wavelengths of the absorption peak λ_{max} = 608 nm.

Figure 7.11 shows the evolution of λ_{max} as a function of the height h of the nanoparticles (Fig. 7.11a, where the interparticle distance is fixed at λ = 70 nm and the width at l = 25 nm), as a function of the interparticle distance (Fig. 7.11b, where the height is fixed at h = 15 nm and the width at l = 25 nm), as function of the width l of the nanoparticles (Fig. 7.11c, where the interparticle

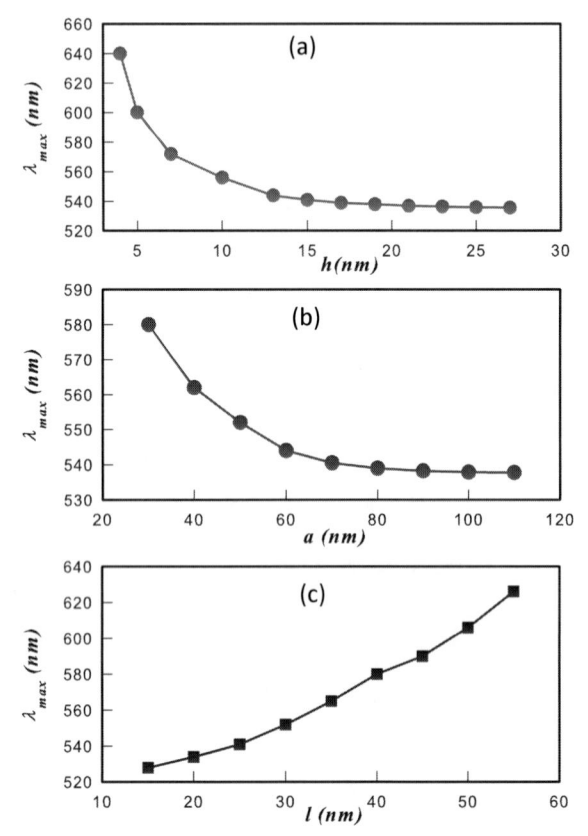

Figure 7.11 (a) Evolution of LSPR λ_{max} peak versus the height h of Au NSs covered with a semi-infinite air medium. The interparticle distance of the Au NSs is fixed at a = 70 nm and the length of the Au NSs at l = 25 nm. (b) Evolution of LSPR λ_{max} peak versus the period a of Au NSs covered with a semi-infinite air medium. The height of the Au NSs is fixed at h = 15 nm and the length of the Au NSs at l = 25 nm. (c) Evolution of LSPR λ_{max} peak versus the length l of Au NSs covered with a semi-infinite air medium. The interparticle distance of the Au NSs is fixed at a = 70 nm and the height of the Au NSs at h = 15 nm.

distance is fixed at λ = 70 nm and the height at h = 15 nm), respectively. We note that a decrease in the height of the Au NSs and thus an increase in the aspect ratio shifts λ_{max} to the red. Absorption spectra with increasing particle diameter and constant height shifts the LSPR λ_{max} to the red (not reported here). In fact, λ_{max} can be tuned continuously from ~400 nm to ~1200 nm by choosing the appropriate nanoparticle aspect ratio (Jensen et al., 2000). Electromagnetic coupling between Au NSs must also be considered in a well-designed experiment as this causes λ_{max} to shift. This coupling can be complicated. To determine the degree of coupling, a theoretical simulation was done in which the max of the LSPR was calculated as a function of the interparticle distance λ. Figure 7.11 shows that coupling is substantial when the nanoparticles are within ~60 nm of one another resulting in a significant red shift by further decreasing the interparticle distance. Our plasmonic interface showed a mean particle distance of about 16 nm and strong coupling is taking place.

7.3.3 Influence of the Thickness of a Diamond Dielectric Overlayer on the LSPR

The periodic structure studied in this paragraph is sketched on Fig. 7.9, where the gold NSs are deposited on the SiO_2 substrate, coated with dielectric films of NCD and covered with air. The experimental LSPR signals of the SiO_2/AuNSs interfaces before and after deposition of different NCD thick films are shown in Fig. (a). The deposition of NCD films of d_{NCD} results in a significant red shift of λ_{max} for all NCD thicknesses. In contrast to the observations on Al_2O_3 coated silver nanotriangles, where the LSPR λ_{max} shift levels off once a saturation point was reached (Szunerits et al., 2008), a different behavior was observed when increasing the thickness of the NCD film. From Fig. (b) it becomes clear that in the case of NCD, a saturation point was first reached at d_{NCD} ≈ 60 nm, which caused a $\Delta\lambda_{max}$ ≈ 110 nm. This is in accordance with predictions that the LSPR shift should saturate at $\Delta\lambda_{max}$ ≈ $m\Delta n$ ≈ 126 nm, where m is the refractive index sensitivity for the SiO_2/Au NSs interface ($m = \delta\lambda_{max}/\delta n$ = 90 nm) and Δn the change in refractive index induced by an adsorbate (Δn = 1.4). However, after the saturation point, a blue shift in λ_{max} was observed between 60 < d_{NCD} < 120 nm, followed by a red shift until d_{NCD} ≈ 200 nm.

These results are in accordance with other reports in the literature demonstrating long-range refractive index sensing on plasmonic nanostructures. The group of Kall et al. (Rindzevicius et al., 2007) studied the short and long range effects on ordered Au nanodisks an nanoholes coated with multilayers composed of 22-tricosenoic acid (n = 1.53) of coating thicknesses up to 340 nm. We have shown a similar behavior on random gold nanostructures deposited on glass with SiO_x and Si_3N_4 coatings having thicknesses between 0–300 nm (Jensen et al., 2000; Szunerits et al., 2008). The SiO_2/Au NSs/NCD hybrid interface is hence another LSPR configuration allowing for long-range sensing. Compared to the oscillating behavior of glass/Au NSs/SiO_x with n = 1.48 (Szunerits et al., 2008) and of glass/Au/Si_3N_4 (Galopin et al., 2009), the SiO_2/Au NSs/NCD hybrid interface shows a steeper linear red shift until $d_{NCD} \approx$ 60 nm.

The experimentally obtained shift in λ_{max} with increasing thicknesses of NCD overlayers (Fig. 7.12b) was compared to theoretically calculated λ_{max} shifts as reported recently (Jensen et al., 2000). Figure 7.12b shows that the experimentally observed oscillation behavior is corroborated by the calculated LSPR curves. The oscillation of λ_{max} extends until 1000 nm with a periodicity of $P \approx$ 140 nm and with increasing amplitude at d_{NCD} > 700 nm. The resonances (corresponding to the maximum of the amplitude of the oscillations) and the anti-resonances (corresponding to the minimum of the amplitude of the oscillations) are similar to the classical Fabry–Pérot cavity. Moreover, let us note also that the periodicity of the oscillation can be calculated by the following equation: $P = \lambda_{max}/2n_2$, where n_2 is the refractive index of the diamond [$P(NCD)$= 655/(2 × 2.4)= 136.5 nm (see Fig. 7.12b)]. The oscillation is anharmonic, that is, the plasmon shift is faster in the blue-shifting regions than in the red-shifting regions.

SiO_2/Au NSs/NCD/Air structure of varying layer thickness d. We obtained (not reported here) the same results when the structure is periodic, of period a = 70 nm, its absorption has been calculated with the FDTD method described previously. As a reference system, we have shown that the absorption spectrum of an SiO_2/Au NSs/NCD structure (no air above the diamond), which exhibits a LSPR resonance at λ_{max} = 655 nm (Saison et al., 2012; Akjouj et al., 2013). No oscillation of the LSPR wavelength occurs in this case as there is no NCD/air interface. However, when the gold NSs are placed inside a NCD layer of finite thickness, the LSPR is shifted as shown in Fig. 7.12b. This oscillation results from the interaction between the NSs and the system of stationary waves established inside the linear Fabry–Pérot cavity formed by the two SiO$_2$/NCD and NCD/air interfaces. In other words, when the thickness d_{NCD} increases, the wavelength for which a maximum of the interference pattern inside the layer occurs at the particle location is blue-shifted due to its FP-cavity mode nature. As a consequence, the resonance wavelength λ_{max} oscillates around the "average value" λ_{max}, determined by the system without air. These observations are consistent with the fact that the λ_{max} oscillation period P obeys the equation used in the previous paragraph. P corresponds to the resonance condition of linear Fabry–Pérot cavity modes.

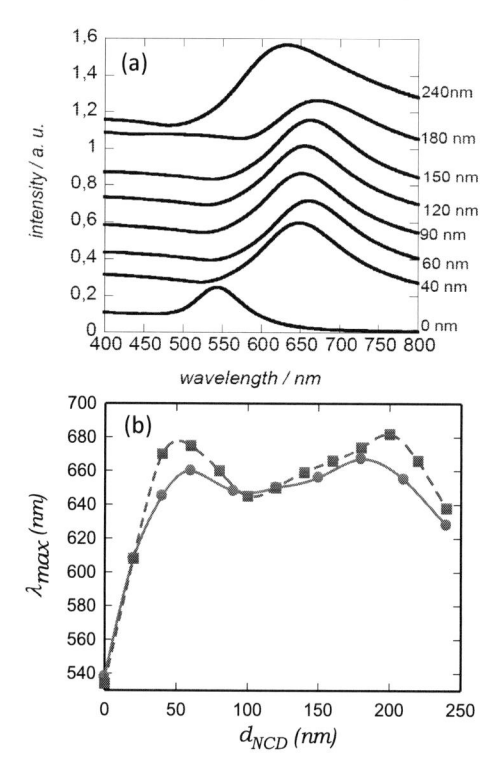

Figure 7.12 (a) UV/Vis transmission spectra of glass/Au NSs coated with 40 nm, 60 nm, 90 nm, 120 nm, 150 nm, 180 nm, 240 nm of NCD in air with a = 70 nm, h = 15 nm, l = 25 nm. (b) Evolution of λ_{max} with the thickness of the NCD overlayer; Comparison of experimental (red circles) and theoretical (green squares) data (see Szunerits et al., 2010).

From these observations, we can build a simple analytical model. Let us consider a plane wave in normal incidence on the silica layer, sent from the indiumtin-oxide (ITO) substrate. The electrical field modulus at the ITO/SiO$_x$ interface is continuous and reads without NSs:

$$E_0(\lambda, d) = E_i t_{12} \left(\frac{1 + 2r_{23}\cos(\phi) + (r_{23})^2}{1 - 2r_{23}r_{21}\cos(\phi) + (r_{23}r_{21})^2} \right)^{1/2}$$

where E_i is the amplitude of the incoming planewave electric field, and with

$$t_{12} = \frac{2n_1}{n_1 + n_2} \tag{7.10}$$

$$r_{ij} = \frac{n_i - n_j}{n_i + n_j} \tag{7.11}$$

$$\phi = \frac{4\pi n_2 d}{\lambda} \tag{7.12}$$

where t_{12} is the transmission coefficient at the ITO/SiO$_x$ interface, r_{21} and r_{23} are respectively the reflection coefficients at SiO$_x$/ITO and SiO$_x$/Air interfaces and ϕ is the phase difference after one back and forth of the light inside the silica layer. Hence, we can use the following expression to model the absorption spectrum of the gold NSs ("th" meaning *theory*):

$$A_{th}(\lambda, d) = E_0^2(\lambda, d) \times L(\lambda) \tag{7.13}$$

with

$$L(\lambda) = \frac{K^2 H}{(\lambda - \lambda_{eff})^2 + K^2}$$

where $L(\lambda)$ is an *ad hoc* fitting function of the Au-NSs plasmon response inside the SiO$_2$/NCD matrix (without air) as a Lorentzian function centered on λ_{max} = 655 nm, of arbitrary amplitude H and full width at half maximum (FWHM) equal to $2K$ = 200 nm (λ_{max} and K were extracted from FDTD comparison. A good agreement between the λ_{max} evolution extracted from the analytical model and from the FDTD simulation results are observed in our results (Saison et al., 2012; Akjouj et al., 2013).

The last figure of this part, Fig. 7.13, gives the evolution of the amplitude, the wavelength, and the periodicity of the oscillation as a function of the geometrical parameter of the Au NSs. An increase in the height of the Au NSs from 10 to 20 nm with a = 70 nm and l = 25 nm shifts the oscillation curves to the blue (Fig. 7.13a) with an increase of the amplitude of the oscillation. Figure 7.13b shows the evolution of oscillation curves for l = 20, 25 and 30 nm with h = 15 nm and λ = 70 nm. The figure shows that the decrease of the length of the Au NSs shifts the oscillation curves to the blue with an increase of the amplitude of the oscillation. In Fig. 7.13c the increase of the lattice parameter "a" from 50 to 90 nm with l = 25 nm and h = 15 nm shifts the oscillation curves to the blue, with an increase of the amplitude of the oscillation and a decrease of the periodicity of the oscillation. In conclusion, this theoretical study shows that to increase the amplitude of the oscillation, it is necessary to increase the values of the height h and of the lattice parameter a and to reduce the value of the length l of the Au NSs. Note also that the periodicity of the oscillation varies only as a function of the lattice parameter of the Au NSs.

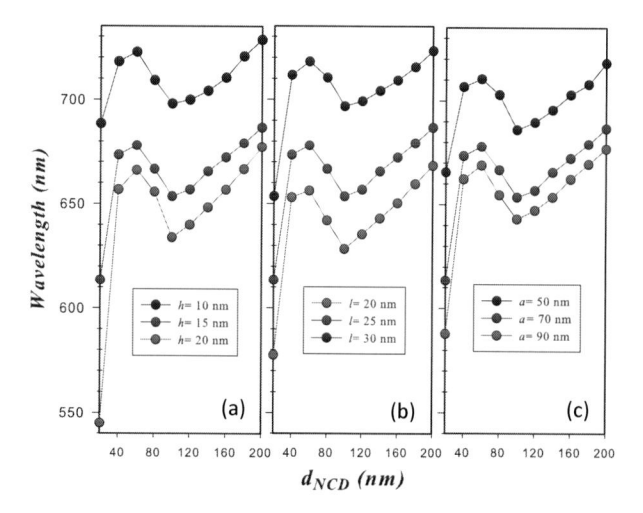

Figure 7.13 Evolution of LSPR peak with change in NCD film thickness d. (a) a = 70 nm, l = 25 nm and h = 10, 15, 20 nm, (b) a = 70 nm, h = 15 nm, and l = 20, 25, 30 nm, (c) l = 25 nm, h = 15 nm and a = 50, 70, 90 nm.

7.3.4 Conclusion

We have studied the optical properties of short range ordered gold nanoparticles as a function of overlayer thickness. The Lorentz–Drude model was used to calculate the optical signal of localized surface plasmon resonances (LSPR) comprising a random array of gold nanostrucutres (Au NSs) on glass and coated with differently thick dielectrics. Independent of the dielectric chosen, diamond (NCD), no saturation of the LSPR signal was observed, but rather an oscillation behavior that extends over several hundreds of nanometer from the glass/Au NSs interface. To validate the model, experimentally obtained LSPR spectra of the novel interfaces were compared to theoretical ones. We are currently investigating interfaces with even higher refractive indices to see if a general conclusion can be drawn. The interest in such multilayer interfaces is the possibility of sensitive long-range sensing with important implications in biological studies.

References

Akjouj, A., Leveque, G., Szunerits, S., Pennec, Y., Djafari-Rouhani, B., Boukherroub, R., and Dobrzynski L. (2013). Nanometal plasmon polaritons, *Surf. Sci. Rep.*, **68**, 1–67.

Berenger, J. P. (1994). A perfectly Matched Layer for the Absorption of Electromagnetic Waves, *J. Comput. Phys.*, **114**, 185–200.

Bohren, C. F., and Huffmann, D. R. (1998). *Absorption and Scattering of Light by Small Particles*, Wiley and Sons.

Faraday, M. (1857). The Bakerian lecture: Experimental relations of gold (and other metals) to light, *Phil. Trans. R. Soc. Lond.*, **147**, 145–181.

Fleischmann, M., Hendra, P.-J., and Mequilla, A.-J. (1974). Raman spectra of pyridine adsorbed at a silver electrode, *Chem. Phys. Lett.*, **26**, 163–166.

Galopin, E., Noual, A., Niedziolka-Jönsson, J., Jönsson-Niedziolka, M., Akjouj, A., Pennec, Y., Djafari-Rouhani, B., Boukherroub, R., and Szunerits, S. (2009). Short-and long-range sensing using plasmonic nanostrucures: Experimental and theoretical studies, *J. Phys. Chem. C*, **113**, 15921–15927.

Gans, R. (1912). Uber die Form ultramikroskopischer Goldteilchen, *Ann. Phys.* **342**, 881–900.

Jensen, T. R., Malinsky, M. D., Haynes, C. L., and Van Duyne, R. P. (2000). Nanosphere lithography: Tunable localized surface plasmon resonance spectra of silver nanoparticles, *J. Phys. Chem. B,* **104**, 10549–10556.

Johnson, P. B., and Christy, R. W. (1972). Optical constants of noble metals, *Phys. Rev. B,* **6**, 12, 4370–4379.

Lévêque, G., and Martin, O. J. F. (2008). Narrow-band multiresonant plasmon nanostructure for the coherent control of light: An optical analog of the xylophone, *Phys. Rev. Lett.,* **100**, p. 117402.

Lévêque, G., and Quidant, R. (2008). Channeling light along a chain of near-field coupled gold nanoparticles near a metallic film, *Opt. Express,* **16**, 22029–22038.

Martin, O. J. F. (2003). Plasmon resonances in nanowires with a non-regular cross-section, *Springer Verlag Series on Topics in Applied Physics,* **88**, 183–210.

Martin, O. J. F., and Piller, N. B. (1998). Electromagnetic scattering in polarization background, *Phys. Rev. E,* **58**, 3909–3915.

Mie, G. (1908). Beiträge zur Optik trüber Medien, speziell kolloidaler Metallösungen, *Ann. Phys.,* **330**, 377–445.

Mubeen, S., Zhang, S., Kim, N., Lee, S., Kra, S., Xu, H., and Moskovits, M. (2012). Plasmonic properties of gold nanoparticles separated from a gold mirror by an ultrathin oxide, *Nano Lett.,* **12**, 2088–2094.

Novotny, L., van Hulst, N. (2011). Antennas for light, *Nat. Photon.,* **5**, 83–90.

Paulus, M., and Martin, O. J. F. (2001). Light propagation and scattering in stratified media: a Greens tensor approach, *J. Opt. Soc. Am. A,* **18**, 854–861.

Rakic, A. D., Djuristic, A. B., Elazar, J. M., and Majewski, M. L. (1998). Analysis of optical channel cross talk for free-space optical interconnects in the presence of higher-order transverse modes, *Appl. Opt.,* **37**, 5271–5283.

Rindzevicius, T.; Alaverdyan, Y.; Käll, M.; Murray, W. A.; Barnes, W. L. (2007). Long-range refractive index sensing using plasmonic nanostructures, *J. Phys. Chem. C,* **111**, 11806–11810.

Saison O., Leveque G., Akjouj A., Pennec Y., Djafari-Rouhani B., Boukherroub R., and Szunerits S. (2012). Plasmonic nanoparticles array for high-sensitivity sensing: A theoretical investigation, *J. Phys. Chem. C,* **116**, 17819–17827.

Saison-Francioso, O., Lévêque, G., Akjouj, A., Pennec, Y., Djafari-Rouhani, B., Boukherroub, R., and Szunerits, S. (2013). Search of extremely sensitive near–infrared plasmonic interfaces: a theoretical study, *Plasmonics,* **8**, 1691–1698.

Szunerits S., Ghodbane S., Niedziolka-Jönsson J., Galopin E., Klauser F., Akjouj A., Pennec Y., Djafari-Rouhani B., Boukherroub R., and Steinmuller-Nethl D. (2010). Development and characterization of a diamond-based localized surface plasmon resonance interface, *J. Phys. Chem. C*, **114**, 3346–3353.

Szunerits, S., Das, M. R., and Boukherroub, R. (2008). Short-and long-range sensing on gold nanostructures, deposited on glass, coated with silicon oxide films of different thicknesses, *J. Phys. Chem. C*, **112**, 8239–8243.

Taflove, A. (1995). *Computational Electrodynamics: The Finite-Difference Time-Domain Method*, Norwood, MA: Artech House.

Underwood, S., and Mulvaney, P. (1994). Effect of the solution refractive index on the color of gold colloids, *Langmuir*, **10**, 3427–3430.

Yaghjian, A. D. (1980). Electric dyadic Green's functions in the source region, *Proc. IEEE*, **68**, 2, 248–263.

Yee, K. S. (1966). Numerical solution of initial boundary value problems involving Maxwell's equations in isotropic media, *IEEE Trans. Antennas Propagat.*, **14**, 302–307.

Zhang, S., Bao, K., Halas, N. J., Xu, H., and Nordlander, P. (2011). Substrate-induced Fano resonances of a plasmonic nanocube: A route to increased-sensitivity localized surface plasmon resonance sensors revealed, *Nano Lett.*, **11**, 1657–1663.

Chapter 8

Advances in the Fabrication of Plasmonic Nanostructures: Plasmonics Going Down to the Nanoscale

Thomas Maurer

Université de Technologie de Troyes,
Laboratoire de Nanotechnologie et d'Instrumentation Optique–Institut Charles Delaunay, CNRS UMR 6281, 12 rue Marie Curie,
CS 42060, F-10004 Troyes Cedex, France

thomas.maurer@utt.fr

8.1 Introduction

One of the major breakthroughs in solid-state physics in the past 30 years has been the miniaturization of materials down to the nanometer scale. The progress that has been made in the past decades has been so tremendous that the expression of "technological revolution" is often employed to qualify the field of nanotechnology. It is generally admitted that the term "nano" refers to materials whose at least one dimension is of the order of the nanometer. However, this definition seems trivial if one considers some natural nanoparticles such as the ones rejected by the volcanic eruptions and that do not present any specific properties.

Introduction to Plasmonics: Advances and Applications
Edited by Sabine Szunerits and Rabah Boukherroub
Copyright © 2015 Pan Stanford Publishing Pte. Ltd.
ISBN 978-981-4613-12-5 (Hardcover), 978-981-4613-13-2 (eBook)
www.panstanford.com

Precisely, a nano-object is so called when its properties differ from the bulk ones. The size reduction indeed makes appear a threshold below which the physical and chemical properties of the nanomaterial vary from the bulk ones. The modification of the material properties in fact stems from the drastic increase of the ratio between the surface and the volume of the material. Below the size threshold, the investigated properties become size-dependent. The size threshold of course depends on both the nature of the material and of the investigated properties and will differ depending if we pay attention to mechanical, chemical, optical, or magnetic features of nanoparticles. In the beginning of the 1980s, the field of plasmonics emerged when Nylander, Liedberg, and Lind demonstrated that surface plasmon resonance (SPR) of plasmonic thin films ($\approx$50 nm) could lead to gas sensing [1]. Plasmonic thin films were extensively studied during the last three decades [2]. Therefore, they will not be included in the scope of the present chapter. More recently, a turning point was marked in the development of plasmonics via the progress in nanofabrication. Indeed, nanostructuration of noble metals allowed inducing interaction between light and the collective coherent oscillation of the metal free electrons without the need of prism or complex setup [3]. Such effect has been involved through centuries by artisans fabricating coloring glasses or church windows without understanding that they were synthesizing gold or copper nanoparticles [4]. The possibility of controlling the size and shape of metal nanoparticles boosted the field of plasmonics at the end of the 1990s and made it very attractive both from a fundamental and applied point of view.

In nanofabrication, there is no technique prevailing over the others. The targeted applications and the plasmonic structure size or organization will drive the choice of the fabrication process. Two approaches have been developed to synthesize nano-objects: the top-down and the bottom-up ones. The top-down approach consists of taking and etching a massive material down to the desired shape and size. The processes associated to it take advantage of their high reproducibility and find some immediate applications in the industry. After many improvements, this approach nowadays allows to produce structures as small as 20 nm in size. Semiconductor companies even project reaching the 14 nm node technology via lithography or etching techniques [5].

However, it is still challenging to obtain structures smaller than 20 nm with well-defined edges over large areas [6]. In order to go below this threshold, synthesis routes based on the *bottom-up* approach have been developed for about 15 years. The idea consists of assembling atoms to fabricate nano-objects. The emergence of such techniques helped for getting large batch of gold nanoparticles, in particular via colloidal synthesis [7]. For the moment, large-scale applications are still limited with this approach. Indeed, a key challenge remains the development of efficient ways for organizing colloids [8, 9] especially when nanoparticles are not magnetic [10–12]. Therefore, new strategies that combine top-down and bottom-up strategies start emerging to drive the self-organization of plasmonic nanoparticles.

This chapter focuses on nanofabrication related to plasmonic structures. It is divided into three distinct parts: (i) top-down techniques, (ii) bottom-up processes, and (iii) routes combining the top-down and bottom-up strategies. The conclusion provides perspectives in the field of nanofabrication for plasmonics.

8.2 Top-Down Techniques: A Mask-Based Process

Top-down techniques are mainly based on the use of masks (see Fig. 8.1). Indeed, from the mask fabrication, nanostructuration is then possible via dry etching of pre-evaporated metal film, mask removal after metal film deposition, or electroplating. Therefore, the fabrication of the masks is a crucial step in top-down nanostructuration and will be discussed in this paragraph related to the field of plasmonics.

8.2.1 Conventional Lithography Techniques: Photolithography and Particle Beam Lithography

The empiric and famous Moore's law is the epitome of the miniaturization success achieved by the microelectronics industry. This success is based on the technological development of two lithography techniques: (i) photolithography and (ii) scanned beam lithography. In both cases, resins are patterned to form masks before nanostructuration of surfaces by either etching or metal deposition.

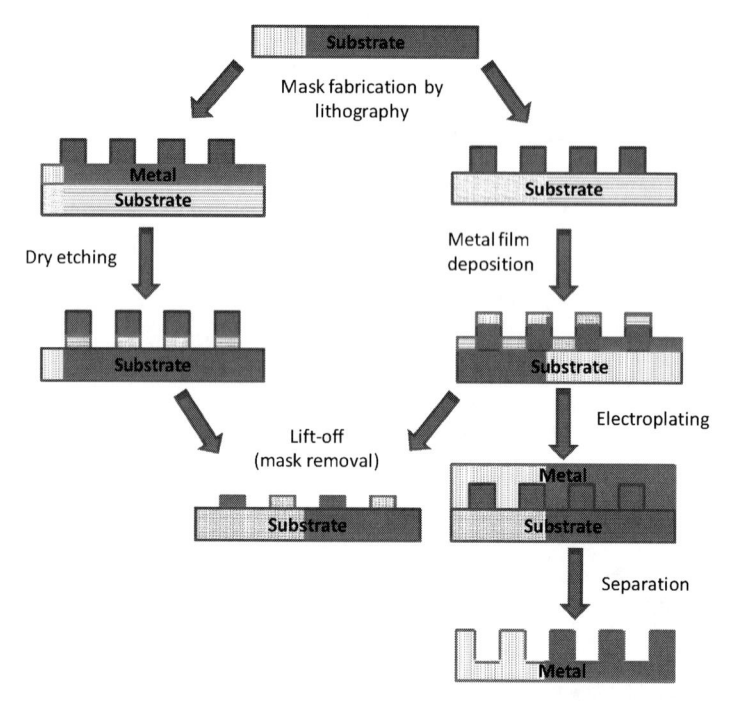

Figure 8.1 Schematic illustration of nanofabrication by top-down techniques. The first step consists of fabricating masks via either resin patterning or self-assembly process. Then dry etching, lift-off after metal evaporation or electroplating leads to nanostructuration of metallic surfaces.

8.2.1.1 Photolithography

Photolithography, the most adapted method for mass production, consists of projecting light through a chromium mask pre-patterned onto a quartz plate in order to expose photoresist-coated substrates. Then, immersion into solvents will lead to surface patterning of the resin-coated substrate by removing the exposed (positive photoresist) or unexposed (negative photoresist) regions. This technique encountered great success in microelectronics and can provide sub-50 nm patterns. However, achieving such performance starts to be very expensive in time, money, and efforts. Indeed, resolution is defined as the possibility to distinguish two diffraction spots coming from two points that are angularly close. A diffraction spot obtained from a circular hole looks like an Airy disc so that these two spots are discernible for distances superior to the angular radius, r, which is the distance central maximum of the first spot intensity profile and the first minimum of the second intensity profile (see Fig. 8.2). Therefore, the instrumental resolution is related to the Rayleigh criteria, [13] providing the angular radius:

$$r = 1.22 \frac{\lambda}{NA}, \qquad (8.1)$$

where λ is the illumination wavelength and NA the numerical aperture of the lens. To increase the instrumental resolution power, there are thus three main possibilities: (i) decreasing the wavelength of light sources down to the deep-UV (193 nm and even 157 nm thanks to F2 lasers) and (ii) increasing the numerical aperture by working in media with higher refractive index as for immersion lithography [14] or (iii) using near-field techniques to break the diffraction limit like the evanescent near-field optical lithography (ENFOL) [15] or the plasmonic nanolithography [16].

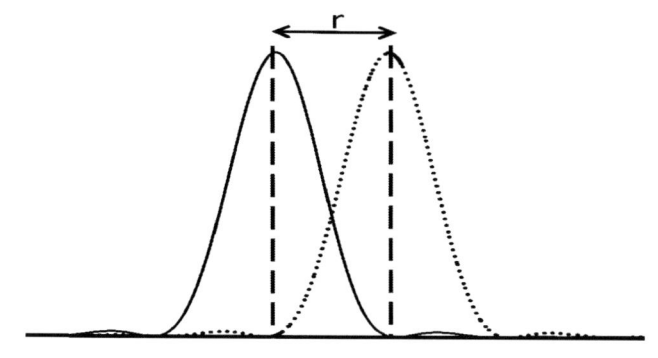

Figure 8.2 Intensity profiles of two diffraction spots discernible from each other. The intenisty profiles are proportional to Airy functions $(J_1(x)/x)^2$, where J_1 is the first-order Bessel function.

In conclusion, photolithography is well adapted for structures larger than 100 nm since instrumental resolution is roughly limited to $\lambda/2$. Therefore, this technique suits better to investigation of plasmon resonances located in the infrared range [17]. That is why particle beam lithography techniques such as electron beam lithography (EBL) and focused ion beam (FIB) lithography have witnessed great success for studying plasmon resonances of nanostructures in the visible range and will now be discussed.

8.2.1.2 Particle beam lithography

Particle beam lithography belongs to the family of scanned beam lithography, which also includes laser beam lithography. Laser beam lithography provides 250 nm resolution and is often used in the industry to fabricate photolithography masks at the micron scale. However, in order to reach sub-50 nm resolution, scanned beam lithography techniques irradiating resins with particles are required. Such techniques are relatively slow and expensive compared to photolithography, but allow breaking the diffraction limit. The two main classes of particle beam lithography are EBL and FIB lithography. They are respectively based on the use of accelerated electrons and ions. These techniques suit perfectly for academic research, especially when the aim is to investigate anisotropic nanostructures (see Fig. 8.3). To increase the resolution, it is necessary to decrease the particle beam diameter and thus to

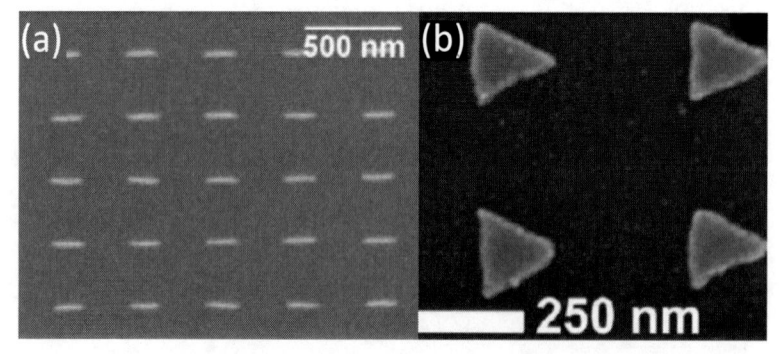

Figure 8.3 (a) array of gold nanorods fabricated by electron beam lithography (EBL). Reprinted from *Synth. Metals*, **139**(3), Grand, J., Kostcheev, S., Bijeon, J.-L., Lamy de la Chapelle, M., Adam, P.-M., Rumyantseva, A., Lérondel, G., Royer, P., Optimization of SERS-active substrates for near-field Raman spectroscopy, 621–624. Copyright (2003), with permission from Elsevier. (b) Gold nanotriangles obtained via EBL. Reprinted with permission from Awada, C., Popescu, T., Douillard, L., Charra, F., Perron, A., Yockell-Lelièvre, H., Baudrion, A.-L., Adam, P. M., Bachelot, R. (2012). Selective excitation of plasmon resonances of single Au triangles by polarization-dependent light excitation, *J. Phys. Chem. C.*, **116**, 14591–14598. Copyright (2012) American Chemical Society.

work with low beam current. This implies preliminary tests for controlling the exposing dose. More practical details can be found in the review paper written by Gilles Lérondel, Sergeï Kostcheev, and Jérôme Plain [18] and will thus not be further discussed here. However, note that a focused particle beam allows patterning photoresist and also direct writing onto metal films or substrates (see Fig. 8.3c).

8.2.2 Advanced Lithography Techniques: Masks Coming from Researcher Imagination

In the past decades, one of the key challenges to improve lithography process has been the development of new strategies to fabricate mask. Efforts have been devoted to decrease the pattern size, increase the complexity of mask geometry, avoid chemical treatment or the use of resin or reduce the cost of fabrication. This section aims to introduce recent development among lithography techniques.

8.2.2.1 Multilevel laser interference lithography

Laser interference lithography belongs to the photolithography techniques and became very attractive for fabricating grating structures over large areas due to a high spatial coherence. As explained in Section 8.2.1, the resolution power is limited by $\lambda/2n$. Therefore, a solution consists of working in deep-UV under immersion (see Fig. 8.4a). Such techniques allow fabricating gratings with sub-50 nm periods (see Figs. 8.4A,B) but their use is limited by their complexity, cost and small exposure area [22]. Nevertheless, recently Chang and coworkers proposed an efficient alternative named multilevel interference lithography, consisting of multiple expositions with π phase-offset for each step (see Figs. 8.4C,D). The fabrication of Au gratings via laser interference lithography is particularly adapted for SPR substrates development [23]. Note that other photolithography techniques developed in the CMOS industry, such as spacer patterning technology [24], could be applied to plasmonic nanofabrication in the next future.

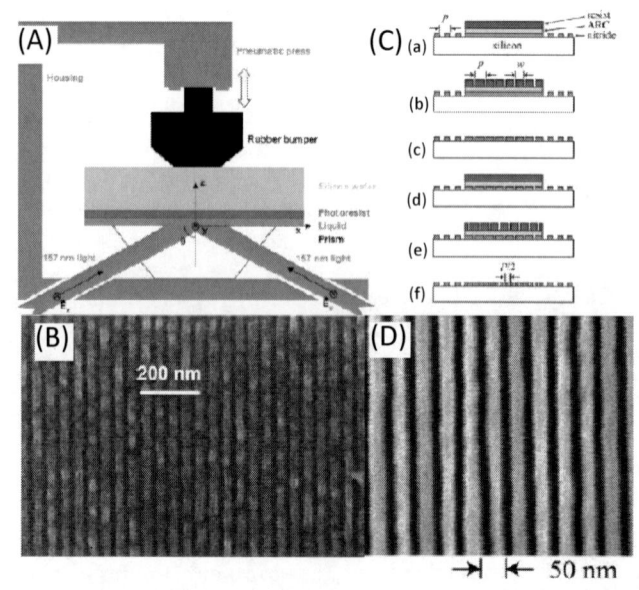

Figure 8.4 (A) Schematic illustration of the immersion laser interference lithography output portion, (B) SEM image of 44 nm pitch patterned photoresist using interference lithography (Reproduced with permission from Bloomstein, T. M., Marchant, M. F., Deneault, S., Hardy, D. E., Rothschild, M. (2006). 22 nm immersion interference lithography, *Opt. Express*, **14**, 6434–6443.), (C) Fabrication process diagram for multilevel interference lithography (Reproduced with permission from Chang, C.-H., Zhao, Y., Heilmann, R. K., Schattenburg, M. L. (2008). Fabrication of 50 nm period gratings with multilevel interference lithography, *Opt. Lett.*, **33**, 1572–1574.): (a) A reference grating is patterned in the outer substrate region, while the center region is spin-coated with anti-reflection coating (ARC) and photoresist, (b) after aligning to the reference grating, the first grating level with a period $p = 200$ nm is patterned, (c) the pattern is transferred into nitride, (d) after spin-coating ARC and photoresist, (e) the second grating level is exposed at an additional π phase-offset, (f) the grating is pattern transferred, resulting in a nitride grating with a period of $p/2 = 100$ nm, (D) 50 nm period grating fabricated by overlaying four 200 nm period grating levels.

8.2.2.2 Nanostencil lithography

One of the main disadvantages of electron-beam or focused ion-beam lithography comes from their low throughput. A way to overcome this limit consists of preparing suspended masks, which will be "stuck" onto substrates and act as stencils. That is why this shadow-mask technique is named nanostencil lithography.

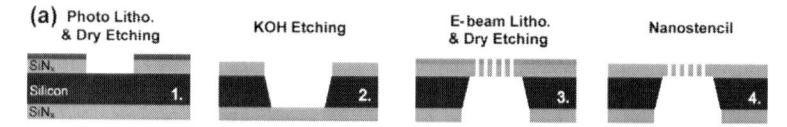

Figure 8.5 (1 and 2) Fabrication of suspended membrane via photolithography and etching of Si/SiN$_X$ films. Note that SiN$_X$ is robust enough to be used as free-standing membrane. (3 and 4) Patterning of the suspended membrane via EBL and dry-etching to form shadow masks. Reprinted with permission from Aksu, S., Yanik, A. A., Adato, R., Artar, A., Huang, M., Altug, H. (2011). Nanostencil lithography for high-throughput fabrication of infrared plasmonic sensors, *Proc. SPIE*, 8031. Copyright (2011) American Chemical Society.

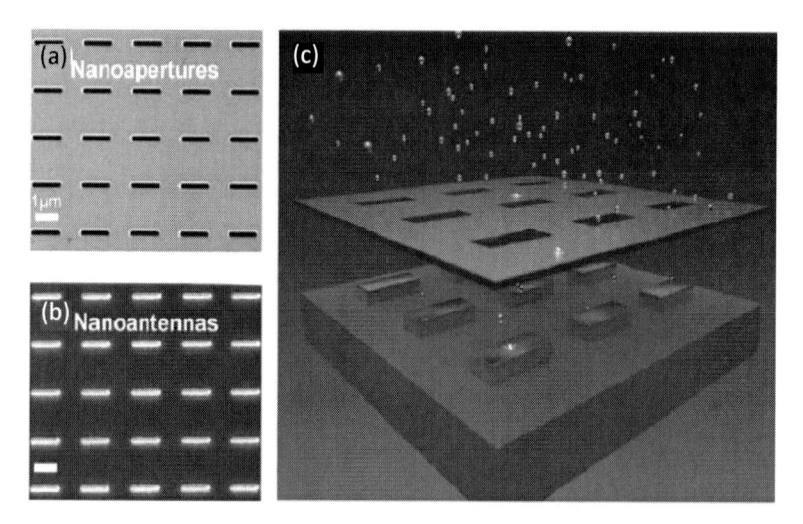

Figure 8.6 (a) A shadow mask acting as nanostencils, (b) nano-antennas made after gold deposition via the shadow mask presented in Fig. 8.6a, and (c) Illustration of nano-antennas fabrication via plasmonic metal deposition via the nanostencils. Reprinted with permission from Aksu, S., Yanik, A. A., Adato, R., Artar, A., Huang, M., Altug, H. (2011). Nanostencil lithography for high-throughput fabrication of infrared plasmonic sensors, *Proc. SPIE*, 8031. Copyright (2011) American Chemical Society.

A pre-patterned mask allows depositing material, which may be metal, dielectric, or organic. Recently, Aksu and coworkers promoted this process for optical nano-antenna fabrication [26, 27]. Nanostencil lithography is composed of three main steps: (i) fabrication of the suspended membrane (see Fig. 8.5(1) and 8.5(2)), (ii) patterning of the nanostencil (see Fig. 8.5(3) and 8.5(4)) and (iii) plasmonic metal deposition (see Fig. 8.6). Another advantage of nanostencil lithography is the absence of lift-off step. This prevents from pre-evaporating adhesion layer and thus from damping plasmon resonances.

8.2.2.3 Self-assembly techniques for mask fabrication: nanosphere lithorgaphy (NSL) and block copolymer lithography (BCL)

One of the most promising perspectives for mask fabrication over large areas at low cost is certainly the use of bottom-up process. A well-known technique, referred to as nanosphere lithography

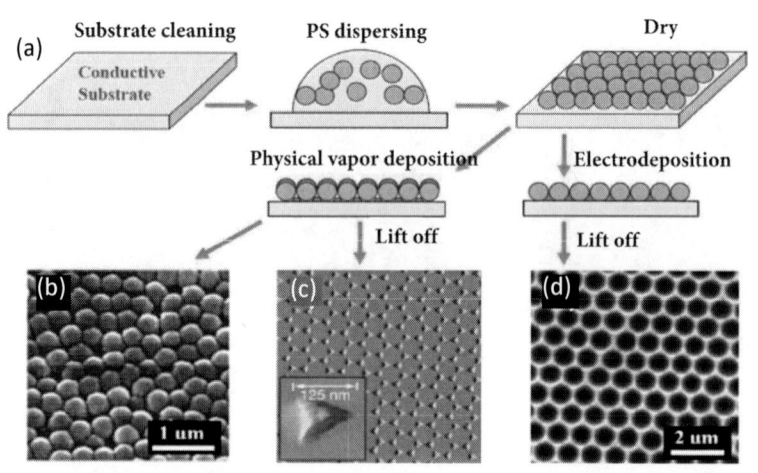

Figure 8.7 (a) Principle of nanosphere lithography. The process is based on the self-organization of polystyrene (PS) nanospheres. Then physical vapor deposition of a metal layer or electroplating leads to the synthesis of metallic triangles (b and c) or connected rings (d). The final step consists of removing PS nanospheres using adhesive tape. Reproduced from Wu, D.-Y., Li, J.-F., Ren, B., Tian, Z.-Q. (2008). Electrochemical surface-enhanced Raman spectroscopy of nanostructures, *Chem. Soc. Rev.*, **37**(5), with permission of The Royal Society of Chemistry.

(NSL) and described in Fig. 8.7, was initially imagined by Hulteen and Van Duyne in 1995 [28] and has since then offered great promise for applications in the field of plasmonics and especially for biosensors development [29].

This nanofabrication process finds potential applications in localized surface plasmon (LSP) [30], surface-enhanced Raman spectroscopy [31] and wavelength-scanned surface-enhanced Raman excitation spectroscopy (WS SERES) sensors. Figure 8.8 highlights how silver nanotriangles prepared via nanosphere lithography are good transducers for Alzheimer's biosensors [32].

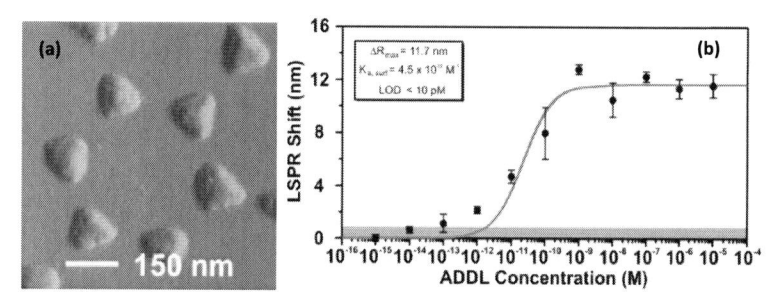

Figure 8.8 (a) Ag nanoparticles (width 90 nm, height 25 nm) made by nanosphere lithography on mica substrates, (b) Detection of the antigen amyloid-β derived diffusible ligand (ADDL)—a biomarker suspected to play a central role in the Alzheimer's disease, thanks to the shift of the Ag LSP resonance [32].

Note that recent developments extended NSL to sparse colloidal lithography (SCL) and hole-mask colloidal lithography (HCL) [33]. In the former technique, nanospheres are sparsely dispersed onto substrates. Hole-mask colloidal lithography allows a better versatility by using a resin sacrificial layer. This technique recently led to the fabrication of 3D chiral plasmonic structures [34].

Another lithography technique based on self-assembly masks is block copolymer lithography. The key idea consists of preparing copolymer masks composed of cylindrical domains whose polymer can be preferably etched compared to the surrounding matrix polymer. Among adequate copolymers, PS-PMMA copolymer is a good choice since PMMA can be much faster etched than PS either via reactive ion etching (RIE) [36–38] or via UV-exposure [38, 39]. Etching of PMMA domains therefore leads to the formation of hole arrays over large areas before metal film evaporation. The main limitation of such process is the difficulty in obtaining deep holes. Indeed, the lift-off procedure requires a depth of about three times the metal thickness that is intended to be deposited. However, RIE and UV-exposure are not fully anisotropic and holes may join together during too long processes. However, metal nanoparticles with 5 to 40 nm thickness may be today achieved when the process is well controlled [35, 39] and lie at the root of plasmonic biosensors with good performances (see Fig. 8.9 for process principles) [35]. Note that such process can lead to large-scale production of biosensors and is thus perfectly adapted to industry.

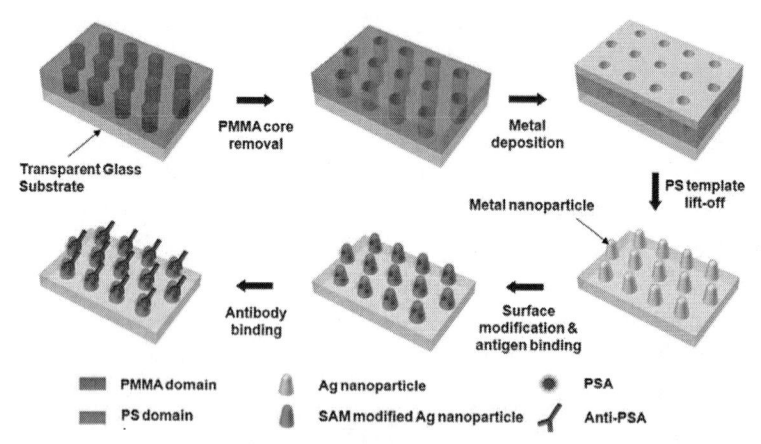

Figure 8.9 Block copolymer lithography (BCL) applied to the fabrication of LSPR biosensors. Reproduced from Shin, D. O., Jeong, J.-R., Han, T. H., Koo, C. M., Park, H.-J., Lim, Y. T., Kim, S. O. (2010). A plasmonic biosensor array by block copolymer lithography, *J. Mater. Chem.*, **20**, 7241–7247, with permission of The Royal Society of Chemistry.

8.2.3 Direct Writing

Direct writing techniques are time-saving since they do not require different steps such as resin coating, exposure, development, metal deposition and lift-off. Among these one-step techniques, one of the most promising is focused ion-beam etching (FIB), which was mentioned in Section 8.2.3. Such a technique allows material milling and finds numerous applications for direct plasmonic nanoparticle fabrication [40]. In fact, direct writing still more concerns the semiconductor industry. In this section, we will mention direct writing routes that have started to be used for plasmonic nanofabrication.

8.2.3.1 Particle beam–induced etching and particle beam–induced deposition

Focused particle beam is a well-adapted technique for direct writing at the nanoscale. Apart from focused ion beam (see Fig. 8.3c), focused electron beam is a powerful process for either material etching or deposition. It has been well detailed by Randolph and coworkers [41]. Such techniques have been recently extended to plasmonic materials [42] and may be used for plasmonic waveguides or gratings fabrication (see Fig. 8.10). Contrary to FIB, focused electron beam still requires investigations to control fabrication parameters such as the choice of gas precursors and their partial pressure [43]. For the moment, FIB etching remains the most used direct writing route for plasmonics, especially due to its high resolution.

Figure 8.10 Au nanodots gratings fabricated using EBID. Reproduced with permission from Dhawan, A., Gerhold, M., Madison, A., Fowlkes, J., Russell, P. E., Vo-Dinh, T., Leonard, D. N. (2009). Fabrication of nanodot plasmonic waveguide structures using FIB milling and electron beam-induced deposition, *Scanning*, **31**, 139–146. Copyright Wiley Periodicals, Inc.

8.2.3.2 Laser ablation

Laser ablation appeared as an original method for producing metallic nanoparticles [44]. In particular, this process is well adapted for laser-assisted nanostructuring of aluminum (Al) by laser ablation in liquids [45]. The advantage of such process is to provide Al nanoparticles with high purity and to limit oxidation when the air dissolved in solvents is outgased (see Fig. 8.11).

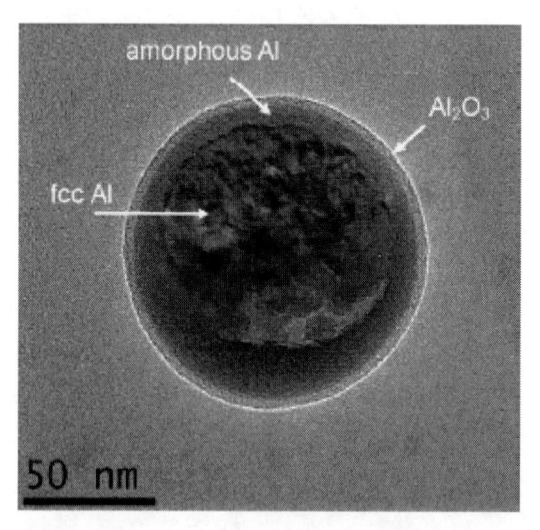

Figure 8.11 TEM image of Al nanoparticle obtained from laser-assisted nanostructuration of aluminum. Reprinted from *Chem. Phys. Lett.*, **501**, Viau, G., Collière, V. L., Shafeev, L.-M. G. A., Internal structure of Al hollow nanoparticles generated by laser ablation in liquid ethanol, 419–422, Copyright (2011), with permission from Elsevier.

8.2.3.3 3D laser lithography

Progress made in femtosecond laser nanofabrication led to the development of 3D laser lithography based on two-photon polymerization [46, 47]. Indeed, such techniques enable nano- and microfabrication beyond the diffraction limit. Figure 8.12, extracted from the Nanoscribe Web page (www.nanoscribe.de), describes 3D laser lithography setup. This technique provides both microsized structures (see Fig. 8.13) and nanosized masks for imprint lithography [48]. In particular, such 3D structuration has been appealing in the photonic crystal community. Applications in plasmonics start appearing. Indeed, surface plasmon polariton can carry information along metallic waveguides on a dielectric substrate, but also along dielectric structures on metal surfaces such as bends or splitters like in Fig. 8.13b. Besides, new strategies consisted of doping resins with plasmonic nanoparticles should lead to the development of 3D nanostructures in the next future [49].

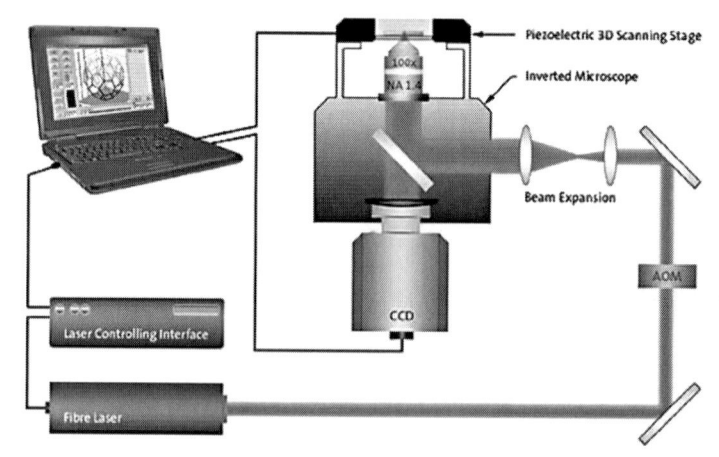

Figure 8.12 Description of the 3D laser lithography setup, extracted from Nanoscribe Web page (www.nanoscribe.de). A piezoelectric 3D scanning stage allows inducing photopolymerization in the three directions and building of 3D structures.

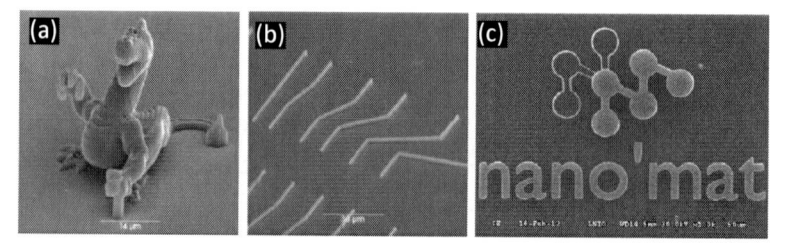

Figure 8.13 SEM images of (a) a micro-scale dragon [50], (b) *bends and splitter structures* [50] and (c) the Nano'mat logo (http://www.nanomat.eu/fr/materiel/nanoscribe.html, structure made by S. Jradi) fabricated by two-photon polymerization. (a) and (b) reproduced with permission from Ostendorf, A., Chichkov, B. N. (2006). Two-photon polymerization: A new approach to micromachining, *Photon. Spectra*, 1–7.

8.2.4 Printing, Replica Molding and Embossing

The previous sections highlighted both lithography and direct-writing techniques. Lithography presents two major drawbacks for industrial applications. First, the lift-off step requires an adhesion layer, which is often metallic and lies at the root of plasmon damping [51]. Second, all the required steps involved in lithography techniques–resin coating, patterning, development, metal deposition and lift-off-lead to low throughput. Direct-writing processes appear to be promising routes to overcome these two limits. Nevertheless, they still exhibit a lack of versatility for mass production. New approaches, coming from the semiconductor industry [52], have therefore been recently extended to plasmonics in order to provide large-scale fabrication tools and can be classified into three main categories: printing, replica molding, and embossing. All these techniques require a mold, which can be hard or soft and is generally made by one of the processes detailed in the previous sections. The difference comes from how the mold may be used to fabricate nanostructures. Printing involves a material transfer from the mold to the substrate. Replica molding includes a process for which a liquid is injected between the mold and the substrate and then solidified. As for embossing, it implies to press the mold into the substrate surface. Few examples of plasmonic nanostructures obtained using these techniques will now be given.

8.2.4.1 Printing

Printing processes involve the use of elastomeric masks generally made with PDMS. Among the printing category techniques, microcontact printing (µCP) is a versatile and large-scale patterning technique of surfaces with self-assembled monolayers (SAMs) such as biomolecules, colloids, or polymers as evidenced in the references included in the review written by Gates and coworkers [52]. When SAM patterns are transferred onto plasmonic film, selective etching of the plasmonic film leads to plasmonic nanostructures fabrication over large areas (see Fig. 8.14). Another way for fabricating plasmonic materials via µCP consists of patterning polymer containing gold nanorods as recently evidenced [53]. Moreover, in the continuation of µCP development, nanotransfer printing (nPT) appears to be a promising way for plasmonic nanostructures fabrication. Indeed, nPT consists of coating the elastomeric stamp with metal and directly transferring it from the mold to the substrates. Therefore, no etching procedure is required to pattern plasmonic metal structures [54].

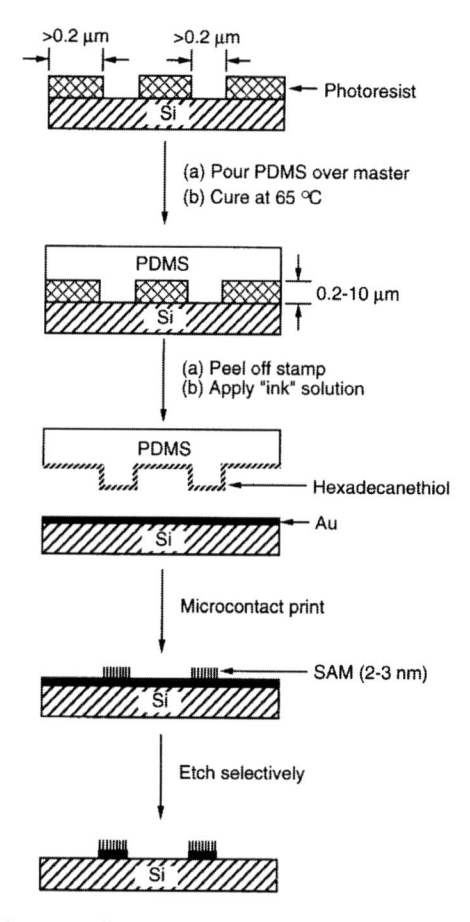

Figure 8.14 Schematic illustration of the µCP procedure leading to gold nanostructures fabrication. Reproduced from Zhao, X.-M., Xia, Y., Whitesides, M. (1997). Soft lithographic methods for nanofabrication, *J. Mater. Chem.*, **7**, 1069–1074, with permission of The Royal Society of Chemistry.

8.2.4.2 Replica molding

Replica molding has been mostly employed to fabricate hard or soft molds. Replica molding comprises three steps: (i) patterning a master via conventional techniques, (ii) transferring the pattern to PDMS by replica molding, and (iii) fabricating a replica of the original master by solidifying a liquid in the PDMS mold. Such process is well adapted for polymer molding; however, original routes using replica molding have been recently developed to pattern plasmonic nanostructures over large areas. For instance, Zhao and coworkers first sputtered gold thin layer on the PDMS stamp before contacting the stamp onto a high-energy surface in order to remove the gold film in contact with the substrate. Then the remaining gold film on the PDMS stamp is transferred to a molded polymer [56]. Another original method, presented in Fig. 8.15, was proposed by Lipomi and coworkers. It consists of depositing gold film onto nanopost before sectioning with ultramicrotome [57].

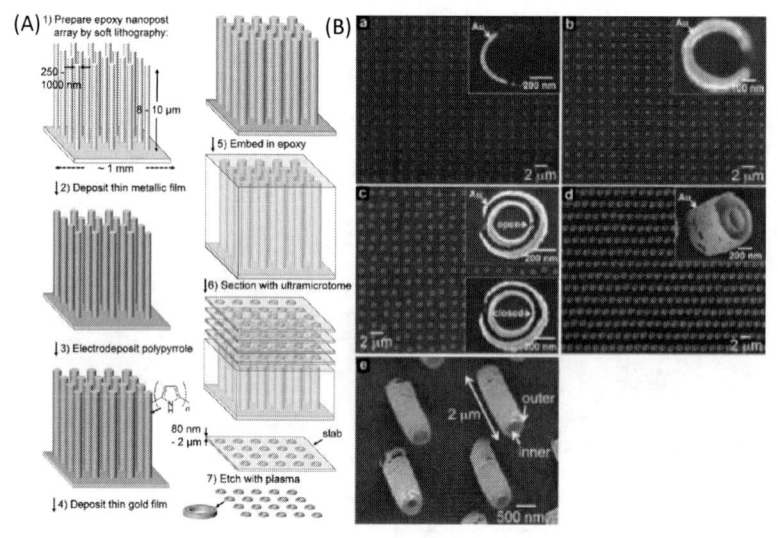

Figure 8.15 (A) Schematic illustration of the process proposed by Lipomi and coworkers consisting of evaporating gold on nanoposts before sectioning them with ultramicrotome (Reprinted with permission from Lipomi, D. J., Kats, M. A., Kim, P., Kang, S. H., Aizenberg, J., Capasso, F., Whitesides, G. M. (2010). Fabrication and replication of arrays of single- or multicomponent nanostructures by replica molding and mechanical sectioning, *ACS Nano*, **4**, 4017–4026. Copyright (2010) American Chemical Society). (B) SEM images of the gold structures obtained with this process: (a) gold crescents, (b) gold split rings, concentric split rings, (d) high-aspect-ratio concentric rings and very-high-aspect-ratio coaxial cylinders of gold.

8.2.4.3 Embossing

Many embossing techniques have been investigated in the semiconductor industry and have been detailed in the review written by Gates and coworkers [52]. The aim of this section is not to address a list of all the related techniques, but to illustrate how one of the most well-known techniques, nano-imprint lithography (NIL), has been applied to plasmonic structures fabrication.

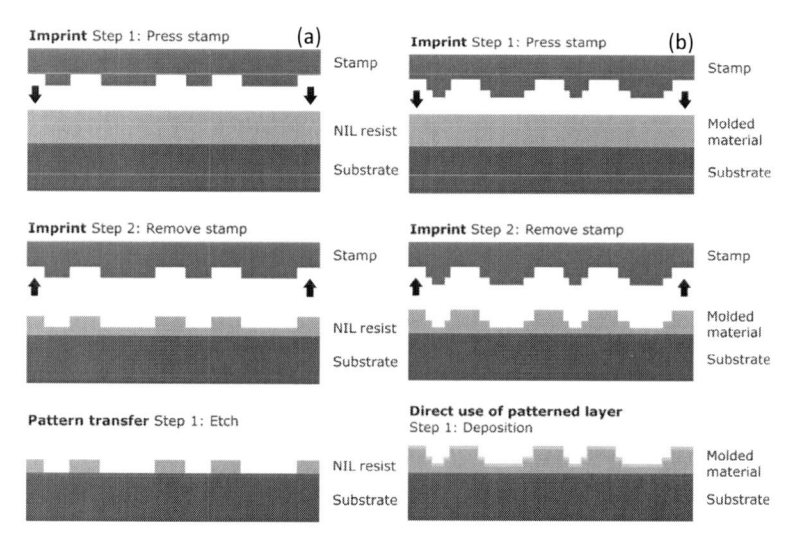

Figure 8.16 Schematic illustrations of the two major NIL processes exposed in Boltasseva's review. (a) the originally proposed NIL process based on mechanical deformation of resists to provide conventional lithography masks, (b) a process where the provided mask will directly be the pattern layer after deposition process as illustrated here. From [58]. © IOP Publishing. Reproduced by permission of IOP Publishing.

For other top-down techniques, resists are chemically or physically etched to provide masks. In NIL, the masks are formed thanks to mechanical deformations of resists. Therefore, the resolution is not limited by light diffraction or particle beam scattering and offers a good way for large-scale patterning of surfaces. Two major processes (among others) have been developed as detailed in the NIL applied to plasmonic components written by Boltasseva [58] and exposed in Fig. 8.16. The original NIL process, developed by S. Y. Chou and coworkers [48, 59] and initially named "hot embossing", is based on thermoplastic deformation of the resist using a hard mold (see Fig. 8.16a). The as-prepared mask then plays the role of conventional lithography masks to reproduce the pattern. The second possibility consists of directly using the patterned layer for metal deposition (see Fig. 8.16b). Since its development in the late 1990s, NIL has found many applications for plasmonics components due to the possibility of large-scale patterning leading to plasmonic crystals [60]. Therefore, NIL appears as a promising alternative for metamaterials fabrication in the next future.

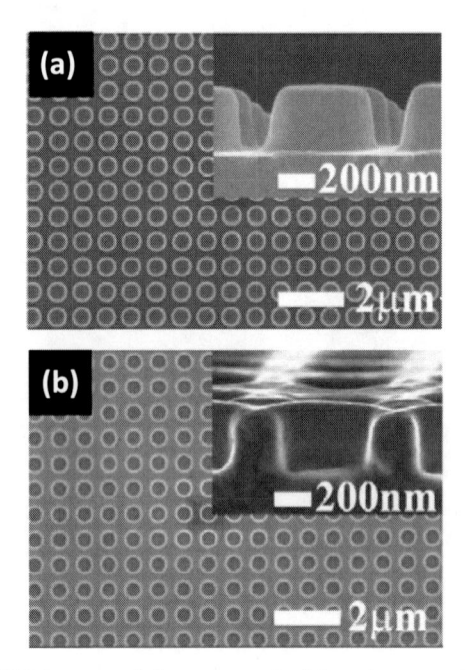

Figure 8.17 SEM images of Au nanoposts (a) and nanowells (b) via NIL using the process methodology presented in Fig. 8.16b. From [60]. © IOP Publishing. Reproduced by permission of IOP Publishing.

8.2.5 Conclusion about the Top-Down Strategy

The top-down strategy for plasmonic nanofabrication has encountered the same great success as in the semiconductor industry. Its advantages are mainly the control of the size and shape dispersion and the pattern reproducibility. In this chapter part, not all the top-down methods have been detailed. The objective was not to establish an exhaustive list. For instance, edge lithography or scanning probe lithography (SPL) has not been mentioned. Details about these techniques can be found in other book chapters or review [61]. The aim of this part of the chapter was to show the reader how from photolithography, progress has been made to first break the light diffraction limit and then favor both size reduction and large-scale production. Indeed, if FIB and EBL are very good laboratory techniques, their use is not really adapted for plasmonic sensors or metamaterials fabrication for which large-scale production is required. This is how emerged printing, molding, and embossing techniques, among which NIL appears as one of the most appealing candidate for further applications in the fields of plasmonics.

8.3 Bottom-Up Techniques: Atom by Atom Building

Like for an artist sculpting rocks, there is a threshold below which the resolution is limited by the pencil itself. That is why scientists have been investigating bottom-up routes for the past 40 years. The bottom-up strategy consists of building particles from the bricks of matter, that is to say molecules or atoms. However, if the bottom-up approach provides plasmonic nanoparticles with tunable size and geometry, efforts have to be focused onto organization of these structures either in dielectric matrix or onto substrates.

8.3.1 The Bottom-Up Strategy

In this approach, chemistry plays a key role as kinetic processes and thermodynamics are involved. Bottom-up processes can be divided into three main routes: physical, electrochemical, and chemical ways.

8.3.1.1 Physical route

Among plasmonic systems, metal island films have been intensively investigated for more than 30 years since the discovery, in 1974, that they may lead to strong enhancement of Raman scattering signal when molecules are absorbed on them [62]. Since that date, these nanostructured plasmonic films have shown great promise to detect biomolecules [63] and many efforts have been devoted to increase the versatility of their synthesis. Indeed, the vacuum evaporation followed by thermal annealing of thin plasmonic films on glass substrates has been considered as a promising option because of its cost-effectiveness. Moreover, it has been proven that gold film annealing at high temperature may lead to a good control of nanoparticle mean size and therefore to the tunability of the LSP resonance [64]. Recently, other protocols like integration of metal masks to create a "shadowing effect" during the evaporation, or repetition of several evaporation cycles have paved the way for large-scale fabrication and applications of such systems as biosensors (see Fig. 8.18) [65, 66].

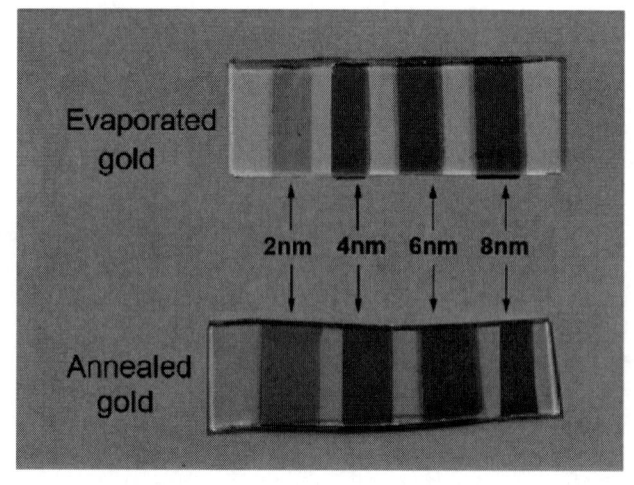

Figure 8.18 Photographs of samples obtained from two different protocols: (top) simultaneous fabrication of a controlled thickness gradient of gold nanoparticles on single glass substrate by using metal staples to produce an artificial glass substrate/gold interface, (bottom) four cycles of gold evaporation (2 nm film for each cycle) leading to four zones of 2, 4, 6, and 8 nm thickness. The area with the wanted thickness is protected by scotch tape. Reprinted Jia, K., Bijeon, J. L., Adam, P. M., Ionescu, R. E. (2012). Sensitive localized surface plasmon resonance multiplexing protocols, *Anal. Chem.*, **84**, 8020–8027. Copyright (2012) American Chemical Society. Note that evaporation of gold thin film followed by thermal annealing has been successfully extended to other plasmonic metals such as aluminum [67].

8.3.1.2 Electrochemical route

A large fraction of nanofabrication methods is based on the chemical reduction of Au or Ag salts. Electrochemical plating has been extensively investigated since the 1990s. Nanoscale template is, however, required to control the nanoparticle geometry, and the synthesis of anisotropic structures such as Au nanorods becomes possible [68–70]. Noncentrosymmetric Au sphere pairs have even been synthesized thanks to an accurate control of the plating conditions [71]. The nanoparticles prepared by electroplating are in solution and efforts have thus to be made to disperse them into a dielectric medium. Among the different studies, one promising example for applications was provided by Dirix and coworkers who fabricated color filters with a single oriented polyethylene film containing monodisperse silver nanoparticles [72].

8.3.1.3 Chemical route

Among all the routes toward plasmonic nanoparticles fabrication, colloidal synthesis probably appears as the most promising solution [73] since it provides not only nanospheres [74, 75] but also anisotropic particles like nanorods [76] or nanoprisms [77]. Knowing that plasmonic properties are highly sensitive to nanoparticles shape, the possibility of shape tuning during colloidal synthesis makes this route powerful among all the existing fabrication methods [78]. Indeed, playing on both the size and shape of the plasmonic nanoparticles synthesized by wet chemistry allows to tune the LSP resonance through the whole visible range (see Fig. 8.19). We will not detail the mechanisms involved in colloidal synthesis since many reviews have been published [73] and many methods have been developed in the past 60 years. Probably the easiest way of synthesizing spherical Au nanoparticles in the 10–100 nm range is the Turkevich method [79–81], which is based on $HAuCl_4$ gold salts dispersion in water and reduction in a boiling sodium citrate solution. This provides 10 nm Au seeds, which then can be grown in a second step [74]. The second well-known process is known as the Brust–Schiffrin method and relies on the $HAuCl_4$ salts reduction in a two-phase solution. Gold salts are dispersed in water and transferred to the toluene phase by means of tetraoctylammonium bromide (TOAB), which plays the role of transfer agent [82]. Many other ways have been developed, especially reducing metal slats by organic [83] or polyol [84, 85] solvents. Since then, efforts have been focused onto anisotropic nanoparticles synthesis because of the possibility to tune the LSP resonance by playing with the rods' aspect ratio [70]. Among the different strategies to synthesize Au nanorods, a seed-mediated growth method with the aid of cetyltrimethyl-ammonium bromide (CTAB) has been intensively investigated and provides narrow size distribution [86, 87]. The associated mechanism, proposed by Murphy and coworkers [88],

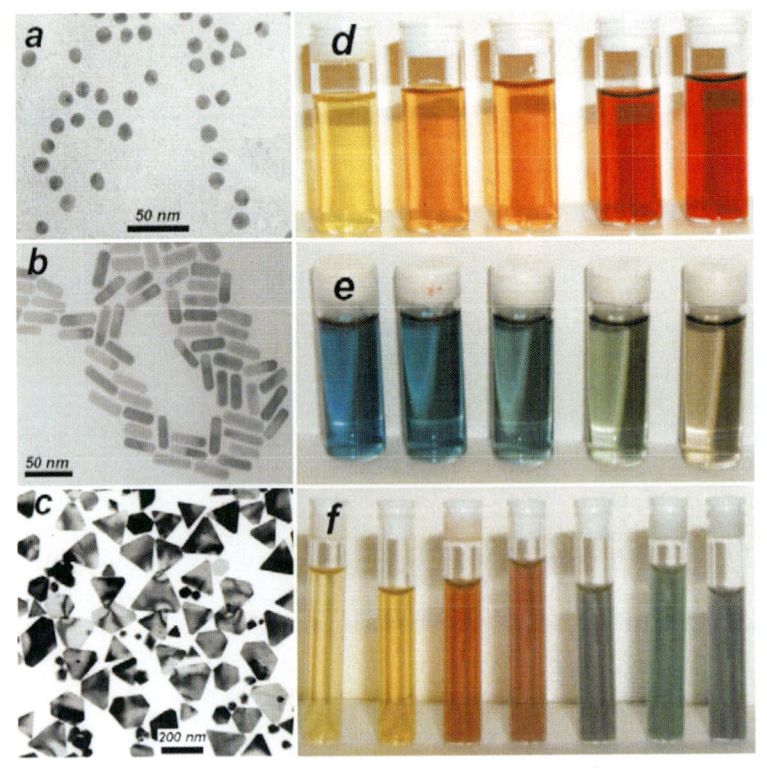

Figure 8.19 Left: Transmission electron micrographs of Au nanospheres and nanorods (a,b) and Ag nanoprisms (c, mostly truncated triangles) formed using citrate reduction, seeded growth, and DMF reduction, respectively. Right: Photographs of colloidal dispersions of AuAg alloy nanoparticles with increasing Au concentration (d), Au nanorods of increasing aspect ratio (e), and Ag nanoprisms with increasing lateral size (f). Reprinted from *Mater. Today*, **7**, Liz-Marzan, L., Nanometals: Formation and color, 26–31, Copyright (2004), with permission from Elsevier.

suggests that the CTAB plays the role of structure-directing agent (see Fig. 8.20a). Note that the CTAB surfactant is also suitable for Ag nanorods synthesis (see Figs. 8.20b,c). To obtain longer Au nanowires, a three-step synthesis process is generally required [89, 90]. Indeed, to avoid self-reduction of gold salts into large spheres, after about 1 min of growth, a small portion of the solution is extracted and injected in a new growth solution. Again, after about 1 min, a portion of the second solution is extracted and injected in a new growth solution. Then several hours are spent to obtain long Au nanowires (see Fig. 8.21).

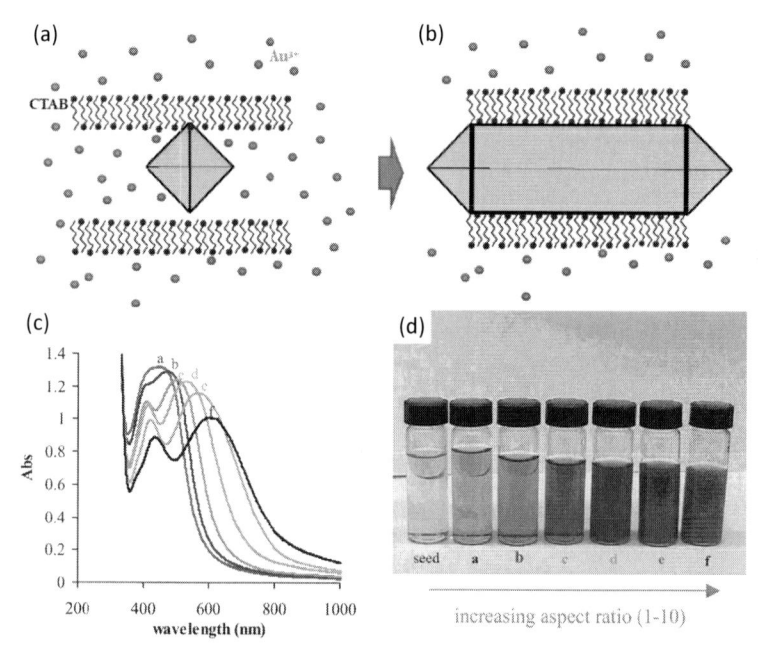

Figure 8.20 (a) Anisotropic growth of the surfactant-directed metal nanorod growth proposed by Murphy and coworkers [87, 88]: the single crystalline seed particles have facets that are differentially blocked by surfactant. Subsequent addition of metal ions and weak reducing agent lead to metallic growth at the exposed particle faces. (b) Visible absorption spectra of Ag nanowires with various aspect ratios from 1 to 10. (c) Corresponding aqueous solutions of silver nanoparticles. (b) and (c) reproduced with permission from Murphy, C. J., Jana, N. R. (2002). Controlling the aspect ratio of inorganic nanorods and nanowires, *Adv. Mater.*, **14**, 80–82. Copyright WILEY-VCH Verlag GmbH.

Figure 8.21 Long Au nanowires synthesized via a three-step process [89, 90].

However, we would like to draw the reader's attention to the fact that many developments have been led allowing to play with not only the shape of particles but also the involved materials. Indeed, a large zoology of core–shell nanoparticles have been synthesized with various targeted applications. For instance, bimetallic AgAu core–shell nanoparticles exhibit high potential for sensing applications [91]. Coating magnetic nanoparticles with gold layer could offer strong sensitivity enhancement for SPR sensors as demonstrated by Halas and coworkers [92] and could also provide applications in the biomedical field taking advantage of both their plasmonic and magnetic properties. Another interesting core–hell system comes from gold–silica and silica–gold core–shells. Successively, Liz-Marzan and coworkers and Halas and coworkers pioneered the synthesis of such core–shell nanoparticles with various dimensions. For example, doping the silica shell with fluorescent molecules could find a technological potential in SPASER (surface plasmon amplification by stimulated emission of radiation) systems proposed by Bergman and Stockman [93]. Finally, wet chemistry now offers a wide range of plasmonic nanoparticle types.

8.3.2 Self-Organization, the Next Challenge of Plasmonics

If the chemical route appears as the most promising way toward nanofabrication for plasmonics, the related challenge is definitively the development of nanoparticles organization processes. Many applications require depositing and organizing nanoparticles onto substrates. Therefore, developing efficient ways for depositing nanoparticles at specific positions has been considered a key challenge in the field of plasmonics. Dry coating routes have been intensively investigated and all the related details are exposed in the review written by Shchukin and Bimberg [94]. In this section, we will focus onto routes involving plasmonic nanoparticles synthesized via chemical routes. Roughly, self-organization techniques can be divided into two classes: laboratory self-assembly techniques and mass-production wet coating processes.

8.3.2.1 Laboratory self-assembly techniques

We will briefly discuss the dominant self-assembly techniques today [95]. Other techniques such as dip and spray coating and electrophoretic deposition (EPD) [96] are also techniques that can lead to self-assembly of nanoparticles. Note that spray coating appears nowadays as a promising route toward large-scale application [97], in particular for the development of paintings or glass coating.

8.3.2.1.1 *Surface functionalization, a path to self-assembled monolayers*

In plasmonics, surface functionalization is a key feature either for driving self-organization of plasmonic nanostructures or for developing plasmonic biosensors. Concerning the development of plasmonic biosensors, surface functionalization is mandatory to graft bioreceptors that will allow molecule sensing. Szunerits and coworkers wrote a detailed review article about surface functionalization strategies in the framework of SPR biosensors [98]. These strategies can be transposed to surface functionalization of LSPR or SERS sensors.

As for plasmonic nanoparticles self-organization, two strategies can be endeavored, either surface functionalization of substrates [99] or surface functionalization of the plasmonic nanostructures themselves [100]. In both cases, the main route for surface functionalization is the use of SAM of molecules with functional groups. Another route, not discussed here, is the physisorption of polymers whose details can be found in other review or book chapters [61, 98].

In this section, we would like to draw the reader's attention to the concept of SAM. Indeed, dispersing metallic nanoparticles onto substrates often leads to the aggregation of the latter and thus non-uniformity of physical properties onto the whole surface. This is of great importance for plasmonics since the LSP resonance wavelength can strongly be shifted by interaction between neighboring nanoparticles [38, 101].

In order to self-assemble monolayers of plasmonic nanoparticles onto substrates, bi-functional molecules are required showing a great affinity for both the substrate and the nanoparticle materials. One well-known functional group remains thiol function since it provides strong affinity for gold and other plasmonic materials such as silver and copper. A review written by Whitesides and coworkers offer all the details about functionalization process with thiol groups depending on the material [102]. In the case of alkanethiols for example, the thiol group plays the role of ligand (or head group), which makes the molecule anchor to the metal surface by modifying electronic states and stabilizing surface atoms. The alkane chain acts as a spacer while at the other side of the molecule, a terminal function group allows grafting the targeted molecules or nanoparticles [102]. Of course, chemistry offers great variety of functional groups with high affinity depending on the materials that come into play. Thiols have great affinity for metals such as gold, silver, and copper. Other useful functional groups for plasmonics are amines and silanes. Indeed, amines also exhibit strong affinity for gold while silanes are often required to functionalize SiO_2 substrates or nanoparticles [99]. This is how aminosilane molecules, such as 3-aminopropyltriethoxysilane, are involved in the synthesis of Au@SiO_2 core–shell nanoparticles [103].

8.3.2.1.2 *The Langmuir–Blodgett technique*

The Langmuir–Blodgett technique consists of transferring monolayers of nanoparticles onto substrates. The nanoparticles are initially dispersed onto a sub-phase solvent in a Langmuir–Blodgett trough when the transfer is granted from compression of the nanoparticle thin film onto the substrate (see Fig. 8.22a). Recent development by chemically bridging neighboring nanoparticles by bifunctional linkers at the air/water interface allowed to significantly increase mechanical stability of gold nanoparticles deposited onto substrates via the Langmuir–Blodgett technique [104] (see Fig. 8.22b).

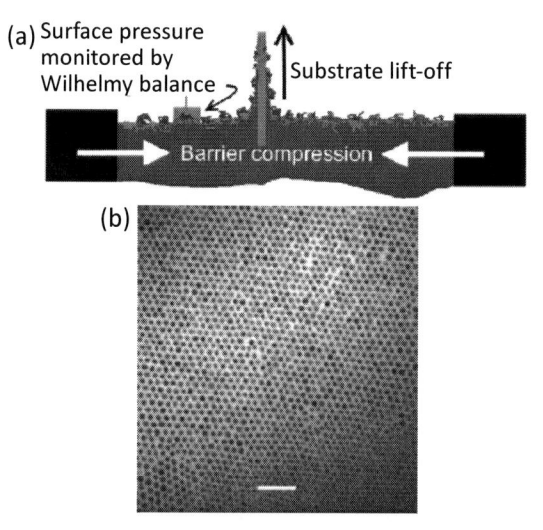

Figure 8.22 (a) Illustration of the Langmuir–Blodgett technique for molecule or nanoparticle transfer onto substrates [105]. (b) TEM micrographs of the C_8-Au particles deposited onto a SiO_x-coated Cu grid at varied surface pressures before and after TBBT (4,4′-thiobisbenzenethiol) cross-linking (Reprinted (with permission from Chen, S. (2001). Langmuir-Blodgett fabrication of two-dimensional robust cross-linked nanoparticle assemblies, *Langmuir*, **17**, 2878–2884. Copyright (2001) American Chemical Society.). The scale bar is 33 nm.

8.3.2.1.3 *The layer-by-layer self-assembly technique*

The layer-by-layer (LbL) self-assembly is an electrostatic thin film deposition technique. It is based on the successive deposition of layers with opposite charges, like polyelectrolytes (see Fig. 8.23). Note that this technique can be combined with other deposition routes such as spin-coating, spraying, and photolithography [106]. The LbL technique has been successfully employed in nano-optics for both quantum dots [107] and plasmonic nanoparticles [108] superlattices fabrication. Note that for Au nanoparticles, using such self-assembly technique, Jiang and coworkers evidenced collective plasmonic effects [109]. Many details about the LbL technique can be found in the review article written by Ariga and coworkers [106].

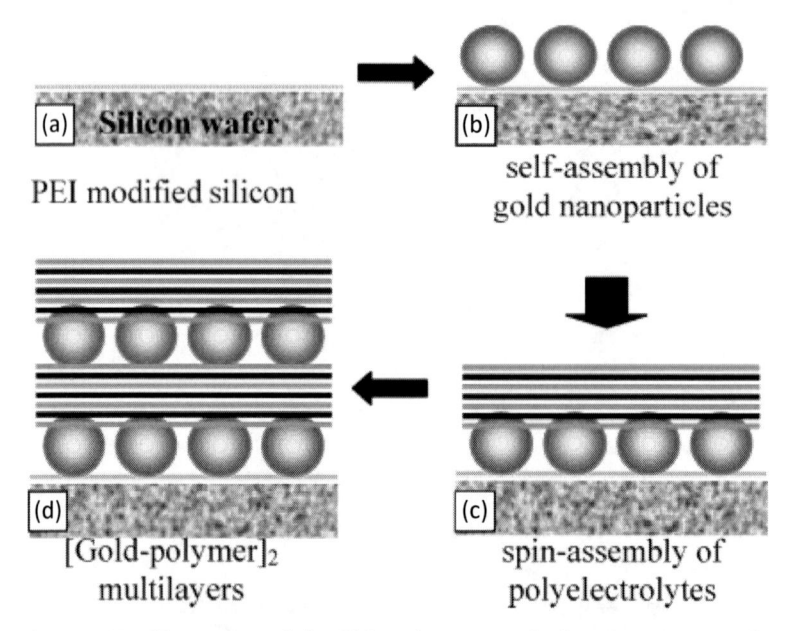

Figure 8.23 Illustration of the LbL technique applied to Au nanoparticle superlattices fabrication. Reprinted with permission from Jiang, C., Markutsya, S., Tsukruk, V. V. (2004). Collective and individual plasmon resonances in nanoparticle films obtained by spinassisted layer-by-layer assembly, *Langmuir*, **20**, 882–890. Copyright (2004) American Chemical Society.

8.3.2.1.4 The evaporation-driven self-assembly technique

When no special attention is borne onto nanoparticle interaction and solvent evaporation kinetics, nanoparticles solution evaporation leads to the formation of nanoparticle aggregates. However, it has been shown that the accurate control of evaporation kinetics and particle interactions with the liquid–air interface offers the possibility to fabricate long-range nanoparticle crystals [110–112]. Jaeger and coworkers evidenced [112] that the two key parameters for the interfacial self-assembly mechanism are a rapid evaporation, to segregate particles near the liquid–air interface, and an attractive interaction between the particles and the liquid–air interface, to localize them on the interface (see Fig. 8.24).

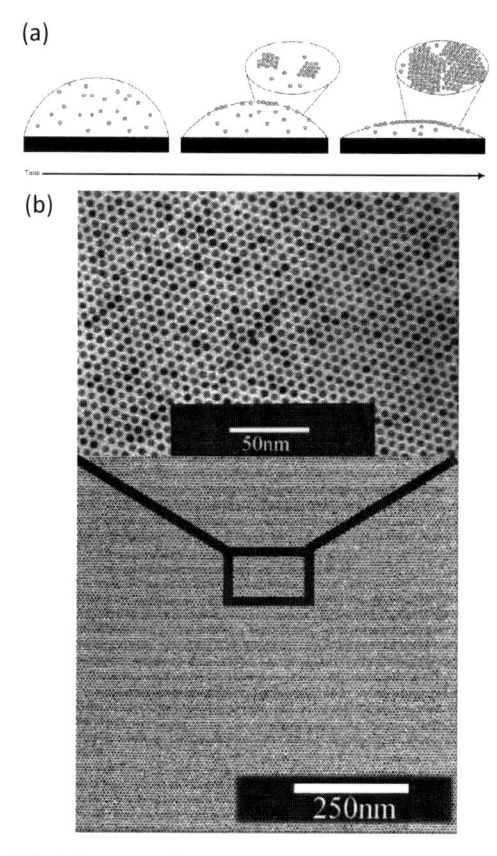

Figure 8.24 (A) Schematic diagram of the self-assembly process during the early stages of drying, showing how nanocrystals self-assemble at the liquid–air interface (Reprinted by permission from Macmillan Publishers Ltd: Nature, Bigioni, T. P., Lin, X. M., Nguyen, T. T., Corwin, E. I., Witten, T. A., Jaeger, H. M. Kinetically driven self assembly of highly ordered nanoparticle monolayers, **5**, 265–270, copyright (2006).), (B) TEM image of nanocrystal monolayer of gold nanoparticles (Reprinted with permission from Lin, X. M., Jaeger, H. M., Sorensen, S. M., Klabunde, K. J. (2001). Formation of long-range-ordered nanocrystal superlattices on silicon nitride substrates, *J. Phys. Chem. B*, **105**, 3353–3357. Copyright (2001) American Chemical Society.).

The evaporation-driven self-assembly (EDSA) technique has been recently upgraded in order to produce more complex assemblies of nanoparticles. For instance, Khanal and Zubarev adapted it for the formation of Au nanorods rings [113]. Another promising route, combining EDSA and colloidal lithography, has been proposed by Cremer and coworkers in order to fabricate long-range nanorings of quantum dots [114] or plasmonic nanoparticles [9] as shown in Fig. 8.25.

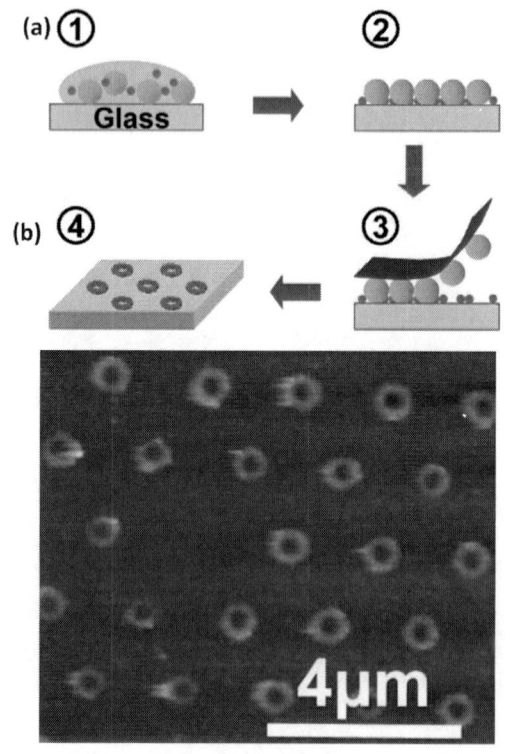

Figure 8.25 (a) Illustration of the evaporation templating procedure employed for forming CdSe nanorings (red particles) on planar substrates using microsphere templates (orange particles) (Reprinted with permission from Chen, J., Liao, W.-S., Chen, X., Yang, T., Wark, S. E., Son, D. H., Batteas, J. D., Cremer, P. S. (2009). Evaporation-induced assembly of quantum dots into nanorings, *ACS Nano*, **3**, 173–180. Copyright (2009) American Chemical Society.) (b) AFM image of Au colloids nanorings obtained by Plain and coworkers using this procedure (Reprinted with permission from Lerond, T., Proust, J., Yockell-Lelièvre, H., Gérard, D., Plain, J. (2011). Self-assembly of metallic nanoparticles into plasmonic rings, *Appl. Phys. Lett.*, **99**, 123110. Copyright [2011], AIP Publishing LLC.).

8.3.2.1.5 Patterning using copolymer templates

Among the most appealing strategies for nanoparticles self-organization, the use of block copolymers have attracted much attention due to their exceptional ability to self-organize into microphase-separated domains with controlled shapes and sizes [37, 115, 116]. Many routes involving block copolymers have been developed: selectively incorporation of inorganic precursors into one block of amphiphilic block copolymers like polystyrene-*b*-poly(2-vinylpyridine) [117, 118] or reduction and stabilization of inorganic precursors like trihydrate tetrachloroaurate ($HAuCl_4 \cdot 3H_2O$) into poly(ethylene oxide)-*b*-poly(propylene oxide) copolymers leading to single step synthesis of gold nanoparticles [119, 120]. It is also possible to extend this process to *PS-b-PMMA* copolymers by adding thiolated PS, which exhibits high affinity for Au salts and AU NPs driving the as-prepared Au NPs into the PS domains [121]. Recently, it has been shown that applying the Langmuir–Blodgett technique to PS-b-PMMA copolymer thin film and thiol-capped Au nanoparticles may lead to an accurate control of nanoparticles self-organization into PS domains [38, 105, 122]. Indeed, the precise location of the NPs, as well as the nature of their assembly, depends on both particle size and nature of the capping ligand. For example, this self-assembly technique can provide long-range ordered Au nanoparticle rings when they are capped with thiolated PS (see Fig. 8.26).

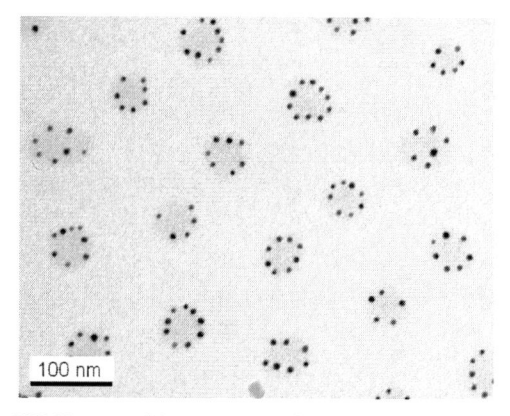

Figure 8.26 TEM image of Au nanoparticles organized in PS domains of PS-b-PMMA copolymer thin films using the Langmuir–Blodgett technique.

8.3.3 Mass Production Using Wet Coating Processes

Nanoparticles synthesized via wet chemistry processes offer great perspectives of applications. However, the key challenge for industry remains the development of large-scale deposition of nanoparticles, especially on continuous substrates [123]. Printing technologies appear as the most promising for plasmonic application transfer to industry [95]. Examples of appealing processes are gap, slot die, roll and gravure coatings. Details can be found in the review written by Yamagushi and coworkers [95].

8.4 Mixing Top-Down and Bottom-Up Routes

A promising alternative for large-scale fabrication is the combination of top-down and bottom-up routes. Indeed, the techniques that have been presented in the earlier sections, can be combined to optimize size or organization control of the nanoparticles. We will not list all the possible combinations in this section but just give a few examples.

8.4.1 Porous Membranes for Ordered Nanowires Growth

A strategy—which has been intensively investigated—in many nanotechnology fields consists of inducing anisotropic growth inside a host material that plays the role of a template. Hexagonal mesoporous silica or alumina membranes are the epitome of such host materials since they provide highly ordered nanopores (see Fig. 8.27a) allowing electrodeposition [124, 125] or chemical growth [126] of various materials like Ag nanowires (see Fig. 8.27b).

8.4.2 Copolymer Template Control of Plasmonic Nanoparticle Synthesis via Thermal Annealing

As previously discussed, thermal annealing of evaporated gold films leads to the formation of Au nanoparticles. However, size and organization of the as-prepared nanoparticles are not so easy to control. On the other hand, copolymer template etching via either RIE or UV-exposure can lead to the formation of ordered nanoholes arrays [38, 39]. Nevertheless, holes are generally not deep enough to provide thick nanoparticles. To allow a good resin removal during the lift-off procedure, depths of about three times the desired height of nanoparticles are required. The combination of both techniques provide a low-temperature process (~200°C) with a good control of Au nanoparticles size dispersion and thus sharper LSP resonance peaks [128]. We can also imagine the deposition of chemically synthesized plasmonic nanoparticles inside as-prepared holes to avoid metal oxidation during the annealing process.

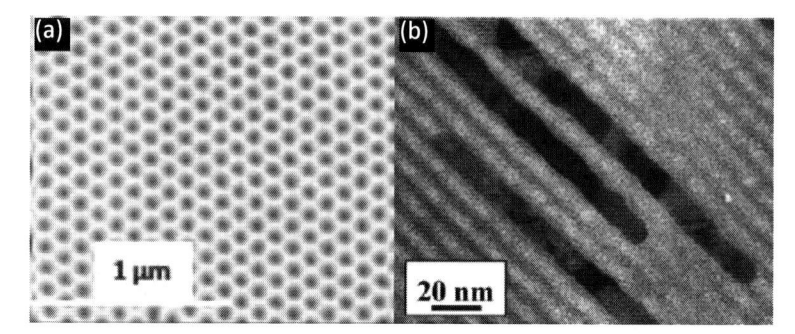

Figure 8.27 (a) SEM image of porous alumina membranes (extracted from [127]), (b) TEM image of Ag nanowires obtained by reduction of AgNO$_3$ by ethylene glycol inside mesoporous silica (Reprinted from Mater. Res. Bull., 38, Piquemal, J.-Y., Viau, G., Beaunier, P., Bozon-Verdurz, P., Fiévet, F., One-step construction of silver nanowires in hexagonal mesoporous silica using the polyol process, 389–394, Copyright (2003), with permission from Elsevier.).

8.4.3 Let's Play Your Imagination

The two techniques mentioned above are just examples of how combining top-down and bottom-up routes to control plasmonic nanoparticle size, organization or both. Many other combinations are possible and we hope that the readers will let run their creativity to conceive original routes for large-scale fabrication of plasmonic structures.

8.5 Conclusion: First, Choose Materials

The aim of this chapter was to give the reader an overview of all the possibilities that may exist for plasmonic material nanostructuration. However, as a conclusion, the reader should remember that the first thing to do in plasmonics is to think about which material is the best addressed to which application. Depending on the targeted application (biosensors, optical nano-antenna, superlenses, light concentrators, waveguides, ... [129]), plasmonic resonance frequency needs to be tuned from UV to a few tens of gigahertz. The control of the size and geometry is therefore not sufficient. The first decision concerns the choice of the material. Historically, plasmonics exploded with metal nanostructuration, which offered the possibility to subwavelength control of light propagation. Indeed, pure metals are good candidates for plasmonic applications since they have large plasma frequencies and high electrical conductivity. They provide negative real permittivity at optical frequencies [130] (see Fig. 8.28a), which is required for plasmonic applications. However, the main limitation for application of pure metals in plasmonics comes from losses (see Fig. 8.28b). One of the mains loss mechanisms at optical frequencies precisely comes from interband transitions. Other mechanisms can come into play like ohmic losses, which are stronger close to the plasmon resonance or scattering processes due to roughness or grain boundaries [131]. Among metals, silver takes advantage of a smaller relaxation time of conduction electrons and appears as the best candidate for application in the visible range [132, 133] due to smaller losses. However, precautions have to be taken due to fast oxidation of Ag nanostructures in air [100, 134]. This is very often a reason of choice of gold for lower near infra-red (NIR) applications even if interband transitions may induce losses at wavelengths around 470–500 nm [132, 135]. Cu [136] and Al [45, 67] have also started to be investigated in spite of larger losses. In particular, Al NPs offer LSP resonances in the UV range and could find specific applications in the solar cell industry [137]. However, both of these metals suffer from rapid oxidation [138, 139].

Other metals may candidate for specific plasmonic applications. Palladium [140] and platinum [141] have been integrated in plasmonic applications rather for their catalytic activity as reminded in West and coworker's review [132]. Very recently, Ni nanostructuration led to the observation of LSP resonances in the 200–500 nm range [142, 143], which makes nickel nanoparticles very appealing for applications in the merging magnetoplasmonics field [144]. Other choices among metals are discussed in the two reviews written by west and coworkers [132] and Lindquist and coworkers [135]. Information about metal alloys is also provided in these reviews.

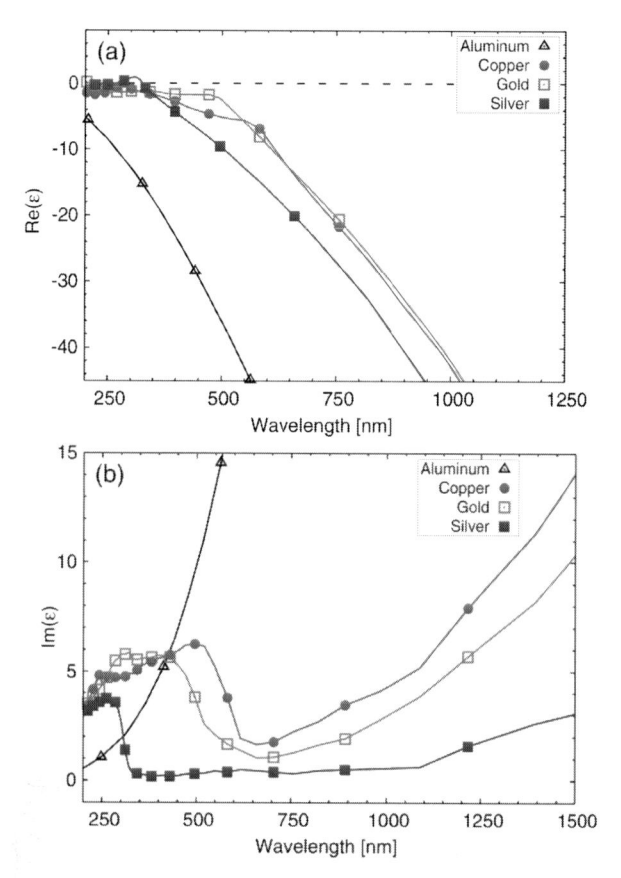

Figure 8.28 (a) Real part and (b) imaginary part of the complex dielectric functions for Al [145], Cu [146], Au [130], and Ag [130]. For plasmonic applications, the real part must be negative. For low loss operation, the imaginary part must be small. Extracted from [135].

However, pure metals or metals alloys tend to behave like perfect conductors in the THz range or above [129]. An experimental and theoretical study on aluminum and stainless steel nanowires recently evidenced the role of the skin effect for surface plasmon polariton (SPP) in this frequency range [147]. Therefore, other materials like semiconductors have been investigated for frequencies above several hundred THz [132]. One of their key advantages is the possibility to tune their properties through carrier doping. However, pushing semiconductor plasmon resonances to the visible range requires a very high doping level and is thus hard to achieve [135]. Among plasmonic materials, wide band gap semiconductors are attractive since they exhibit no interband transition and low losses. All the necessary details are enclosed in the review written by West and coworkers [132]. It may just be underlined that transparent conducting oxides like indium tin oxide (ITO) or doped ZnO have been studied for applications in optoelectronics. Among non-transparent semiconductors, InGaAS, AlInAs, and SiC also offer promising perspectives as plasmonic materials.

We would like rather end this book section by mentioning graphene plasmonics as a promising emerging field. Initial theoretical works predicted graphene as an attractive material for applications in the THz range [148] and has been recently imaged in the IR-range via a-SNOM measurements [149]. Graphene as a plasmonic material takes advantage of strong confinement and relatively long propagation distances [150]. Moreover, the possibility to tune its doping level via electrostatic gating offers many perspectives. Indeed, graphene could suffer from significant losses when non-doped due to interband transitions. However, this could be avoided thanks to high enough level of doping. Jablan and coworkers evidenced that doped graphene should enable low losses for frequencies below 0.2 eV. A promising route concerns graphene nanostructuration in order to push plasmon resonances to the IR and visible ranges [151] or to design graphene quantum dots [152]. Another perspective for graphene applications is its use as substrates for metallic nanostructures [153, 154].

Finally, the message from this book chapter is that plasmonic applications require a smart convergence between the choice of materials and the choice of nanofabrication methods.

Acknowledgments

Financial support of NanoMat (www.nanomat.eu) by the "Ministère de l'enseignement supérieur et de la recherche," the "Conseil régional Champagne-Ardenne," the "Fonds Européen de Développement Régional (FEDER) fund," and the "Conseil général de l'Aube" is acknowledged. Thomas Maurer also thanks the DRRT (Délégation Régionale à la Recherche et à la Technologie) of Champagne-Ardenne, the CNRS via the chair "optical nanosensors," the UTT-EPF strategic program for supporting the CODEN project and the Labex ACTION project (contract ANR-11-LABX-01-01) for financial support.

References

1. Nylander, C., Liedberg, B., Lind, T. (1982). Gas detection by means of surface plasmon resonance, *Sens. Actuators*, **3**, 79–88.
2. Homola, J., Yee, S. S., Gauglitz, G. (1999). Surface plasmon resonance sensors: Review, *Sens. Actuators B: Chem.*, **54**, 3–15.
3. Jain, P. K., Huang, X., El-Sayed, I. H., El-Sayed, M. A. (2008). Noble metals on the nanoscale: Optical and photothermal properties and some applications in imaging, sensing, biology, and medicine, *Acc. Chem. Res.*, **41**, 1578–1586.
4. Hunt, L. B. (1976). The true story of purple of cassius, *Gold Bull.*, **9**, 134–139.
5. Kaneko, A., Yagishita, A., Yahashi, K., Kubota, T., Omura, M., Matsuo, K., Mizushima, I., Okano, K., Kawasaki, H., Inaba, S., Izumida, T., Kanemura, T., Aoki, N., Ishimura, K., Kanemura, T., Aoki, N., Ishimura, K., Ishiuchi, H., Suguro, K., Eguchi, K., Tsunashima, Y. (2005). Sidewall transfer process and selective gate sidewall spacer formation technology for sub-15 nm finFET with elevated source/drain extension. *Electron Devices Meeting, 2005. IEDM Technical Digest. IEEE International*, 844–847.
6. Pelton, M., Aizpurua, J., Bryant, G. (2008). Metal-nanoparticle plasmonics, *Laser Photon. Rev.*, **2**, 136–159.
7. Sun, Y., Xia, Y. (2002). Shape-controlled synthesis of gold and silver nanoparticles, *Science*, **298**, 2176–2179.
8. Li, J.-R., Lusker, K. L., Yu, J. J., Garno, J. C. (2009). Engineering the spatial selectivity of surfaces at the nanoscale using particle lithography

combined with vapor deposition of organosilanes, *ACS Nano*, **3**, 2023–2035.

9. Lerond, T., Proust, J., Yockell-Lelièvre, H., Gérard, D., Plain, J. (2011). Self-assembly of metallic nanoparticles into plasmonic rings, *Appl. Phys. Lett.*, **99**, 123110.

10. Maurer, T., Ott, F., Chaboussant, G., Soumare, Y., Piquemal, J.-Y., Viau, G. (2007). Magnetic nanowires as permanent magnet materials, *Appl. Phys. Lett.*, **91**, 172501.

11. Robbes, A. S., Cousin, F., Meneau, F., Dalmas, F., Boué, F., Jestin, J. (2011). Nanocomposite materials with controlled anisotropic reinforcement triggered by magnetic self-assembly, *Macromolecules*, **44**, 8858–8865.

12. Kim, J. M., Ge, J., Kim, J., Choi, S., Lee, H., Lee, H., Park, W., Yin, Y., Kwon, S. (2009). Structural colour printing using a magnetically tunable and lithographically fixable photonic crystal, *Nat. Photon.*, **3**, 534–540.

13. Levinson H. J. (2010). In *Principles of Lithography*. 3rd ed., Chapters 2 and 10. Society of Photo-Optical Instrumentation Engineers; Bellingham, WA.

14. Switkes, M., Rothschild, M. (2001). Immersion lithography at 157 nm, *J. Vacuum Sci. Technol. B*, **19**, 2353–2356.

15. Alkaisi, M. M., Blaikie, R. J., McNab, S. J., Cheung, R., Cumming, D. R. S. (1999). Sub-diffraction-limited patterning using evanescent near-field optical lithography, *Appl. Phys. Lett.*, **75**, 3560–3562.

16. Srituravanich, W., Fang, N., Sun, C., Luo, Q., Zhang, X. (2004). Plasmonic nanolithography, *Nano Lett.*, **4**, 1085–1088.

17. Felts, J. R., Law, S., Roberts, C. M., Podolskiy, V., Wasserman, D. M., King, W. P. (2013). Near-field infrared absorption of plasmonic semiconductor microparticles studied using atomic force microscope infrared spectroscopy, *Appl. Phys. Lett.*, **105**, 152110.

18. Lérondel, G., Kostcheev, S., Plain, J. (2012). Nanofabrication for plasmonics, in: *Plasmonics* (Bonod, S. E. A. N., ed.), Springer-Verlag Berlin Heidelberg, pp. 269–315.

19. Grand, J., Kostcheev, S., Bijeon, J. L., Lamy de la Chapelle, M., Adam, P. M., Rumyantseva, A., Lérondel, G., Royer, P. (2003). Optimization of SERS substrates for near-fiel Raman spectroscopy, *Synthetic Metals*, **139**, 621–624.

20. Awada, C., Popescu, T., Douillard, L., Charra, F., Perron, A., Yockell-Lelièvre, H., Baudrion, A.-L., Adam, P. M., Bachelot, R. (2012). Selective excitation of plasmon resonances of single Au triangles

by polarization-dependent light excitation, *J. Phys. Chem. C.*, **116**, 14591–14598.

21. Zhu, S., Zhou, W. (2012). Plasmonic properties of two-dimensional metallic nanoholes fabricated by focused ion beam lithography, *J. Nanoparticle Res.*, **14**, 1–6.

22. Bloomstein, T. M., Marchant, M. F., Deneault, S., Hardy, D. E., Rothschild, M. (2006). 22-nm immersion interference lithography, *Opt. Express*, **14**, 6434–6443.

23. Arriola, A., Rodriguez, A., Perez, N., Tavera, T. Withford, J., Fuerbach, A., Olaizola, S. M. (2012). Fabrication of high quality sub-micron Au gratings over large areas with pulsed laser interference lithography for SPR sensors, *Opt. Mater. Express*, **2**, 1571–1579.

24. Choi, Y.-K., King, T.-J. (2002). A spacer patterning technology for nanoscale CMOS, *IEEE Trans. Electron Devices*, **49**, 436–441.

25. Chang, C.-H., Zhao, Y., Heilmann, R. K., Schattenburg, M. L. (2008). Fabrication of 50 nm period gratings with multilevel interference lithography, *Opt. Lett.*, **33**, 1572–1574.

26. Aksu, S., Huang, M., Artar, A., Adato, R., Yanik, A. A., Altug, H. (2011). High resolution large area nanopatterning for plasmonics and metamaterials with nanostencil lithography, in: *Laser Science* XXVII (F. I. O. 2011, Ed.), OSA San Jose, California United States.

27. Aksu, S., Yanik, A. A., Adato, R., Artar, A., Huang, M., Altug, H. (2011). Nanostencil lithography for high-throughput fabrication of infrared plasmonic sensors, *Proc. SPIE*, 8031.

28. Hulteen, J. C., Van Duyne, R. P. (1995). Nanosphere lithography: A materials general fabrication process for periodic particle array surfaces, *J. Vacuum Sci. Technol. A*, **13**, 1553–1558.

29. Liao, H., Nehl, C., Hafner, J. H. (2006). Biomedical applications of plasmon resonant metal nanoparticles, *Nanomedecine*, **1**, 201–208.

30. Zhang, X., Whitney, A. V., Zhao, J., Hicks, E. M., Van Duyne, R. P. (2006). Advances in contemporary nanosphere lithographic techniques, *J. Nanosci. Nanotechnol.*, **6**, 1–15.

31. Camden, J. P., Dieringer, J. A., Zhao, J., Van Duyne, R. P. (2008). Controlled plasmonic nanostructures for surface-enhanced spectroscopy and sensing, *Acc. Chem. Res.*, **41**, 1653–1661.

32. Haes, A. J., Chang, L., Klein, W. L., Van Duyne, R. P. (2005). Detection of a biomarker for Alzheimer's disease from synthetic and clinical samples using a nanoscale optical biosensor, *J. Am. Chem. Soc.*, **127**, 2264–2271.

33. Fredriksson, H., Alaverdyan, Y., Dmitriev, A., Langhammer, C., Sutherland, D. S., Zäch, M., Kasemo, B. (2007). Hole–mask colloidal lithography, *Adv. Mater.*, **19**, 4297–4302.
34. Frank, B., Yin, X., Schäferling, M., Zhao, J., Hein, S. M., Braun, P. V., Giessen, H. (2010). Large-area 3D chiral plasmonic structures, *ACS Nano*, **7**, 6321–6329.
35. Shin, D. O., Jeong, J.-R., Han, T. H., Koo, C. M., Park, H.-J., Lim, Y. T., Kim, S. O. (2010). A plasmonic biosensor array by block copolymer lithography, *J. Mater. Chem.*, **20**, 7241–7247.
36. Asakawa, K., Fujimoto, A. (2005). Fabrication of subwavelength structure for improvement in light-extraction efficiency of light-emitting devices using a self-assembled pattern of block copolymer, *Appl. Opt.*, **44**, 7475–7482.
37. Asakawa, K., Hiraoka, T. (2002). Nanopatterning with microdomains of block copolymers using reactive-Ion etching selectivity, *Japanese J. Appl. Phys.*, **41**, 6112–6118.
38. Maurer, T., Sarrazin, A., Plaud, A., Béal, J., Nicolas, R., Lamarre, S. S., Proust, J., Nomenyo, K., Herro, Z., Kazan, M., Lerondel, G., Plain, J., Adam, P. M., Ritcey, A. M. (2013). Strategies for self-organization of Au nanoparticles assisted by copolymer templates, *Plasmonics: Metallic Nanostructures and Their Optical Properties XI, Proceedings of the SPIE Society San Diego*. San Diego, August 24–29, 2013.
39. Kang, G. B., Kim, S.-I., Kim, Y. T., Park, J. H. (2009). Fabrication of metal nano dot dry etching mask using block copolymer thin film, *Curr. Appl. Phys.*, **9**, 82–84.
40. Einsle, J. F. B., Dickson, J.-S. W., Zayats, A. V. (2011). Hybrid FIB milling strategy for the fabrication of plasmonic nanostructures on semiconductor substrates, *Nanoscale Res. Lett.*, **6**, 1–5.
41. Randolph, S. J., Fowlkes, J. D., Rack, P. D. (2006). Focused, nanoscale electron-beam-induced deposition and etching, *Crit. Rev. Solid State Mater. Sci.*, **31**, 55–89.
42. Dhawan, A., Gerhold, M., Madison, A., Fowlkes, J., Russell, P. E., Vo-Dinh, T., Leonard, D. N. (2009). Fabrication of nanodot plasmonic waveguide structures using FIB milling and electron beam-induced deposition, *Scanning*, **31**, 139–146.
43. Lobo, C. J., Toth, M., Wagner, R., Thiel, B. L., Lysaght, M. (2008). High resolution radially symmetric nanostructures from simultaneous electron beam induced etching and deposition, *Nanotechnology*, **19**, 1–6.

44. Shafeev, S. A. (2008). Formation of nanoparticles under laser ablation of solids in liquids, in: *Nanoparticles: New Research* (Lombardi, S. L., ed.), Nova Science Publishers Inc., pp. 1–37.
45. Viau, G., Collière, V. L., Shafeev, L.-M. G. A. (2011). Internal structure of Al hollow nanoparticles generated by laser ablation in liquid ethanol, *Chem. Phys. Lett.*, **501**, 419–422.
46. Sun, H.-B., Kawata, S. (2004). Two-photon photopolymerization and 3D lithographic microfabrication, *APS*, **170**, 169–273.
47. Maruo, S., Nakamura, O., Kawata, S. (1997). Three-dimensional microfabrication with two-photon-absorbed photopolymerization, *Opt. Lett.*, **22**, 132–134.
48. Chou, S. Y., Krauss, P. R., Renstrom, P. J. (1996). Imprint lithography with 25-nanometer resolution, *Science*, **272**, 85–87.
49. Shukla, S., Furlani, E. P., Vidal, X., Swihart, M. T., Prasad, P. N. (2010). Two-photon lithography of sub-wavelength metallic structures in a polymer matrix, *Adv. Mater.*, **22**, 3695–3699.
50. Ostendorf, A., Chichkov, B. N. (2006). Two-photon polymerization: A new approach to micromachining, *Photon. Spectra*, 1–7. http://www.aerotech.com/media/246109/TwoPhotonPoly.pdf.
51. Vial, A., Laroche, T. (2007). Description of dispersion properties of metals by means of the critical points model and application to the study of resonant structures using the FDTD method, *J. Phys. D: Appl. Phys.*, **40**, 7152.
52. Gates, B. D., Xu, Q., Stewart, M., Ryan, D., Willson, C. G., Whitesides, G. M. (2005). New approaches to nanofabrication: molding, printing, and other techniques, *Chem. Rev.*, **105**, 1171–1196.
53. Roche, P. J. R., Wang, S., Cheung, M. C.-K., Chodavarapu, V. P., Kirk, A. G. (2012). A nanorod polymer micro-array formed by micro-contact printing, *Nanoscale Imaging, Sensing, and Actuation for Biomedical Applications VIII*, Proc. SPIE, San Jose, United States, February 1, 2012. doi:10.1117/12.910142.
54. Bhandari, D., Wells, S. M., Polemi, A., Kravchenko, I. I., Shuford, K. L., Sepaniak, M. J. (2011). Stamping plasmonic nanoarrays on SERS-supporting platforms, *J. Raman Spectrosc.*, **42**, 1916–1924.
55. Zhao, X.-M., Xia, Y., Whitesides, M. (1997). Soft lithographic methods for nano-fabrication, *J. Mater. Chem.*, **7**, 1069–1074.
56. Zhao, Y., Yoon, Y.-K., Choi, S.-O., Wu, X., Liu, Z., Allen, M. G. (2009). Three dimensional metal pattern transfer for replica molded microstructures, *Appl. Phys. Lett.*, **94**, 023301.

57. Lipomi, D. J., Kats, M. A., Kim, P., Kang, S. H., Aizenberg, J., Capasso, F., Whitesides, G. M. (2010). Fabrication and replication of arrays of single- or multicomponent nanostructures by replica molding and mechanical sectioning, *ACS Nano*, **4**, 4017–4026.
58. Boltasseva, A. (2009). Plasmonic components fabrication via nanoimprint, *J. Opt. A: Pure Appl. Opt.*, **11**, 114001.
59. Chou, S. Y., Krauss, P. R., Renstrom, P. J. (1996). Nanoimprint lithography, *J. Vacuum Sci. Technol. B*, **14**, 4129.
60. Truong, T. T., Maria, J., Yao, J., Stewart, M. E., Lee, T.-W., Gray, S. K., Nuzzo, R. G., Rogers, J. A. (2009). Nanoposts plasmonic crystals, *Nanotechnology*, **20**, 434011.
61. Lerondel, G., Kostcheev, S., Plain, J. (2012). Nanofabrication for plasmonics, in: *Plasmonics* (Bonod, S. E. A. N., ed.), Springer Series in Optical Sciences, pp. 269–316.
62. Fleischmann, M., Hendra, P. J. McQuillan, A. J. (1974). Raman spectra of pyridine adsorbed at a silver electrode, *Chem. Phys. Lett.*, **26**, 163–166.
63. Vo-Dinh, T. (1998). Surface-enhanced Raman spectroscopy using metallic nanostructures, *Trends Anal. Chem.*, **17**, 557–582.
64. Karakouz, T., Tesler, A. B., Bendikov, T. A., Vaskevich, A., Rubinstein, I. (2008). Highly stable localized plasmon transducers obtained by thermal embedding of gold island films on glass, *Adv. Mater.*, **20**, 3893–3899.
65. Jia, K., Bijeon, J. L., Adam, P. M., Ionescu, R. E. (2012). Sensitive localized surface plasmon resonance multiplexing protocols, *Anal. Chem.*, **84**, 8020–8027.
66. Jia, K., Bijeon, J.-L., Adam, P.-M., Ionescu, R. E. (2012). Large scale fabrication of gold nano-structured substrates via high temperature annealing and their direct use for the LSPR detection of atrazine, *Plasmonics*, **8**, 143–151.
67. Martin, J., Proust, J., Gérard, D., Plain, J. (2013). Localized surface plasmon resonances in the ultraviolet from large scale nanostructured aluminum films, *Opt. Mater. Express*, **3**, 954–959.
68. Link, S., Mohamed, M. B., El-Sayed, M. A. (1999). Simulation of the optical absorption spectra of gold nanorods as a function of their aspect ratio and the effect of the dielectric constant, *J. Phys. Chem. B*, **103**, 3073–3077.
69. Ying, Y.-Y., Chang, S.-S., Lee, C.-L., Wang, R.-C. (1997). Gold nanorods: Electrochemical synthesis and optical properties, *J. Phys. Chem. B*, **101**, 6661–6664.

70. Link, S., Mohamed, M. B., El-Sayed, M. A. (1999). Simulation of the optical absorption spectra of gold nanorods as a function of their aspect ratio and the effect of the medium dielectric constant, *J. Phys. Chem. B*, **103**, 3073–3077.

71. Sandrock, M. L., Pibel, C. D. G., Foss, F. M. C. A. (1999). Synthesis and second-harmonic generation of noncentrosymmetricgold nanostructures, *J. Phys. Chem. B*, **103**, 2668–2673.

72. Dirix, Y., Bastiaansen, C., Caseri, W., Smith, P. (1999). Oriented pearl-necklace arrays of metallic nanoparticles in polymers: A new route toward polarization-dependent color filters, *Adv. Mater.*, **11**, 223–227.

73. Liz-Marzan, L. (2004). Nanometals: formation and color, *Mater. Today*, **7**, 26–31.

74. Jana, N. R., Gearheart, L., Murphy, C. J. (2001). Evidence for seed-mediated nucleation in the chemical reduction of gold salts to gold nanoparticles, *Chem. Mater.*, **13**, 2313–2322.

75. Duff, G., Baiker, A., Edwards, P. P. (1993). A new hydrosol of gold clusters. 1. Formation and particle size variation, *Langmuir*, **9**, 2301–2309.

76. Perez-Juste, J., Pastoriza-Santos, I., Liz-Marzan, L. M., Mulvaney, P. (2005). Gold nanorods: synthesis, characterization and applications, *Coordination Chem. Rev.*, **249**, 1870–1901.

77. Pastoriza-Santos, I., Liz-Marzán, L. M. (2002). Synthesis of silver nanoprisms in DMF, *Nano Lett.*, **2**, 903–905.

78. Grzelczak, M., Pérez-Juste, J., Mulvaney, P., Liz-Marzán, L. M. (2008). Shape control in gold nanoparticle synthesis, *Chem. Soc. Rev.*, **37**, 1783–1791.

79. Kimling, J., Maier, M., Okenve, B., Kotaidis, V., Ballot, H., Plech, A. (2006). Turkevich method for gold nanoparticle synthesis revisited, *J. Phys. Chem. B*, **110**, 15700–15707.

80. Turkevich, J. (1985). Colloidal gold. Part I, *Gold Bull.*, **18**, 86–91.

81. Turkevich, J., Stevenson, P. C., Hillier, J. (1951). A study of the nucleation and growth processes in the synthesis of colloidal gold, *Discuss. Faraday Soc.*, **11**, 55.

82. Brust, M., Walker, M., Bethell, D., Schiffrin, D. J., Whyman, R. (1994). Synthesis of thiol-derivatised gold nanoparticles in a two-phase liquid-liquid system, *J. Chem. Soc. Chem. Commun.*, **7**, 801–802. http://pubs.rsc.org/en/Content/ArticleLanding/1994/C3/c39940000801#!divAbstract.

83. Hirai, H., et al. (1979). Preparation of colloidal transition metals in polymers by reduction with alcohols or ethers, *J. Macromol. Sci. Chem.*, **13**, 727.

84. Ducamp-Sanguesa, C., Herrera-Urbina, R., Figlarz, M. (1992). Synthesis and characterization of fine and mono disperse silver particles of uniform shape, *J. Solid State Chem.*, **100**, 272.

85. Fievet, F., Lagier, P., Blin, B., Beaudoin, B., Figlarz, M. (1989). Homogeneous and heterogeneous nucleations in the polyol process for the preparation of micron and submicron size metal particles, *Solid State Ionics*, **32–33**, 198–205.

86. Jiang, X. C., Brioude, A., Pileni, M. P. (2006). Gold nanorods: Limitations on their synthesis and optical properties, *Colloids Surf. A: Physicochem. Eng.*, Aspects, **277**, 201–206.

87. Murphy, C. J., Jana, N. R. (2002). Controlling the aspect ratio of inorganic nanorods and nanowires, *Adv. Mater.*, **14**, 80–82.

88. Murphy, C. J., Sau, T. K., Gole, A. M., Orendorff, C. J., Gao, J., Gou, L., Hunyadi, S. E., Li, T. (2005). Anisotropic metal nanoparticles: Synthesis, assembly, and optical applications, *J. Phys. Chem. B*, **109**, 13857–13870.

89. Chen, H. M., Peng, H.-C., Liu, R.-S., Asakura, K., Lee, C.-L., Lee, J.-L., Hu, S.-F. (2005). Controlling the length and shape of gold nanorods, *J. Phys. Chem. B Lett.*, **109**, 19553–19555.

90. Gao, J., Bender, C. M., Murphy, C. J. (2003). Dependence of the gold nanorod aspect ratio on the nature of the directing surfactant in aqueous solution, *Langmuir*, **19**, 9065.

91. Manivannan, S., Ramaraj, R. (2009). Core–shell Au/Ag nanoparticles embedded in silicate sol–gel network for sensor application towards hydrogen peroxide, *J. Chem. Sci.*, **121**, 735–743.

92. Wang, H., Brandl, D. W., Le, F., Nordlander, P., Halas, N. J. (2006). Nanorice: A hybrid plasmonic nanostructure, *Nano Lett.*, **6**, 827–832.

93. Bergman, D. J., Stockman, M. I. (2003). Surface plasmon amplification by stimulated emission of radiation: Quantum generation of coherent surface plasmons in nanosystems, *Phys. Rev. Lett.*, **90**, 027402.

94. Shchukin, V. A., Bimberg, D. (1999). Spontaneous ordering of nanostructures on crystalm surfaces, *Rev. Modern Phys.*, **71**, 1125–1171.

95. Maenosono, S., Okubo, T., Yamaguchi, Y. (2003). Overview of nanoparticle array formation by wet coating, *J. Nanoparticle Res.*, **5**, 5–15.

96. Teranishi, T., Hosoe, M., Tanaka, T., Miyake, M. (1999). Size control of monodispersed Pt nanoparticles and their 2D organization by electrophoretic deposition, *J. Phys. Chem. B*, **103**, 3818–3827.

97. Wang, W., Cassar, K., Sheard, S. J., Dobson, P. J., Bishop, P., Parkin, I. P., Hurst, S. (2009). Spray deposition of Au/TiO$_2$ composite thin films using preformed nanoparticles, in: *Nanotechnology in Construction 3 Proceedings of the NICOM3* (Zdeněk Bittnar, Bartos, P. J. M., Jiří Němeček, Vít Šmilauer, Jan Zeman, ed.), Springer Berlin Heidelberg, pp. 395–401.

98. Wijaya, E., Lenaerts, C., Maricot, S., Hastanin, J., Habraken, S., Vilcot, J.-P., Boukherroub, R., Szunerits, S. (2011). Surface plasmon resonance-based biosensors: From the development of different SPR structures to novel surface functionalization strategies, *Curr. Opin. Solid State Mater. Sci.*, **15**, 208–224.

99. Schlecht, C. A., Maurer, J. A. (2011). Functionalization of glass substrates: Mechanistic insights into the surface reaction of trialkoxysilanes, *RSC Adv.*, **1**, 1446–1448.

100. Maurer, T., Abdellaoui, N., Gwiazda, A., Adam, P.-M., Vial, A., Bijeon, J.-L., Chaumont, D., Bourezzou, M. (2013). Optical determination and identification of organic shells around nanoparticles: application to silver nanoparticles, *Nano*, **8**, 130016.

101. Tabor, M., Murali, R., Mahmoud, M., El-Sayed, M. A. (2009). On the use of plasmonic nanoparticle pairs as a plasmon ruler: The dependence of the near-field dipole plasmon coupling on nanoparticle size and shape, *J. Phys. Chem. A*, **113**, 1946–1953.

102. Love, J. C., Estroff, L. A., Kriebel, J. K., Nuzzo, R. G., Whitesides, G. M. (2005). Self-assembled monolayers of thiolates on metals as a form of nanotechnology, *Chem. Rev.*, **105**, 1103–1169.

103. Oldenburg, S. J., Averitt, R. D., Westcott, S. L., Halas, N. J. (1998). Nanoengineering of optical resonances, *Chem. Phys. Lett.*, **288**, 243–247.

104. Chen, S. (2001). Langmuir-Blodgett fabrication of two-dimensional robust cross-linked nanoparticle assemblies, *Langmuir*, **17**, 2878–2884.

105. Lamarre, S. S., Sarrazin, A., Proust, J., Yockell-Lelièvre, H., Plain, J. Ritcey, A.-M., Maurer, T. (2013). Optical properties of Au colloids self-organized into rings via copolymer templates, *J. Nanoparticle Res.*, **15**, 1656.

106. Ariga, K., Hill, J. P., Ji, Q. (2007). Layer-by-layer assembly as a versatile bottom-up nanofabrication technique for exploratory research and realistic application, *Phys. Chem. Chem. Phys.*, **9**, 2319–2340.
107. Cassagneau, T., MAllouk, T. E., Fendler, J. H. (1998). Layer-by-layer assembly of thin film zener diodes from conducting polymers and CdSe nanoparticles, *J. Am. Chem. Soc.*, **120**, 7848–7859.
108. Fu, Y., Xu, H., Bai, S., Qiu, D., Sun, J., Wang, Z., Zhang, X. (2002). Fabrication of a stable polyelectrolyte/Au nanoparticles multilayer film, *Macromol. Rapid Commun.*, **23**, 256–259.
109. Jiang, C., Markutsya, S., Tsukruk, V. V. (2004). Collective and individual plasmon resonances in nanoparticle films obtained by spin-assisted layer-by-layer assembly, *Langmuir*, **20**, 882–890.
110. Denkov, N. D., Velev, O. D., Kralchevsky, P. A., Ivanov, I. A., Yoshimura, H., Nahgayama, K. (1993). Two-dimensional crystallization, *Nature*, **361**, 26.
111. Lin, X. M., Jaeger, H. M., Sorensen, S. M., Klabunde, K. J. (2001). Formation of long-range-ordered nanocrystal superlattices on silicon nitride substrates, *J. Phys. Chem. B*, **105**, 3353–3357.
112. Bigioni, T. P., Lin, X. M., Nguyen, T. T., Corwin, E. I., Witten, T. A., Jaeger, H. M. (2006). Kinetically driven self assembly of highly ordered nanoparticle monolayers, *Nature*, **5**, 265–270.
113. Khanal, B. P., Zubarev, E. R. (2007). Rings of nanorods, *Angew. Chem.*, **46**, 2195–2198.
114. Chen, J., Liao, W.-S., Chen, X., Yang, T., Wark, S. E., Son, D. H., Batteas, J. D., Cremer, P. S. (2009). Evaporation-induced assembly of quantum dots into nanorings, *ACS Nano*, **3**, 173–180.
115. Fahmi, A., Pietsch, T., Mendoza, C., Cheval, N. (2009). Functional hybrid materials, *Mater. Today*, **12**, 44–50.
116. Kim, S. H., Misner, M. J., Xu, T., Kimura, M., Russell, T. P. (2004). Highly oriented and ordered arrays from block copolymers via solvent evaporation, *Adv. Mater.*, **16**, 226–231.
117. Spatz, J. P., Mössmer, S., Möller, M. (1996). Mineralization of gold nanoparticles in a block copolymer microemulsion, *Chem. Eur. J.*, **2**, 1552–1555.
118. Mössmer, S., Spatz, J. P., Möller, M., Aberle, T., Schmidt, J., Burchard, W. (2000). Solution behavior of poly(styrene)-block-poly(2-vinylpyridine) micelles containing gold nanoparticles, *Macromolecules*, **33**, 4791–4798.

119. Alexandridis, P. (2011). Gold nanoparticle synthesis, morphology control, and stabilization facilitated by functional polymers, *Chem. Eng. Technol.*, **34**, 15–28.
120. Sakaï, T., Alexandridis, P. (2004). Single-step synthesis and stabilization of metal nanoparticles in aqueous pluronic block copolymer solutions at ambient temperature, *Langmuir*, **20**, 8426–8430.
121. Sarrazin, A., Gontier, A., Plaud, A., Béal, J., Yockell-Lelièvre, H., Bijeon, J.-L., Plain, J., Adam, P.-M., Maurer, T. (2014). Single step synthesis and organization of gold colloids assisted by copolymer templates, *Nanotechnology*, **25**, 225603.
122. Lamarre, S. S., Lemay, C., Labrecque, C., Ritcey, A. M. (2013). Controlled 2D organization of gold nanoparticles in block copolymer monolayers, *Langmuir*, **29**, 10891–10898.
123. Jacobson, J. M., Hubert, B. N., Ridley, B., Nivi, B., Fuller, S. (2001). Nanoparticle-based electrical, chemical, and mechanical structures and methods of making same, US Patent, 6, 294, 401, in: US Patent, 294, 401 (Ed.) US Patent, United States.
124. Maurer, T., Gautrot, S., Ott, F., Chaboussant, G., Zighem, F., Cagnon, L., Fruchart, O. (2014). Ordered arrays of magnetic nanowires investigated by polarized small-angle neutron scattering, *Phys. Rev. B*, accepted.
125. Maurer, T., Zighem, F., Gautrot, S., Ott, F., Chaboussant, G., Cagnon, L., Fruchart, O. (2013). Magnetic nanowires investigated by Polarized SANS, *Phys. Procedia*, **42**, 74–79.
126. Piquemal, J.-Y., Viau, G., Beaunier, P., Bozon-Verdurz, P., Fiévet, F. (2003). One-step construction of silver nanowires in hexagonal mesoporous silica using the polyol process, *Mater. Res. Bull.*, **38**, 389–394.
127. Maurer, T. (2009). *Magnetism of Anisotropic Nano-Objects: Magnetic and Neutron Studies of $Co_{1-x}Ni_x$ Nanowires*, Paris XI, Orsay, France.
128. Plaud, A., Sarrazin, A., Béal, J., Proust, J., Royer, P., Bijeon, J.-L., Plain, J., Adam, P. M., Maurer, T. (2013). Copolymer template control of gold nanoparticle synthesis via thermal annealing, *J. Nanoparticle Res.*, **15**, 1–6.
129. Maier, S. A. (2007). *Plasmonics: Fundamentals and Applications*, Springer Science + Business Media LLC, 233 Spring Street, New York, NY 10013, U S A.
130. Johnson, P. B., Christy, R. W. (1972). Optical constants of the noble metals, *Phys. Rev. B*, **6**, 4370–4379.
131. Kuttge, M., Vesseur, E. J. R., Verhoeven, J., Lezec, H. J., Atwater, H. A., Poman, A. (2008). Loss mechanisms of surface plasmon polaritons

on gold probed by cathodoluminescence imaging spectroscopy, *Appl. Phys. Lett.*, **93**, 113110.

132. West, P. R., Ishii, S., Naik, G. V., Emani, N. K., Shalaev, V. M., Boltasseva, A. (2010). Searching for better plasmonic materials, *Laser Photon. Rev.*, **4**, 795–808.

133. Debarre, A., Jaffiol, R., Julien, C., Tchénio, P. (2004). Raman scattering from single Ag aggregates in presence of EDTA, *Chem. Phys. Lett.*, **386**, 244–247.

134. Qi, H., Alexson, D. A., Glembocki, O. J., Prokes, S. M. (2011). Synthesis and oxidation of silver nanoparticles, quantum dots and nanostructures: Synthesis, characterization, and modeling VIII. Edited by Eyink, K. G., Szmulowicz, F., Huffaker, D. L. *Proc. SPIE*, 7947, article id. 79470Y, 11 pp.

135. Lindquist, N. C., Nagpal, P., McPeak, K. M., Norris, D. J., Oh, S.-H. (2012). Engineering metallic nanostructures for plasmonics and nanophotonics, *Rep. Prog. Phys.*, **75**, 036501.

136. Chan, G. H., Zhao, J., Hicks, E. M., Schatz, G. C., Van Duyne, R. P. (2007). Plasmonic properties of copper nanoparticles fabricated by nanosphere lithography, *Nano Lett.*, **7**, 1947–1952.

137. Hylton, N. P., Li, X. F., Giannini, V., Lee, K.-H., Ekins-Daukes, N. J., Loo, J., Vercruysse, D., Van Dorpe, P., Sodabanlu, H., Maier, S. A. (2013). Loss mitigation in plasmonic solar cells: aluminium nanoparticles for broadband photocurrent enhancements in GaAs photodiodes, *Sci. Rep.*, **3**, 2874.

138. Kim, J. H., Ehrman, S. H., Germer, T. A. (2004). Influence of particle oxide coating on light scattering by submicron metal particles on silicon wafers, *Appl. Phys. Lett.*, **84**, 1278–1280.

139. Rai, A., Park, K., Zhou, L, Zachariah, M. R. (2006). Understanding the mechanism of aluminium nanoparticle oxidation, *Combustion Theory Model.*, **10**, 843–859.

140. Tobiska, P., Hugon, O., Trouillet, A., Gagnaire, H. (2001). An integrated optic hydrogen sensor based on SPR on palladium, *Sens. Actuators B-Chem.*, **74**, 168–172.

141. Baldelli, S., Eppler, A. S., Anderson, E., Shen, Y. R., Somorjai, G. A. (2000). Surface enhanced sum frequency generation of carbon monoxide adsorbed on platinum nanoparticle arrays, *J. Chem. Phys.*, **113**, 5432–5438.

142. Nouneh, K., Oyama, M., Diaz, R., Abd-Lefdil, M., Kityk, I. V., Bousmina, M. (2011). Nanoscale synthesis and optical features of metallic nickel nanoparticles by wet chemical approaches, *J. Alloys Compounds*, **509**, 5882–5886.

143. Xiong, Z., Chen, X., Wang, X., Peng, L., Yan, D., Lei, H., Fu, Y., Wu, J., Li, Z., An, X., Wu, W. (2013). Size dependence of plasmon absorption of Ni nanoparticles embedded in $BaTiO_3/SrTiO_3$ superlattices, *Appl. Surf. Sci.*, **268**, 524–528.

144. Bonanni, V., Bonetti, S., Pakizeh, T., Pirzadeh, Z., Chen, J., Nogues, J., Vavssori, P., Hillenbrand, R., Akerman, J., Dmitriev, A. (2011). Designer magnetoplasmonics with nickel nanoferromagnets, *Nano Lett.*, **11**, 5333–5338.

145. Rakic, A. (1995). Algorithm for the determination of intrinsic optical constants of metal films: Application to aluminum, *Appl. Opt.*, **34**, 4755–4767.

146. Palik, E. (1985). *Handbook of Optical Constants*, Academic Press.

147. Wang, K., Mittleman, D. M. (2006). Dispersion of surface plasmon polaritons on metal wires in the terahertz frequency range, *Phys. Rev. Lett.*, **96**, 157401.

148. Rana, F. (2008). Graphene terahertz plasmon oscillators, *IEEE Trans. Nanotechnol.*, **7**, 91–99.

149. Fei, Z., Rodin, A. S., Andreev, G. O., Bao, W., McLeod, A. S., Wagner, M., Zhang, L. M., Zhao, Z., Thiemens, M., Dominguez, G., Fogler, M. M., Castro Neto, A. H., Lau, C. N., Keilmann, F., Basov, D. N. (2012). Gate-tuning of graphene plasmons revealed by infrared nano-imaging, *Nature*, **487**, 82–85.

150. Koppens, F. H. L., Chang, D. E., Garcia de Abajo, F. J. (2011). Graphene plasmonics: A platform for strong lightmatter interactions, *Nano Lett.*, **11**, 3370–3377.

151. Fang, Z., Thongrattanasiri, S., Schlather, A., Liu, Z., Ma, L., Wang, Y., Ajayan, P. M., Nordlander, P., Halas, N. J., García de Abajo, F. J. (2013). Gated tunability and hybridization of localized plasmons in nanostructured graphene, *ACS Nano*, **7**, 2388–2395.

152. Russo, P., Hu, A., Compagnini, G., Zhou, N. Y. (2013). Femtosecond laser ablation of highly oriented pyrolytic graphite: Green route for large-scale production of porous graphene and graphene quantum dots, *Nanoscale*, accepted.

153. Fang, Z., Wang, Y., Liu, Z., Schlather, A., Ajayan, P. M., Koppens, F. H. L., Nordlander, P., Halas, N. J. (2012). Plasmon-induced doping of graphene, *ACS Nano*, **6**, 10222–10228.

154. Maurer, T., Nicolas, R., Leveque, G., Subramanian, P., Proust, J., Béal, J., Schuermans, S., Vilcot, J. P., Herro, Z., Kazan, M., Plain, J., Boukherroub, R., Akjouj, A., Djafari-Rouhani, B., Adam, P. M., Szunerits, S. (2013). Enhancing LSPR sensitivity of Au gratings through graphene coupling to Au film, *Plasmonics*, accepted.

Chapter 9

Colorimetric Sensing Based on Metallic Nanostructures

Daniel Aili[a] and Borja Sepulveda[b]

[a]*Division of Molecular Physics, Department of Physics, Chemistry and Biology, Linköping University, 581 83 Linköping, Sweden*
[b]*ICN2—Institut Catala de Nanociencia i Nanotecnologia, Campus UAB and CSIC—Consejo Superior de Investigaciones Cientificas, ICN2 Building, 08193 Bellaterra, Barcelona, Spain*
daniel.aili@liu.se, borja.sepulveda@cin2.es

9.1 Introduction and Historical Perspective

The interesting optical properties of plasmonic metal nanoparticles have intrigued humans for thousands of years and nanoparticles of gold-, silver-, and copper have been used for staining of glass and pottery since ancient times [1]. The Lycurgus cup, made in Rome in the fourth century AD, is perhaps the most famous historical evidence of the skills of the Roman glassmakers [2]. In addition to the beautiful carvings on the glass, the cup has a unique green luster. The true marvel, however, appears when light is shown through the glass and the cup instead displays a striking ruby red color (Fig. 9.1). The origin of this phenomenon was long debated until

---

Introduction to Plasmonics: Advances and Applications
Edited by Sabine Szunerits and Rabah Boukherroub
Copyright © 2015 Pan Stanford Publishing Pte. Ltd.
ISBN 978-981-4613-12-5 (Hardcover), 978-981-4613-13-2 (eBook)
www.panstanford.com

carful examinations using electron microscopy revealed that gold and silver nanoparticles were embedded in the glass [3].

The art of producing colloidal gold dispersed in aqueous media, so-called drinkable gold or *Aurum Potabile*, was developed far later and as consequence of the discovery of *aqua regia* in the eighth century A.D. by the Arab alchemist, Jabir ibn Hayyan [4]. The ability of dissolving metallic bulk gold using a mixture of nitric and hydrochloric acid is an important step for obtaining the gold salt required for the synthesis of gold nanoparticles. The first mentioning of colloidal gold and its application to medicine can be found in a book from 1618 with the title *"Panacea Aurea-Auro Potabile"* by Francisci Antonii [5]. Antonii describes detailed procedures for synthesis of colloidal gold and their uses for treatment of various diseases. In 1676, the German chemist Johann Kunckels published his observation of the properties of *Auro Potabilie* and came to the conclusion that the pinkish solution must contain metallic gold but in such small fragments that it could not be seen by the human eye [6]. The first scientific report scrutinizing the properties of colloidal gold was published by Michael Faraday in 1857 [7]. Faraday came to the same conclusion as Kunckel did, that the red color must emanate from suspended particles of metallic gold albeit *"the state of division of these particles must be extreme; they have not as yet been seen by any power of the microscope"* [7].

Figure 9.1 The Lycurgus Cup, Late Roman, 4th century AD, probably made in Rome. In reflected light (left) and in transmitted light (right). © Trustees of the British Museum.

There was for a long time a strong belief in the magical curative properties of colloidal gold, [6] mainly spurred by the reference of *Aurum Potabile* as the elixir of life by some alchemists. The first scientifically based medical application of colloidal gold was reported by Lange in 1912 and was a colorimetric sensor for diagnosis of syphilis [8]. This early bioassay exploited the dramatic optical shift that is obtained when gold nanoparticles aggregate, which changes the color of a suspension from ruby red to bluish or purple. The assay was based on a previous finding made by Zsigmondy that a solution of proteins could prevent gold nanoparticles from aggregating. Lange discovered that gold nanoparticles aggregated when added to diluted and salinated cerebrospinal fluid from patients suffering from syphilis, but not in healthy patients.

The art of producing and using plasmonic nanoparticles for sensing has thus been known for centuries. There has, however, been a tremendous development in the field since the mid 1990s driven by the improvements in methods for synthesis and characterization of nanoparticles along with increasing knowledge about nanoparticle surface chemistry and functionalization. The majority of colorimetric sensors are based on spherical gold nanoparticles (AuNPs). There are several reasons for this bias toward AuNPs, perhaps most obviously because they absorb light in the visible wavelength range, which makes sense if the aim is to develop a colorimetric sensor. AuNPs are also relatively easy to synthesize, stable in the sense that they do not oxidize (like silver nanoparticles) and they can easily be chemically functionalized using gold–thiol chemistry. This chapter will thus focus on the properties and applications of gold nanoparticles for colorimetric sensors.

AuNP-based colorimetric sensors have the advantage over other biosensor technologies in that they transduce molecular interactions by an optical signal that in many cases can be observed by the naked eye or at least by a simple and cheap spectrophotometer. They are thus best suited for applications where a more complex and expensive biosensor is not required, not affordable, or not practical to use, such as in assays for home use or in primary care. Colorimetric assays may also provide additional information not necessarily accessible by traditional biosensors or assays and can in many cases also be adapted for high throughput drug screening.

9.2 Synthesis of Gold Nanoparticles

The methods for synthesis of AuNPs has been refined and improved over the years and in addition to spherical nanoparticles in sizes ranging from a few nanometers up to about 250 nm, protocols for obtaining other shapes, including cubes, rods and prisms, have also been developed. A more elaborate description of methods for synthesis of gold nanoparticles can be found elsewhere [9].

Briefly, spherical AuNPs, in the size range 10–100 nm, are typically prepared by reduction of chloroauric acid ($HAuCl_4$) using sodium citrate in water. This method was first described by Turkevich [10] and later modified by Frens [11] and results in relatively monodisperse particles that are electrostatically stabilized by physisorbed citrate. Gold nanorods (AuNRs) can be obtained by a seed-mediated growth process based on the reduction of gold salt in the presence of small quantities of silver ions and the surfactant cetyltrimethylammonium bromide (CTAB) [12].

Particles smaller than 5 nm are usually prepared by the method developed by Brust and Schiffrin [13]. $AuCl_4^-$ is transferred to toluene from an aqueous solution using tetraoctylammonium bromide as a phase-transfer reagent, and reduced with aqueous sodium borohydride ($NaBH_4$) in the presence of dodecanethiol. The resulting monolayer protected clusters (MPC) are typically in the size range 2–6 nm. Strategies for using other thiols have also been developed as well as methods for exchange of ligands in order to obtain MPCs suitable for a variety of applications [9].

9.3 Optical Properties of Gold Nanoparticles

Suspensions of spherical AuNPs have a ruby red color with a distinct extinction band in the visible wavelength range. The color emanates from the collective electron oscillations of near free electrons at the particle surface, so-called localized surface plasmon resonance (LSPR) [9].

The position of the LSPR band depends on a number of factors, including size and shape of the nanoparticles, but also on the refractive index (RI) of the surrounding medium. Changes in the RI consequently induce a measurable shift of the LSPR band. The extent of the shift is typically not very large as the RI sensitivity of spherical AuNPs are <100 nm per RI unit. Significantly larger spectral shifts are obtained upon particle aggregation resulting in a change in color from red to purple, clearly visible to the naked eyes. The magnitude of the red shift is dependent on the size of the nanoparticles, density, and internal structure of the aggregates (Fig. 9.2). Both possibilities to induce spectral shifts, i.e., by changes in RI or by changes in colloidal stability, are exploited in AuNP-based biosensors. Sufficiently small AuNPs (<3 nm) can, however, no longer support surface plasmons but can instead demonstrate fluorescent properties [14, 15].

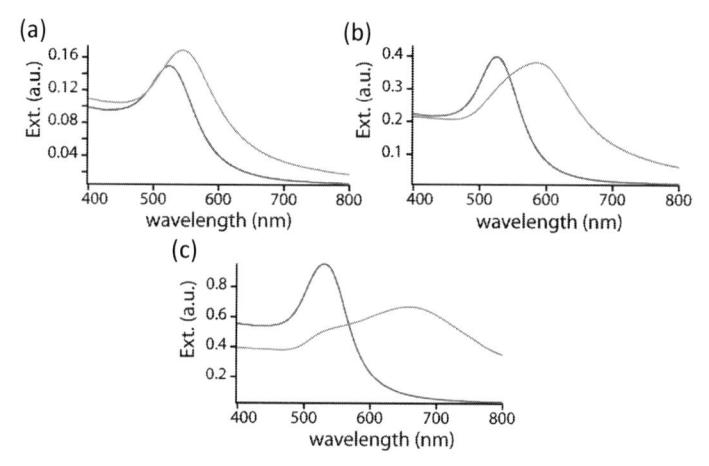

Figure 9.2 Aggregation of AuNPs gives rise to a pronounced red shift and broadening of the LSPR band. UV-vis spectra of (a) 20 nm, (b) 40 nm and (c) 100 nm diameter AuNPs before (red) and after (blue) aggregation). The distances between the aggregated nanoparticles are roughly the same for all sizes. Adapted with permission from ref. [16]. Copyright (2011) American Chemical Society.

9.4 Colloidal Stability and Surface Chemistry of Gold Nanoparticles

Colloidal particles are always subject to Brownian motion and most often also hydrodynamic forces, and will thus frequently undergo collisions. If the repulsive interactions are stronger than the attractive interactions, the particles will separate again after contact whereas aggregation will be the outcome if the opposite is true. In colloidal science, the term *stability* is often used to describe how well particles are dispersed in the continuous phase and can resist aggregation. Whether a colloidal system is stable will depend on the balance of repulsive and attractive forces acting between the particles.

There are a number of forces that play a role in the interaction between colloids such as van der Waals forces, electrostatic interactions, solvation (hydration), and osmotic and entropic interactions related to macromolecular adsorption [17]. At long range (typically <100 nm) the interactions between particles in liquid are dominated by electrostatic, steric-polymer and van der Waals forces, whereas at shorter range (<1–3 nm) solvation and other types of steric forces tend to be more dominant [18].

In colorimetric sensors, biospecific interactions are often employed to modulate the stability of the nanoparticles. The interactions can be manifested in numerous ways but typically involve specific recognition between two or more biomolecules where at least one of the species is immobilized on the nanoparticle surface. The recognition event can result either in binding or dissociation of species or in an enzymatic modification of any of the involved molecules, with the intention that the process should change the colloidal stability of the system, resulting in a colorimetric response. Examples illustrating this process will be highlighted and discussed in more detail in Section 9.5.

9.4.1 Surface Functionalization

A well-defined surface chemistry is a requirement in order to be able to tune the stability of colloids and to provide specific recognition and avoid undesirable interactions related to, e.g., non-specific protein binding. Gold nanoparticles have the advantage of providing a possibility for a robust and flexible surface chemistry owing to the high affinity between sulfur and gold. The gold–thiolate bond has a binding energy of 170 kJ/mol, which is close to the strength of the gold–gold bond [19, 20]. Gold–thiol chemistry using alkanethiols and other thiol- and disulfide-containing molecules have been extensively employed for surface modification and formation of self-assembled monolayers (SAMs) on planar gold surfaces [20].

The principles for assembly of monolayers on nanoparticles are about the same as on flat 2D-surfaces. There are, however, a few differences between a particle and a planar surface and some practical issues that that have to be taken into consideration. One important aspect is that on the surface of dispersed particles, molecules or ions that promote particle stability are already adsorbed. Removal and replacement of these stabilizers can cause an irreversible aggregation of the particles. Furthermore, the surface of a nanoparticle is finite in size and often presents a lot of defects, and edges and corners between the various facets. It may also present variable crystal facets and not only the (111) lattice plane commonly used for two-dimensional SAMs. In particular, the number of edges and corners increases as the size of the particles decreases, whereas the number of (111) facets on gold nanoparticles increases with increasing particle diameter [21]. On smaller particles ($r < \sim 5$ nm) the curvature of the surface can also have a substantial influence on the conformation of adsorbed species [22, 23]. On planar surfaces, the adsorbate can occupy a volume in space with a cylindrical geometry, whereas on surfaces with high curvature the available space is conical. The smaller the particles the larger will the conical segment be, which decreases the stabilizing lateral interactions between adsorbed molecules and allow for a more flexible conformation. A molecule protruding 2 nm from the surface of a 13 nm particle can thus occupy a ~30% larger volume than the corresponding molecule immobilized on a planar substrate. The smaller the particles the larger will the difference in accessible volume be. Furthermore, the higher curvature on spherical particles may result in a higher number of adsorbate molecules per metal surface atom, from the bulk value of one thiol per three gold surface atoms to roughly one thiol per two gold surface atoms on clusters 2 nm in diameter [22, 24].

9.5 Molecular Recognition for Modulation of Nanoparticle Stability

The colorimetric assay developed by Lange in 1912 exploited the aggregation of gold nanoparticles caused by unspecific interactions that also could vary significantly between different samples [25]. An assay intended for diagnostic applications must, however, produce a predictable and reproducible response that is proportional to the concentration or activity of a particular component without interference from other substances. The recognition event should thus ideally be very specific and at the same time unspecific interactions must be prevented, which can be a great challenge.

In order to obtain a substantial colorimetric response, the gold nanoparticles must form rather dense aggregates. As a rule of thumb, the separation should be smaller than the particle diameter (Fig. 9.3). Considering that AuNPs typically used in colorimetric assays are about 20 nm in diameter many biomolecular receptors cannot be utilized as recognition elements. IgG antibodies, for example, are about 14.5 × 8.5 × 4.0 nm in size, and when homogenously grafted to AuNPs the smallest separation between aggregated nanoparticles would be about 20 nm, and then the size of the antigens is not even taken into account. Because of their sheer size, antibodies are thus not suitable recognition molecules in AuNP-based colorimetric assays but much more appropriate for other assay formats, such as flocculation assays, enzyme linked immunosorbent assays (ELISA) and in lateral flow devices.

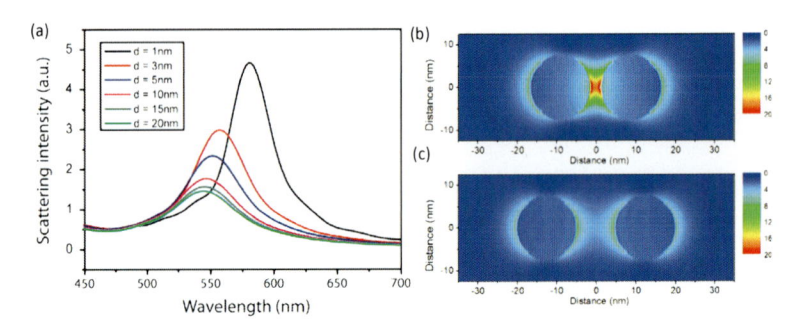

Figure 9.3 The plasmonic coupling is highly dependent on the separation (d) between individual nanoparticles. (a) FTDT simulations of a dimer of 10 nm AuNPs showing a distinct red shift in the scattering component for $d < 10$ nm. Corresponding field plots for AuNP dimers with $d = 3$ (b) nm and $d = 10$ nm (c).

A problem in the development of AuNP-based colorimetric assay is thus finding or designing recognition molecules that are relatively small. In addition to being small, the molecules must also of course provide specific recognition. Recognition should also result in a change in colloidal stability, leading to either aggregation or redispersion of the nanoparticles. Numerous recognition molecules and assay strategies have been proposed and described in the literature for detection and characterization of nucleic acids, drug molecules, heavy metal ions, protein receptors, and a large variety of enzymes [26–28]. The common denominator is that recognition causes either an increase or a decrease in colloidal stability. The latter strategy is by far the most common because of the inherent tendency of nanoparticles to aggregate and because aggregation is often not reversible. In the following sections, a number of recent examples of colorimetric assays will be described, illustrating different methods to link biorecognition to changes in nanoparticle stability.

9.5.1 Cross-Linking Assays

The first specific colorimetric AuNP-based assay was described in 1996 by Chad Mirkin and coworkers, using hybridization of oligonucleotides to modulate nanoparticle stability [29]. In this seminal paper, two batches of 13 nm citrate stabilized AuNPs were separately modified using two different 28-base oligonucleotides both having a terminal-SH moiety. The ssDNA modified AuNPs aggregated extensively upon addition of a strand that was complementary to both immobilized strands (Fig. 9.4). The relatively small size of the DNA oligonucleotides and the specific recognition mediated by hybridization resulted in formation of dense aggregates and a significant colorimetric shift. The aggregation was reversible and heating the sample above the DNA melting temperature (T_m) caused redispersion. Systems formed as a result of molecular self-assembly often display a large degree of cooperativity, which in this particular case was manifested as a much sharper melting transition than for the corresponding free DNA strands. Addition of non-complementary strands did not result in aggregation whereas strands with one or two single base pair miss-matches triggered aggregation, but with significantly lower melting temperatures [30].

The DNA-hybridization-mediated assembly of AuNPs is a prime example of a colorimetric cross-linking assay, where specific molecular interactions are exploited to induce a bridging aggregation of the nanoparticles.

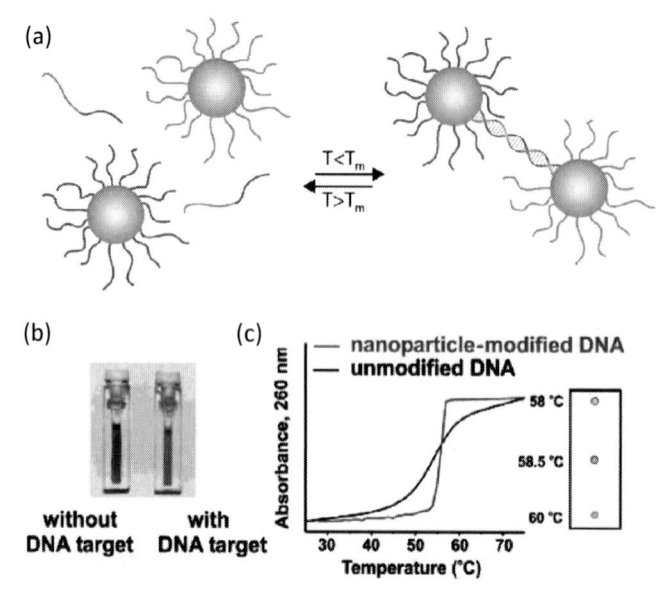

Figure 9.4 AuNPs modified with mercaptoalkyloligonucletiodes aggregate in the presence of complementary DNA (a). The resulting color change from red to purple is easily observed by the naked eye or using UV-vis spectroscopy (b). The melting of DNA on AuNP, and subsequent redispersion of aggregates, shows a sharper temperature dependence than the corresponding DNA in solution (c). Adapted with permission from ref. [27]. Copyright (2005) American Chemical Society.

In a similar approach, but intended for detection of mercury ions, two sets of AuNPs were functionalized with complementary single stranded DNA. When the AuNPs were combined aggregates were formed without the need of a third strand [31]. By deliberately introducing a thymidine–thymidine (T-T) mismatch in the otherwise complementary strands the T_m was reduced. The highly toxic solvated mercury ion (Hg^{2+}) is known to coordinate selectively to a T-T mismatch. In the presence of Hg^{2+} T_m increased linearly with the Hg^{2+} concentration, from about 46 to 56°C, as a consequence of the increase in stability of the DNA duplex caused by the coordination of the ions to the T-T mismatch [31].

In addition to DNA, another group of biomolecules that are extensively utilized in colorimetric AuNP-based assays are polypeptides. Just as oligonucleotides they are relatively small, typically less than 50 amino acids, and can also be produced synthetically using solid-phase synthesis strategies. The compositional resemblance to proteins provides means to mimic recognition sites for proteases and other enzymes normally targeting proteins.

Guarise et al. designed a peptide substrate for proteases that was flanked by Cys residues on both sides of the recognition sequence [32]. The intact peptide caused aggregation of citrate stabilized AuNPs since the peptide simultaneously could bind to two separate nanoparticles. The cross-linking was abolished when the peptide was first exposed to the target protease. In addition to the simple and convenient route for triggering aggregation, the peptides did not suffer from sterical constrains caused by immobilization on a nanoparticle and were hence easily accessible to the protease. A peptide can still be cleaved by proteases when immobilized as long as the peptide is sufficiently large or is tethered to the nanoparticle by a flexible linker. In a recent paper, Peng et al. demonstrated a protease assay based on AuNPs functionalized with a polypeptide containing two protease recognition sites located at either side of a central Cys residue [33]. The thiol-containing Cys tethered the polypeptide to the AuNPs that were electrosterically stabilized by the Glu-rich polypeptide. Exposure to the target protease (matrilysin) reduced the stability of the nanoparticles significantly as a large fraction of the peptide was excised, which caused aggregation in a protease concentration dependent manner.

Peptides can also be designed to have structural features resembling proteins, but in contrast to proteins polypeptides tend to be structurally very robust and can have tunable assembly, oligomerization and folding properties. Peptides that are random coil as monomers, but dimerize and fold into either coiled-coils or helix-loop-helix four-helix bundles have been demonstrated to provide numerous ways for precise tuning of Au NP stability. Since the recognition of a target protein and modulation of nanoparticle stability essentially are two different molecular events they can in fact be provided by two different molecules. The decoupling of recognition and stability functionalities makes the design of receptors more straightforward and can also enable use of a wider variety of receptors or substrates.

Along this line, Aili et al. have demonstrated a generic strategy for colorimetric detection of proteins using AuNPs modified by two different polypeptides, one optimized for target recognition and the other for stability modulation (Fig. 9.5) [34]. The peptide employed for triggering and controlling the AuNP aggregation was a 42 amino acid residue helix-loop-helix polypeptide (JR2EC) designed to homodimerize and fold into a four-helix bundle in the presence of mM concentrations of Zn^{2+} at physiological pH [35]. The peptide was immobilized on AuNPs via a Cys residue in the loop region, and the addition of Zn^{2+} to the suspension caused a reversible nanoparticle aggregation as a result of folding. Co-immobilization of a peptide-based synthetic receptor did not alter the aggregation characteristics of the nanoparticles unless exposed to a sample containing the analyte protein. Binding of the protein to the receptor resulted in a marked reduction in the aggregation because the protein sterically prevented the dimerization event. Two completely different peptide-based receptors were evaluated and that were designed to bind human carbonic anhydrase II and a Fab fragment, respectively, showing similar results.

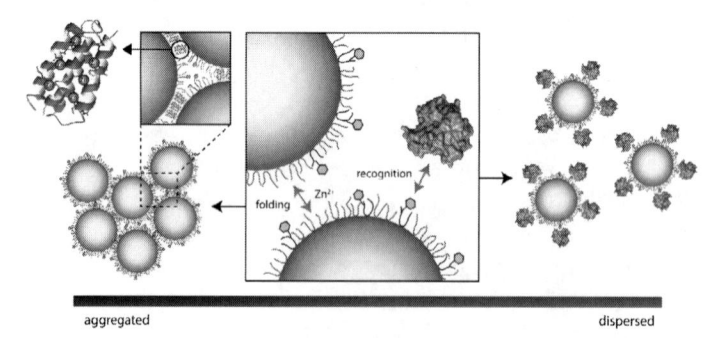

Figure 9.5 AuNPs were functionalized with two different peptides for separate recognition of a target analyte and stability modulation by a Zn^{2+}-triggered folding mediated aggregation, respectively. In the absence of analyte the addition of Zn^{2+} caused extensive aggregation (left). Analyte binding sterically prevented formation of dense aggregates (right). Reprinted with permission from ref. [34]. Copyright (2009) Wiley-VCH Verlag GmbH & Co. KGaA.

In addition, to enable use of a wider variety of receptors, the detection of certain types of enzymes, such as phospholipases, can require or benefit significantly from using separate molecular entities for recognition and for triggering nanoparticle cross-linking. Phospholipase A_2 (PLA_2) is a surface-active enzyme that hydrolyzes glycero-phospholipids in lipid bilayers and the enzymatic activity is very sensitive to the phase state and packing of the lipids in the bilayer. A colorimetric assay specifically designed to account for the substrate requirements of phospholipases has been described based on the release of polypeptides encapsulated in liposomes [36]. The liposomes acted as a substrate for PLA_2, whereas the polypeptides were designed to specifically cross-link AuNPs modified with complementary polypeptides. The digestion of the lipids by PLA_2 consequently triggered the release of the encapsulated polypeptides resulting in aggregation and a substantial red shift of the plasmon band (Fig. 9.6).

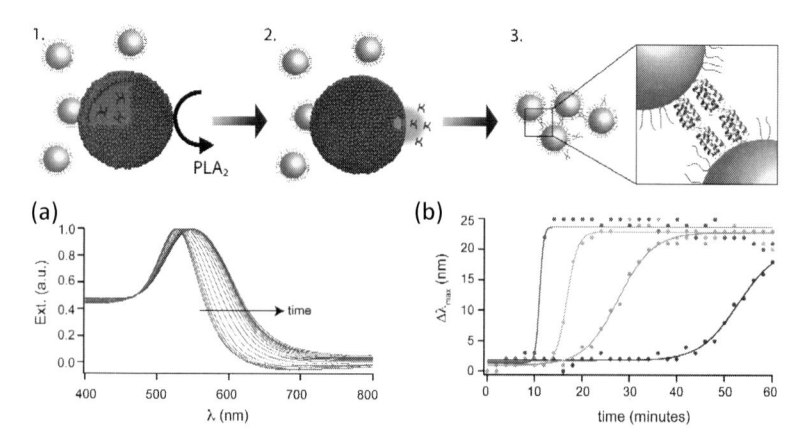

Figure 9.6 A colorimetric assay for Phospholipase A_2 was realized using liposomes encapsulating a divalent polypeptide designed to associate with complementary polypeptides on AuNPs. The encapsulated polypeptide was released as a result of PLA_2 hydrolyzing the lipids, causing bridging aggregation of the AuNPs as a result of the heteroassociation and folding of the polypeptides and a significant red shift of the LSPR band (a). The assay showed a typical lag-burst behavior with a concentration dependent lag- time. (b) Response for 7 nM (red), 3.5 nM (orange), 1.4 nM (green), and 700 pM (blue) PLA2 at 37°C. Adapted with permission from ref. [36]. Copyright (2011) American Chemical Society.

9.5.2 Redispersion Assays

Redispersion of aggregated nanoparticles as a result of a biospecific recognition event is significantly more complicated than the other way around. Nanoparticle aggregation caused by supramolecular interactions is typically highly cooperative and multivalent, and redispersion will thus require numerous interactions to be broken, either simultaneously or in a step-wise manner if the interactions do not reform once broken.

Stevens and coworkers demonstrated the possibility to redisperse loosely aggregated peptide functionalized AuNPs using proteases [37]. The peptides consisted of a short recognition sequence flanked by a terminal Cys residue and an N-(fluorenyl-9-methoxycarbonyl) (Fmoc) group at the other end. When immobilized on AuNPs the particles formed loosely associated aggregates because of $\pi-\pi$ stacking and hydrophobic interactions. The hydrolysis of the peptide by the target proteases removed the Fmoc moiety leaving the peptide with a terminal NH_3^+-group, thus shifting attractive interactions toward repulsive and consequently a gradual redispersion accompanied by a colorimetric blue shift (Fig. 9.7).

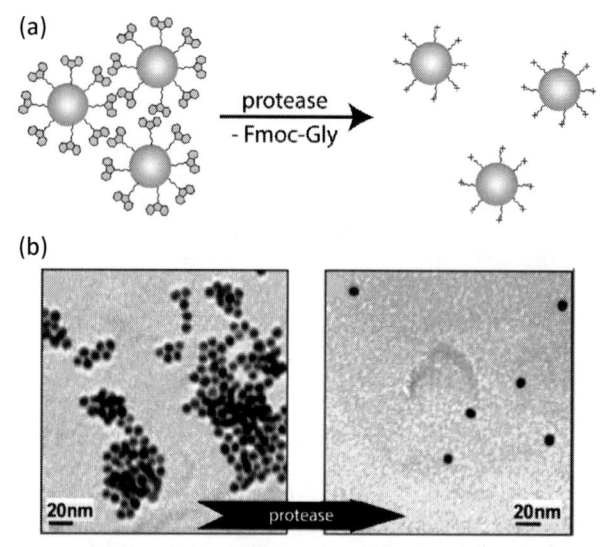

Figure 9.7 AuNPs modified with Fmoc terminated peptides that can be cleaved by proteases redisperse as a result of hydrolysis, which change the color of the suspensions from bluish to red. (a) Schematic illustration of the assay strategy and (b) transmission electron micrographs of nanoparticles before and after exposure to a protease. Adapted with permission from ref. [37]. Copyright (2007) American Chemical Society.

Liu et al. developed a redispersion assay for colorimetric detection of Pb(II) ions by exploiting the ability of a Pb(II) dependent DNAzyme to selectively associate with, and hydrolytically cleave, a substrate strand [38]. The substrate strand was designed to also be complementary to a 5′-thio-modified 12-mer DNA attached to 13 nm gold nanoparticles. Addition of the DNAzyme and the substrate to the DNA modified AuNPs caused aggregation of the nanoparticles. Addition of Pb(II) activated the DNAzyme and the substrate was cleaved and the AuNPs were redispersed (Fig. 9.8).

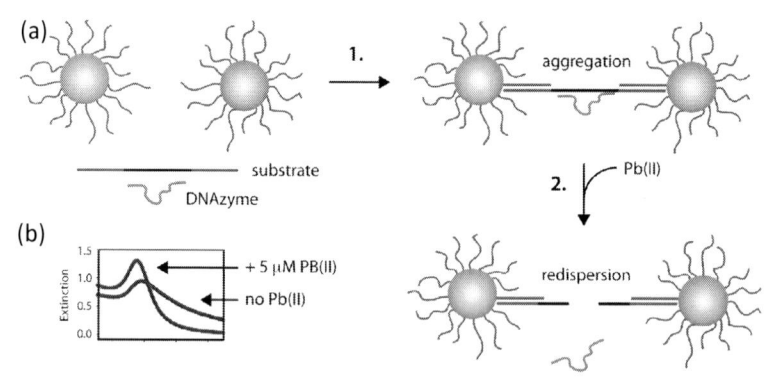

Figure 9.8 (a) AuNPs modified with short single stranded DNA aggregated in the presence of substrate strand (1) that in addition to sequences complementary to the immobilized DNA also contained a recognition sequence for a Pb(II) dependent DNAzyme. Addition of Pb(II) (2) leads to DNAzyme-mediated hydrolysis of the substrate strand and a redispersion of the nanoparticles. (b) UV-vis spectra obtained before (blue) and after addition of 5 μM Pb(II). Adapted with permission from ref. [38]. Copyright (2003) American Chemical Society.

Lin et al. exploited another approach for colorimetric sensing of Pb(II), using a decoupled recognition and stability modulation strategy [39]. AuNPs were modified with the thiolated crown ether ([15]crown-5)$CH_2O(CH_2)_4SH$, and thioctic acid (TA), that provided analyte recognition and control over AuNP aggregation, respectively. The nanoparticles aggregated rapidly in a solution of methanol containing less than 20% water, because of the formation of interparticle hydrogen bonds between the carboxylic acids of TA. Addition of μM concentrations of Pb(II) immediately redispersed the aggregated AuNPs followed by a dramatic spectral blue shift. The redispersion was found to be a consequence of the association of PB(II) to the crown ether, which leads to an increase in zeta potential. Zeta potential reflects the electrostatic potential near the nanoparticle surface and particles with sufficiently high (> +25 mV) or low (< −25 mV) zeta potentials are electrostatically stable. In the absence of Pb(II) the zeta potential was less than 10 mV, whereas in the presence of a few µM Pb(II) the value increased to about 30 mV, which explains the sudden redispersion of the nanoparticles. Other cations could also bind to the same crown ether but the change in zeta potential was less pronounced, even at significantly higher concentrations.

9.5.3 Non-Cross-Linking Assays

The commonly used method for synthesis of AuNPs developed by Turkevich and Frens yields nanoparticles stabilized by physisorbed citrate, [10, 11] that aggregate extensively and irreversibly at elevated ionic strength. Adsorption of other molecular species on the nanoparticle surface may, however, drastically improve the stability of AuNPs and prevent unspecific aggregation. Rothberg et al. [40] showed that single stranded DNA (ssDNA) adsorbed on AuNPs, which improved their stability and the nanoparticles remained dispersed even when exposed to 0.3 mM NaCl. In the absence of ssDNA, the AuNPs aggregated extensively at this salt concentration. Double stranded DNA (dsDNA) was found not to adsorb on the nanoparticle surface to the same extent and did not stabilize the AuNPs at this relatively high ionic strength. Based on this observation a simple assay for DNA hybridization was developed. Exposing ssDNA to a complementary strand triggered aggregation of AuNPs after addition of NaCl because of the hybridization and formation of dsDNA. Non-complementary strands, containing a single base-pair mismatch, did not result in aggregation since enough ssDNA to stabilize the AuNPs was present in the sample. The assay could thus detect specific interactions of molecular species without any of them being immobilized on the nanoparticle surface. The use of unspecific interactions that are linked to a specific recognition event, to trigger nanoparticle aggregation is quite common and is usually referred to as non-cross-linking assays.

A similar but more generic approach, was described by Plaxco and coworkers, [41] based on the observation that dsDNA was also capable of stabilizing AuNPs at low ionic strength. The authors employed a conjugated polymer (CP), capable of binding ssDNA but not dsDNA or otherwise folded DNA, to sequester non-hybridized or unfolded DNA aptamers. AuNPs thus aggregated when subjected to ssDNA and the CP whereas samples containing dsDNA remained stable. Aptamers are oligonucleotides developed to bind a specific target molecule with high affinity and selectivity. Aptamers can be modified to transition from a largely unordered single stranded conformation to a compact folded state upon analyte binding, where the later was found to have poor affinity for the CP. In the absence of analyte, the nanoparticles aggregated because of the association of the unordered aptamer and the CP. In the presence of analyte, the aptamer stabilized the AuNPs and prevented aggregation.

The molecular species of interest must thus not necessarily be immobilized on the AuNPs in order to provide mechanisms for changing the colloidal stability as a result of a specific molecular recognition event. Immobilization may even impose sterical constraints that affect the interactions and assembly of molecules, e.g., by restricting the accessibility of a substrate to an enzyme. Assays where the interactions of interests occur in solution rather on the nanoparticle surface can hence turn out to be more accurate. On the other hand, the system can potentially be more sensitive to contaminating species and unspecific binding to the unprotected gold nanoparticle surface.

9.6 Outlook and Challenges

The development of novel colorimetric sensor strategies based on the triggered aggregation or redispersion of AuNPs has been tremendous, both in terms of number of papers published but also with respect to the numerous creative assays that has been realized. In this chapter, a few examples have been highlighted, that demonstrate the possibilities to utilize molecular recognition for tuning the aggregation of AuNPs. They all have in common that aggregation or redispersion of the nanoparticles produces a colorimetric response that provides information about the interactions causing the change in colloidal stability. The chapter has mainly been focused on discussing the actual mechanisms typically utilized to trigger these changes rather than for what purpose the assays are developed. The rapid development of novel bioassays is of course spurred by the great need for rapid and sensitive bioanalytical tools for applications ranging from diagnostics and drug development to water quality monitoring and detection of pathogens and illegal substances. The requirements on an assay in terms of sensitivity and robustness vary widely for different applications but in general, the main challenge is to be able to selectively and reproducibly detect a small amount of analytes in a complex matrix. Not only must the affinity and selectivity of the recognition elements be extraordinary, the transduction mechanism should also not be affected by unspecific interactions. There is most likely no general solution to these challenges, but rather something that has to be dealt with differently for each sensor strategy and for each type of application. A few emerging technologies based on plasmonic signal amplification have recently been proposed that at least address the problems related to detection of low concentrations of analytes. These technologies combine traditional antibody-based detection (ELISA) with enzymatic processes that affect the resonance conditions of plasmonic nanostructures to yield a colorimetric response [42, 43].

9.6.1 Assays with Reversed Sensitivity and Plasmonic ELISA

In an attempt to push the detection limits of plasmonic assays to ultra low concentrations of analytes, Stevens and coworkers have developed a concept that provides a larger response when the concentration of the target molecule decreases [42]. Instead of relying on changes in colloidal stability to yield a colorimetric response, an enzymatic reaction was exploited that controlled the nucleation of silver nanocrystals on a plasmonic star-shaped gold nanoparticle. The enzyme, glucose oxidase (GOx), generates hydrogen peroxide that at low concentration reduced and deposited silver in the form of a thin film at the surface of the gold nanostar, resulting in significant blue shift of the LSPR band. At higher concentration of hydrogen peroxide, formation of silver nanocrystals was instead favored and a smaller LSPR shift was obtained. A generic sensor concept was realized by combining the gold nanostars with a sandwich ELISA using GOx labeled secondary antibodies. The concentration of hydrogen peroxide was thus proportional to the concentration of analyte whereas the resulting LSPR response was inversely proportional to the analyte concentration. A low concentration of analyte resulted in a larger colorimetric response than when analyte was present at high concentrations. Using this approach, the cancer biomarker prostate-specific antigen (PSA) was detected at concentrations down to as little as 10^{-18} g ml^{-1} in whole serum.

When instead using the enzymatic conversion of hydrogen peroxide to water by catalase, the nucleation and growth of gold nanoparticles from a gold salt solution could be tuned with very high precision [43]. With increasing concentration of hydrogen peroxide, in a very narrow interval around 120 mM, a transition from generating dispersed quasi-spherical AuNPs to aggregated and polydisperse AuNP was observed. The colors of the suspensions thus acquired either a red or blue tonality, reflecting the concentration of hydrogen peroxide. A generic colorimetric assay strategy with remarkable sensitivity was demonstrated by combining this mechanism for sensing of hydrogen peroxide concentration with an ELISA using catalase-labeled secondary antibodies. Based on this technique, prostate specific antigen (PSA) and HIV-1 capsid antigen p24 were both detected in whole serum at concentrations of 10^{-18} g ml^{-1}, which is well below the detection limits of current gold standard tests.

9.6.2 Assays for the Future

A great challenge in the development of bioassays and biosensors in general, is to accomplish technologies that not only provide a sensitive, selective and robust detection, but that also can do so in complex matrices such as blood or serum. Most assays developed so far have not passed the stage of demonstrating proof-of-concept because of the difficulties in translating a technology from the lab bench to an actual product that would work under non-ideal conditions and when handled by non-experts. Despite the many hurdles, intensive efforts by researchers and entrepreneurs start to show results as several plasmonic colorimetric assays are being commercialized. Albeit having a long history of applications in medicine, dating back to the 17th century, the applications and use of gold nanoparticles is currently undergoing a renaissance that most likely will provide researchers and medical practitioners with cheap, rapid and accurate diagnostic and bioanalytical tools.

Acknowledgment

DA acknowledges the financial support from Linköping University, the Swedish Research Council (VR), and the Swedish Foundation for Strategic Research (SSF).

References

1. Colomban, P. (2009). The use of metal nanoparticles to produce yellow, red and iridescent colour, from bronze age to present times in lustre pottery and glass: Solid state chemistry, spectroscopy and nanostructure. *J. Nano Res.*, **8**, 109–132.
2. Freestone, I., et al. (2007). The lycurgus cup—A Roman nanotechnology. *Gold Bull.*, **40**, 270–277.
3. Barber, D. J., Freestone, I. C. (1990). An investigation of the origin of the color of the lycurgus cup by analytical transmission electron-microscopy. *Archaeometry*, **32**, 33–45.
4. Arvizo, R. R., et al. (2012). Intrinsic therapeutic applications of noble metal nanoparticles: past, present and future. *Chem. Soc. Rev.*, **41**, 2943–2970.
5. Antonii, F., *Panacea Aurea-Auro Potabile* (1618). Bibliopolio frobenimo, Hamburg.
6. Hauser, E. A. (1952). Aurum potabile. *J. Chem. Edu.*, **29**, 456.
7. Faraday, M. (1857). The Bakerian Lecture: Experimental relations of gold (and other metals) to Light. *Philos. T. R. Soc. Lond.*, **147**, 145–181.
8. Lange, C. F. A. (1912). Über die Ausflockung von Goldsol durch Liquor cerebrospinalis. *Berliner klinische Wochenschrift*, **49**, 897–901.
9. Daniel, M. C., Astruc, D. (2004). Gold nanoparticles: Assembly, supramolecular chemistry, quantum-size-related properties, and applications toward biology, catalysis, and nanotechnology. *Chem. Rev.*, **104**, 293–346.
10. Turkevich, J., Stevenson P. C., and Hillier, J. (1951). A study of the nucleation and growth processes in the synthesis of colloidal gold. *Disc. Farady Soc.*, **11**, 55–75.

11. Frens, G. (1973). Controlled nucleation for the regulation of the particle size in monodisperse gold suspensions. *Nature (London) Phys. Sci.*, **241**, 20–22.
12. Nikoobakht, B., El-Sayed, M. A. (2001). Evidence for bilayer assembly of cationic surfactants on the surface of gold nanorods. *Langmuir*, **17**, 6368–6374.
13. Brust, M., et al. (1994). Synthesis of thiol-derivatised gold nanoparticles in a two-phase liquid-liquid system. *J. Chem. Soc. Chem. Commun.*, 801–802.
14. Wilcoxon, J. P., et al. (1998). Photoluminescence from nanosize gold clusters. *J. Chem. Phys.*, **108**, 9137–9143.
15. Schaeffer, N., et al. (2008). Fluorescent or not? Size-dependent fluorescence switching for polymer-stabilized gold clusters in the 1.1–1.7 nm size range. *Chem. Commun.*, 3986–3988.
16. Aili, D., et al. (2011). Polypeptide folding-mediated tuning of the optical and structural properties of gold nanoparticle assemblies. *Nano Lett.*, **11**, 5564–5573.
17. Stuart, M. A. C., et al. (1991). Adsorption of ions, polyelectrolytes and proteins. *Adv. Colloid Interfaces.*, **34**, 477–535.
18. Israelachvili, J. N. (1992). *Intermolecular and Surface Forces*, 2nd ed. (London: Academic Press).
19. Hakkinen, H. (2012). The gold-sulfur interface at the nanoscale. *Nat. Chem.*, **4**, 443–455.
20. Ulman, A. (1996). Formation and structure of self-assembled monolayers. *Chem. Rev.*, **96**, 1533–1554.
21. van Hardefeld, R., Hartog, F. (1969). Statistics of surface atoms and surface sites on metal crystals. *Surf. Sci.*, **15**, 189–230.
22. Terrill, R. H., et al. (1995). Monolayers in three dimensions: NMR, SAXS, thermal, and electron hopping studies of alkanethiol stabilized gold clusters. *J. Am. Chem. Soc.*, **117**, 12537–12548.
23. Weeraman, C., et al. (2006). Effect of nanoscale geometry on molecular conformation: Vibrational sum-frequency generation of alkanethiols on gold nanoparticles. *J. Am. Chem. Soc.*, **128**, 14244–14245.
24. Leff, D. V., et al. (1995). Thermodynamic control of gold nanocrystal size–experiment and theory. *J. Phys. Chem.*, **99**, 7036–7041.
25. Bernsohn, J., Borman, E. K. (1947). Proteins in the colloidal gold reaction. *J. Clin. Invest.*, **26**(5), 1026–1030.

26. Saha, K., et al. (2012). Gold nanoparticles in chemical and biological sensing. *Chem. Rev.*, **112**, 2739–2779.
27. Rosi, N. L., Mirkin, C. A. (2005). Nanostructures in biodiagnostics. *Chem. Rev.*, **105**, 1547–1562.
28. Aili, D., Stevens, M. M. (2010). Bioresponsive peptide-inorganic hybrid nanomaterials. *Chem. Soc. Rev.*, **39**, 3358–3370.
29. Mirkin, C. A., et al. (1996). A DNA-based method for rationally assembling nanoparticles into macroscopic materials. *Nature*, **382**, 607–609.
30. Elghanian, R., et al. (1997). Selective colorimetric detection of polynucleotides based on the distance-dependent optical properties of gold nanoparticles. *Science*, **277**, 1078–1081.
31. Lee, J.-S., Han, M. S., Mirkin, C. A. (2007). Colorimetric detection of mercuric ion (Hg^{2+}) in aqueous media using DNA-functionalized gold nanoparticles. *Angew. Chem. Int. Ed.*, **46**, 4093–4096.
32. Guarise, C., et al. (2006). Gold nanoparticles-based protease assay. *Proc. Nat. Acad. Sci. U. S. A.*, **103**, 3978–3982.
33. Chen, P., et al. (2013). Peptide functionalized gold nanoparticles for colorimetric detection of matrilysin (MMP-7) activity. *Nanoscale*, **5**, 8973–8976.
34. Aili, D., et al. (2009). Colorimetric protein sensing by controlled assembly of gold nanoparticles functionalized with synthetic receptors. *Small*, **5**, 2445–2452.
35. Aili, D., et al. (2008). Folding induced assembly of polypeptide decorated gold nanoparticles. *J. Am. Chem. Soc.*, **130**, 5780–5788.
36. Aili, D., et al. (2011). Hybrid nanoparticle-liposome detection of phospholipase activity. *Nano Lett.*, **11**, 1401–1405.
37. Laromaine, A., et al. (2007). Protease-triggered dispersion of nanoparticle assemblies. *J. Am. Chem. Soc.*, **129**, 4156–4157.
38. Liu, J., Lu, Y. (2003). A colorimetric lead biosensor using DNAzyme-directed assembly of gold nanoparticles. *J. Am. Chem. Soc.*, **125**, 6642–6643.
39. Lin, S.-Y., Wu, S.-H., Chen, C.-H. (2006). A simple strategy for prompt visual sensing by gold nanoparticles: General applications of interparticle hydrogen bonds. *Angew. Chem. Int. Ed.*, **45**, 4948–4951.
40. Li, H., Rothberg, L. (2004). Colorimetric detection of DNA sequences based on electrostatic interactions with unmodified gold nanoparticles. *Proc. Nat. Acad. Sci. U. S. A.*, **101**, 14036–14039.

41. Xia, F., et al. (2010). Colorimetric detection of DNA, small molecules, proteins, and ions using unmodified gold nanoparticles and conjugated polyelectrolytes. *Proc. Nat. Acad. Sci. U. S. A.*, **107**, 10837–10841.
42. Rodriguez-Lorenzo, L., et al. (2012). Plasmonic nanosensors with inverse sensitivity by means of enzyme-guided crystal growth. *Nat. Mater.*, **11**, 604–607.
43. de la Rica, R., Stevens, M. M. (2012). Plasmonic ELISA for the ultrasensitive detection of disease biomarkers with the naked eye. *Nat. Nanotechnol.*, **7**, 821–824.

Chapter 10

Surface-Enhanced Raman Scattering: Principles and Applications for Single-Molecule Detection

Diego P. dos Santos,[a] Marcia L. A. Temperini,[a] and Alexandre G. Brolo[b]

[a]*Institute of Chemistry, University of São Paulo,
Avenida professor Lineu Prestes 748, CEP 05513-970, São Paulo, Brazil*
[b]*Department of Chemistry, University of Victoria,
3800 Finnerty Road, V8W 3V6, Victoria, Canada*

agbrolo@uvic.ca

10.1 Introduction

Surface-enhanced Raman scattering (SERS) is among the most studied spectroscopic method in nanosciences. The fascination with the "SERS effect" is driven by both scientific curiosity and the possibility of new technologically relevant applications. Surface-enhanced Raman scattering is attractive from the phenomenological point of view because it delivers an unexpected increase in a normally very weak Raman process, allowing the observation of minute amount of adsorbed species. The possibility of controlling the magnitude of this enhancement by tailoring geometrical surface parameters at nanometric level adds another degree of

Introduction to Plasmonics: Advances and Applications
Edited by Sabine Szunerits and Rabah Boukherroub
Copyright © 2015 Pan Stanford Publishing Pte. Ltd.
ISBN 978-981-4613-12-5 (Hardcover), 978-981-4613-13-2 (eBook)
www.panstanford.com

excitement. The control of plasmonic properties challenges the imagination of chemists, physicists, and engineers to create new nanostructures guided by the magnitude of the SERS effect. From the application front, SERS combines exquisite sensitivity and the possibility of single molecular detection to the specificity given by the vibrational molecular fingerprint.

As a hallmark technique in nanosciences, SERS matured with the advent of tools for nanofabrication and sub-wavelength optical measurements. This rapid development of SERS has been reviewed by several authors who focused on specific aspects of the technique, including SERS substrates (metallic nanostructures that support the effect) [1], SERS applications in analytical [2], biological [3] and environmental chemistry [4], SERS theory [5–7], and other modern technologies derived from SERS, such as tip-enhanced Raman scattering (TERS) [8–10].

In this chapter, we will provide a comprehensive overview of the fundaments behind the SERS phenomenon. We will start with a classical description of the normal Raman scattering and emphasize the role of the different mechanisms that contribute to the SERS effect. Then, we will focus on a description of the behavior of single molecules in SERS hot spots. Our ultimate goal is to provide a clear picture of the statistical aspects that dominates at the single-molecule regime, and how these aspects can be explored to provide interesting insights into the SERS mechanism.

10.2 Raman Scattering

In this section, we describe the basic fundamentals of SERS by starting with a simple exposition of Raman scattering [11–13]. The electric field of a given electromagnetic source interacts with the electrons of a target molecule, which changes its electronic density distribution leading to an induced dipole, as described by Eq. 10.1:

$$\vec{P} = \hat{\alpha}\vec{E}, \tag{10.1}$$

where $\hat{\alpha}$ is the molecular polarizability tensor, which is physically associated to the facility for distortion of the electronic density of a molecule. This is a rank 2 tensor, as it transforms the electric field vector (with E_x, E_y, and E_z components) to the induced dipole moment vector $\vec{P}$ (with components P_x, P_y, and P_z):

$$\begin{bmatrix} P_x \\ P_y \\ P_z \end{bmatrix} = \begin{bmatrix} \alpha_{xx} & \alpha_{xy} & \alpha_{xz} \\ \alpha_{yx} & \alpha_{yy} & \alpha_{yz} \\ \alpha_{zx} & \alpha_{zy} & \alpha_{zz} \end{bmatrix} \begin{bmatrix} E_x \\ E_y \\ E_z \end{bmatrix} \qquad (10.2)$$

For instance, the P_x component of the induced dipole moment can be described by $P_x = \alpha_{xx} E_x + \alpha_{xy} E_y + \alpha_{xz} E_z$.

The molecular polarizability is not static, but it depends on the system dynamics, for instance the vibrational dynamics. One can describe the contribution of a given molecular vibration (which can be any of the 3N–6, for non-linear molecules, degrees of freedom, where N is the number of atoms in a molecule) by an expansion series in terms of the vibrational mode coordinate, Q_j, of that particular vibrational mode:

$$\hat{\alpha}(Q_j) = \hat{\alpha} + \left(\frac{\partial \hat{\alpha}}{\partial Q_j}\right)_0 Q_j(t) + \left(\frac{\partial^2 \hat{\alpha}}{\partial Q_j^2}\right)_0 Q_j^2(t) + \cdots \qquad (10.3)$$

Molecular vibrations constitute of small atomic displacements from the equilibrium positions, which allow the disregard, in a first approximation, of the higher order terms (2, 3, and so on). Therefore, we can approximately describe the polarizability dependence to the vibrational normal mode coordinate as an expansion truncated at the first order term ($Q_j(t)$, electrical harmonicity approximation).

Once the molecular electronic density perturbation is caused by a time-varying electric field, $\vec{E}(t)$, we can write

$$\vec{E}(t) = \text{Re}[\vec{E}_0 \exp(-i\omega t)], \qquad (10.4)$$

where $\vec{E}_0$ is the amplitude vector for the incident electric field and ω is the frequency (in Hz) of the incident radiation. If we assume $Q_j(t)$ in Eq. 10.3 to behave as a harmonic oscillator (mechanical harmonicity approximation), which is a valid approximation for small values of Q_j, we can write

$$Q_j(t) = \text{Re}[Q_j^0 \exp(-i\omega_j t)], \qquad (10.5)$$

where Q_j^0 is the maximum amplitude for the normal mode coordinate and ω_j corresponds to the vibrational frequency. Taking into account that $\exp(ix) = \cos(x) + i\sin(x)$, $\exp(-ix) = \cos(x) - i\sin(x)$, and the fact that $\mathrm{Re}[\exp(ix)] = \cos(x)$, an alternative description for the vibrational normal mode coordinate can be written as

$$Q_j(t) = \frac{1}{2}Q_j^0[\exp(-i\omega_j t) + \exp(i\omega_j t)] \tag{10.6}$$

The substitution of Eqs. 10.6, 10.4, and 10.3 into Eq. 10.1 leads to the calculation of the induced dipole moment. In the above formalism, the electric field and the molecular polarizability are described by functions that oscillate in time; hence, the induced dipole moment also oscillates in time according to

$$\vec{P}(t) = \hat{\alpha}\vec{E}_0 \,\mathrm{Re}[\exp(-i\omega t)] + \frac{1}{2}\left(\frac{\partial \hat{\alpha}}{\partial Q_j}\right)_0 Q_j^0 \vec{E}_0 \tag{10.7}$$

$$\{\mathrm{Re}[\exp(-i(\omega + \omega_j)t)] + \mathrm{Re}[\exp(-i(\omega - \omega_j)t)]\}$$

In Eq. 10.7, the time dependence for the induced dipole moment is described by three terms, each of which has a characteristic oscillation frequency that correspond to ω, $\omega + \omega_j$, and $\omega - \omega_j$. Given that an oscillating dipole is a source of electromagnetic radiation, the above expression characterizes, in the classical sense, emission of electromagnetic radiation with frequencies ω, $\omega + \omega_j$, and $\omega - \omega_j$. The first of these corresponds to the elastic scattering (or Rayleigh scattering), where the scattered light has the same frequency of the incident radiation. The other two components correspond to inelastic scatterings, or Raman (vibrational) scattering. One of the components of inelastic scattering presents a smaller frequency ($\omega - \omega_j$, therefore smaller energy) than the incident excitation and is called Stokes scattering. The higher frequency component ($\omega + \omega_\varphi$) is called anti-Stokes scattering. In both types of inelastic scattering (anti-Stokes and Stokes), the difference between the frequencies of incident and scattered light corresponds to a vibrational frequency of a particular molecular vibrational mode, j. Therefore, the Raman scattering holds vibrational information of a molecule, which is a molecular fingerprint.

One important aspect of the phenomenological description of Raman scattering provided by Eq. 10.7 is that its intensity (proportional to $|\vec{P}(t)^2|$) depends on the change in the molecular polarizability with the vibrational motion, which translates into a selection rule for Raman scattering. This feature is an important key aspect that differentiates the Raman from elastic (Rayleigh) scattering.

Figure 10.1 shows a typical experimental Raman spectrum of carbon tetrachloride obtained with a HeNe laser at 632.8 nm. The Raman shift, defined as the difference between incident and scattered frequencies, is presented in wavenumber units (cm^{-1}). Therefore, the positive side of the Raman shift axis corresponds to the Stokes scattering and the negative region to the anti-Stokes scattering. It can be observed from Fig. 10.1 that the anti-Stokes and Stokes sides differ only in terms of scattering intensities. This difference in intensities between the branches cannot be explained classically, requiring arguments from quantum mechanics. Figure 10.2 is a diagram that illustrates the interaction between light and molecular quantum states. The horizontal solid lines in Fig. 10.2 represent vibrational energy states of the ground electronic state (HOMO—highest occupied molecular orbital) and the first excited electronic state (LUMO—lowest unoccupied molecular orbital). The incident electromagnetic field causes a perturbation

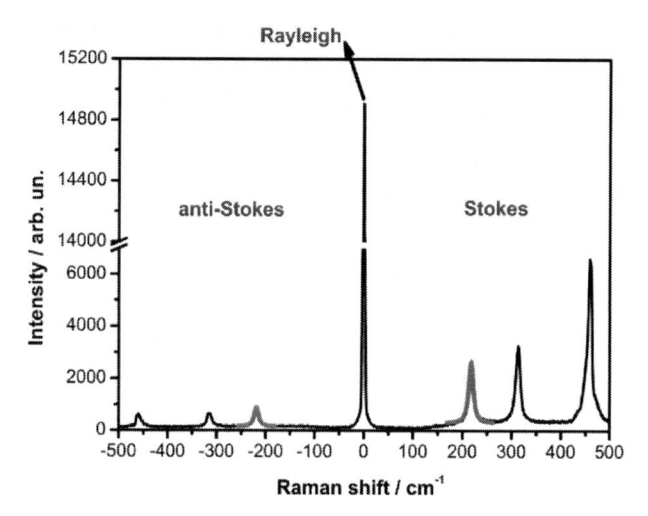

Figure 10.1 Typical Raman spectrum of carbon tetrachloride (CCl$_4$) in the range −500 to 500 cm^{-1}, obtained with excitation laser of wavelength 632.8 nm. The spectrum shows both the anti-Stokes and Stokes regions, as depicted in the figure.

in the molecule (the dashed horizontal lines in Fig. 10.2 represent the perturbed state) such that the scattering process can be elastic (Rayleigh) or inelastic (Raman anti-Stokes and Stokes). The anti-Stokes scattering depends on the population of molecules at excited vibrational states. On the other hand, the Stokes scattering intensity is mostly related to the population of molecules at the vibrational ground state, as described in Fig. 10.2.

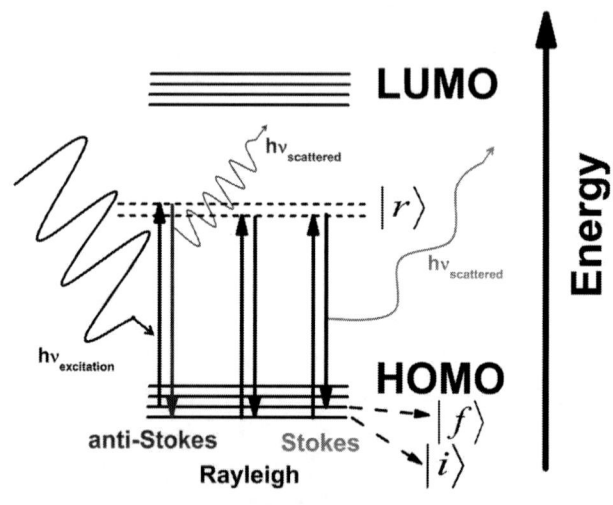

Figure 10.2 Schematic representation of the Raman scattering process. The horizontal solid lines represent vibrational states of the ground (HOMO) and first excited (LUMO) electronic states. The dashed lines represent the perturbed state after molecule/photon interaction. The scattering process is represented in the figure by the vertical arrows and wave-like arrows: in the Raman scattering, the energy of the scattered radiation (represented by the blue and red wave-like arrows) is different from the energy of the incident radiation (black wave-like arrow).

The population of molecules at vibrational excited states depends on the temperature and the energy difference among the vibrational states (characterized by the vibrational frequency[1]). Therefore, it is possible to calculate the expected ratio (ρ) between anti-Stokes and Stokes scattering intensities (I_{AS} and I_S, respectively) for a particular vibrational mode of frequency

[1] In the case of the harmonic oscillator approximation (as used in this context), the vibrational energy states are equally spaced.

(ω_{vib}, in cm^{-1} units) at a given temperature (T, in K unit) using the Boltzmann distribution as indicated in Eq. 10.8 [13–16]:

$$\rho = \frac{I_{AS}}{I_S} = \left[\frac{(\omega_{exc}+\omega_{vib})}{(\omega_{exc}-\omega_{vib})}\right]^4 \cdot \frac{\exp\left(\frac{-hc\omega_{vib}}{k_B T}\right)}{\left(1+\exp\left(\frac{-hc\omega_{vib}}{k_B T}\right)\right)}, \qquad (10.8)$$

where k_B is the Boltzmann constant, ω_{exc} is the frequency of the excitation laser (in cm^{-1}) and c is the speed of light in vacuum (in cm·s^{-1} units). The term inside the bracket (ratio between the frequency of anti-Stokes and Stokes scattered light) raised to the fourth power in Eq. 10.8 corresponds to a correction due to the dependence of the scattering efficiency with the light frequency. The explicit temperature dependence in the anti-Stokes to Stokes ratio (seen in the Boltzmann term in Eq. 10.8) attributes to Raman spectroscopy the ability to be used as a tool to probe the local temperature properties in a given system [17–19].

The above description corresponds to a phenomenological approach to the Raman scattering. The appropriate (or a quantitative) treatment requires the tools of quantum mechanics. For instance, the Raman scattering intensity can be described by

$$I_{Raman} \propto N_{mol}(\omega \pm \omega_j)^4 \, |\vec{E}_0|^2 \sum_{\rho,\sigma}(\hat{\alpha}_R)^2_{\rho\sigma}, \qquad (10.9)$$

where N_{mol} is the number of molecules probed by the electromagnetic radiation whose electric field magnitude is $|\vec{E}_0|$ and $(\hat{\alpha}_R)_{\rho\sigma}$ describes the components of the Raman polarizability tensor, where ρ and σ correspond to x, y, or z directions. In comparison to Eq. 10.7, such a tensor can be qualitatively described by

$$\hat{\alpha}_R = \frac{1}{2}\left(\frac{\partial\hat{\alpha}}{\partial Q_j}\right)_0 Q_j^0$$

The calculation of the Raman intensity for a particular vibration involves the calculation of the components of this tensor. Using second order perturbation theory, it can be shown that the components of this tensor can be described as [13]

$$(\hat{\alpha}_R)_{op} = \frac{1}{\hbar} \sum_{r \neq i,f} \left\{ \frac{\langle \Psi_f | \hat{p}_\rho | \Psi_r \rangle \langle \Psi_r | \hat{p}_\sigma | \Psi_i \rangle}{\omega_{ri} - \omega - i\Gamma} + \frac{\langle \Psi_f | \hat{p}_\sigma | \Psi_r \rangle \langle \Psi_r | \hat{p}_\rho | \Psi_i \rangle}{\omega_{ri} + \omega - i\Gamma} \right\} \quad (10.10)$$

Equation 10.10 is called Kramers–Heisenberg–Dirac (KHD) dispersion equation. In this chapter, we will focus on a qualitative description of this equation. Let us take the first term in the summation of Eq. 10.10. The numerator involves the product of transition moments between two states in the molecule, which constitute, in a qualitative interpretation, as the source of electromagnetic radiation in the same sense as the induced dipole moment in the classical picture. In the denominator we have $\omega_{ri} - \omega$, where ω_{ri} is the frequency associated to the energy of a transition between states i and r (see Fig. 10.2) and ω is the frequency of the incident radiation. Although in the KHD equation the summation is over all the states r of the molecule, if the laser is close or equal to one allowed electronic transition in the molecule (like the HOMO → LUMO transition in Fig. 10.2) the denominator describes a resonance condition and the Raman intensity is enhanced. This phenomenon is called resonance Raman scattering and it constitutes a very important and powerful aspect of Raman spectroscopy [13]. For instance, the intensities of the vibrational modes associated to the chromophore are preferentially enhanced, which is important for the study of electronic transitions (Raman scattering is a vibronic spectroscopy). In conditions away from resonance (ω_{ri} value is very different than ω), Eq. 10.10 shows that the efficiency of the Raman effect is low. In fact, the Raman cross section (typically about 10^{-29} cm^2) is very small if compared to other spectroscopy techniques, such as fluorescence spectroscopy (cross section about 10^{-19} cm^2), for instance [11, 12]. The small Raman cross-section limits the Raman signal detectable from a small amount of a molecule of interest, restricting the applicability of the technique in analytical chemistry. However, this limitation can be overcome by another special Raman enhancement effect observed from molecules adsorbed on certain types (specially gold, silver and copper) of metallic nanostructures called SERS [11, 12, 20]. SERS is a central phenomenon in nanotechnology and plasmonics and allows the detection of a small number of species confined to nanostructured metal surfaces. In the next sections, we will provide an overview of the principles of SERS and its application in single-molecule detection.

10.3 SERS

In 1974, Fleishmann et al. [21] observed very strong Raman signals from pyridine adsorbed onto an electrochemically roughened silver electrode. This effect was first associated to an increased surface area from the roughened surface and, therefore, to the increase in the number of probed molecules, as indicated in Eq. 10.9. Such interpretation of increased surface area was shown by Jeanmaire and Van Duyne [22] and Albretch and Creighton [23] to be inconsistent with the magnitude of the observed enhancement in respect to the normal Raman condition. For example, in the case of pyridine adsorbed onto an electrochemically activated (by oxidation and reduction cycles [ORCs]) silver electrode, the observed Raman signals were about 10^6 times more intense than in the case of a silver surface not subjected to the electrochemical roughening procedure. However, the expected increase in the number of molecules by such activation procedure was predicted to be no more than 10 times. Therefore, the experiments showed that the effect of the surface on the efficiency of the Raman scattered by adsorbed molecules was much more than expected by a mere increase of surface area. Other experimental observations definitely showed that the enhancement, now called SERS, was a new phenomenon. These observations include the fact that the enhancements were only possible from nanostructured (10–100 nm features) surfaces of certain metals, such as Ag, Au, and Cu metals; the magnitude of the observed enhancement was different for different molecules; different enhancement efficiencies were observed for different vibrational modes of a given molecule; experiments in electrochemical conditions showed maxima in the intensity profiles of different vibrational modes as function of the applied potential for some molecules.

Whichever the experimental conditions where SERS signals are observed, it is possible to define an enhancement factor (EF) as [24]

$$\mathrm{EF} = \frac{I_{\mathrm{SERS}}}{\langle I_{\mathrm{Raman}} \rangle} \cdot \frac{N_{\mathrm{Raman}}}{N_{\mathrm{SERS}}} \qquad (10.11)$$

where I_{SERS} and $\langle I_{\mathrm{Raman}} \rangle$ are the observed SERS and normal Raman intensities for a particular vibrational mode, and N_{SERS} and N_{Raman}, are the number of molecules in the SERS and in the normal Raman experiments responsible for the observed signals. The difficulty in the calculation of EF for the various experimental conditions is in the calculation of N_{SERS} and N_{Raman}. In the case of normal Raman, this parameter can be estimated from the optical configuration of the Raman spectrometer, i.e., from the knowledge of the sample volume probed by the incident laser beam. On the other hand, N_{SERS} is a much more difficult parameter to be estimated and it is the main source of problems in the calculation of the experimental SERS EFs.

The theoretical calculation of the EF requires the knowledge of the physics behind the mechanism of Raman scattering enhancement. Moskovits was the first to suggest that the enhancement should be related to collective electronic excitations from the metallic nanostructures [25]. That work preceded a series of papers from different authors that correlated the origin of the enhancement to some type of electromagnetic effect. The excitation of surface plasmons from the metallic nanostructure, as hinted by Moskovits, has been accepted as the dominant electromagnetic contribution in SERS. Meanwhile, another class of mechanisms, championed, for instance, by Otto [26], suggests that the enhancement originates from the chemical interactions in the metal–molecule system that leads to new types of electronic resonances. Nowadays, these two main mechanisms (chemical (CE) and electromagnetic (EM) mechanisms) are still sought to be responsible for the observed enhancement. The CE mechanism is the one related to the formation of a chemical bond between molecule and surface upon adsorption. The interaction between molecule and metal electronic states contributes to changes in the Raman polarizability tensor elements (Eq. 10.10), modifying the scattering intensity. The EM mechanism is related to the effect of the optical properties of the metal surface, i.e., the surface plasmon resonance, on the local properties of a dipole scattering. The SERS

EF is then generally described as contribution from both classes of mechanisms:

$$EF = EF_{CE} \cdot EF_{EM}, \tag{10.12}$$

where EF_{CE} and EF_{EM} are the chemical (CE) and electromagnetic (EM) contributions to the total SERS enhancement factor (EF). It is accepted that the EF_{EM} is dominant over EF_{CE} in the majority of systems. In fact, EF_{EM} is so important to the observed enhancement that SERS and plasmonics benefits from each other: not only plasmonics knowledge is of central importance to SERS, but also some plasmonic properties of nanostructures can be studied by SERS.

In what follows, the main aspects of both the CE and the EM mechanisms will be described, and their relationship to experimental observations in SERS will be discussed.

Chemical mechanism

The chemical mechanism (CE) arises from the change in the elements of the polarizability tensor (Eq. 10.10) upon adsorption. The main source for this change in the polarizability comes from photoinduced charge transfer between molecular energy levels (HOMO and LUMO, or $|I\rangle$ and $|K\rangle$ in Fig. 10.3, respectively) and metal states close to the Fermi level, $|F\rangle$, where the metal presents the highest density of states [27].

The photoinduced charge transfer (CT) contribution to SERS can be interpreted in the same basis as the resonance Raman scattering, discussed previously in this Chapter. In this formalism, first presented by Lombardi et al. [7, 27–29], if the laser energy matches the difference between the molecule and metal states close to the Fermi level, a resonance condition can be established for a metal to molecule ($|F\rangle \to |K\rangle$) or molecule to metal ($|I\rangle \to F\rangle$) transition, which contributes to an enhancement in the polarizability tensor elements in the same way as in the resonance Raman effect.

One of the most important experimental observations supported by such mechanism is the dependence of the SERS intensity profile on the applied potential, observed from experiments using roughened electrodes. Those intensity profiles present maximum intensities, for particular SERS bands, when a given potential is applied and the laser energy is kept constant.

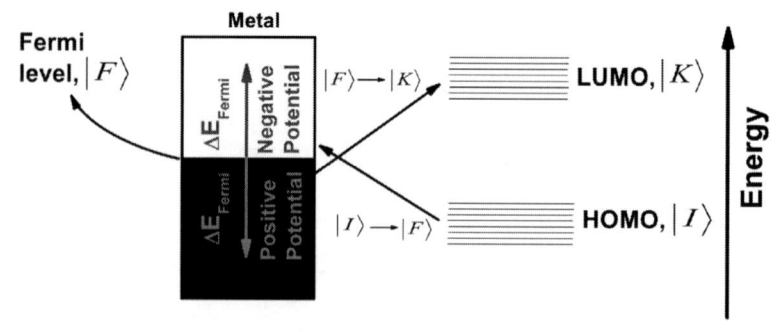

Figure 10.3 Schematic representation of photoinduced electronic transitions in the metal-molecule system. The metal is represented in the figure by a rectangular box, in which, the black area corresponds to the valence band, whereas the white region represents the conduction band. The interface between the valence and conduction bands in the metal corresponds to the Fermi level, $|F\rangle$, whose energy can be changed by the application of electric potential to the metal. Because the Fermi level contains the highest density of states, transitions between molecule and metal occur with higher intensities when involving metal states close to the Fermi level. The figure represents three types of electronic transitions that can occur upon laser excitation in the case of resonance condition: metal to molecule ($|F\rangle \rightarrow |K\rangle$), molecule to metal ($|I\rangle \rightarrow |F\rangle$) and a transition involving the electronic states of the molecule ($|I\rangle \rightarrow |K\rangle$).

The applied potential affects the position of the Fermi level, and therefore, if the metal/molecule transition is permitted, the energy of this transition will match the fixed energy of the excitation laser only for a particular applied potential, which characterizes the resonance condition in the SERS intensity profile [29, 30].

Although the chemical mechanism can account for a number of experimental observations, the calculation of the magnitude of such contribution to the overall enhancement factor (EF), defined in Eq. 10.12, is a very hard task, since it needs, from the theoretical point of view, the consideration of very large number of electronic states. There are some recent developments where calculations of EF_{CE} were obtained for small clusters, where it is demonstrated the possibility of utilizing a hybrid quantum mechanics/classical electrodynamics simulations to calculate the SERS spectra of molecules adsorbed on small nanoparticle surfaces [31, 32].

Electromagnetic mechanism

In the majority of cases, the most important contribution to the SERS enhancement is from the electromagnetic mechanism (EM) mechanism. The origins for this mechanism rely on localized surface plasmon resonances, which will be referred simply as surface plasmon. The coupling of the incident light with the surface plasmon modes of a metal nanoparticle leads to an enhancement of the local electric field experienced by the molecule. The magnitude of the Raman scattering depends on the magnitude of the local electric field, as indicated in Eq. 10.9. Another very important factor that has to be taken into account is that this excitation field creates an induced dipole, which is also the source of the Raman scattering and the presence of a metal surface modifies this dipole emission, which significantly contributes to EF_{EM}, as we shall see in this section [12, 20].

The surface plasmon resonance effect on the SERS enhancement can be calculated by considering the optical response of metal nanoparticles, which is characterized by the metal dielectric function, $\varepsilon(\omega)$, a wavelength-dependent complex function described by Eq. 10.13

$$\varepsilon(\omega) = \varepsilon_1(\omega) + i\varepsilon_2(\omega), \quad (10.13)$$

where $\varepsilon_1(\omega)$ and $\varepsilon_2(\omega)$ are the real and imaginary parts of this function. The knowledge of these functions enables qualitative and quantitative insights about the SERS intensity for a given condition. For instance, considering a metallic nanosphere of dimensions much smaller than the wavelength of the incident electromagnetic radiation as a model system (here we first considered the interaction between the excitation field and the metal nanoparticle), the electric field at a given distance r from the center of the sphere ($E(r)$) can be calculated within the so-called electrostatic approximation [12, 20, 33], as described by the result in Eq. 10.14:

$$\vec{E}(\vec{r}) = |\vec{E}_0|\hat{x} - |\vec{E}_0|\left(\frac{\varepsilon_1(\omega) - \varepsilon_0 + i\varepsilon_2(\omega)}{\varepsilon_1(\omega) + 2\varepsilon_0 + i\varepsilon_2(\omega)}\right)a^3\left[\frac{\hat{x}}{r^3} - \frac{3\hat{x}}{r^5}(x\hat{x} + y\hat{y} + z\hat{z})\right], \quad (10.14)$$

where $\hat{x}$, $\hat{y}$, and $\hat{z}$ are the unit vectors and a is the sphere radius. The second term in the right-hand side of Eq. 10.14 corresponds to the contribution to the local electric field at position r from the field scattered by the plasmon mode, where it can clearly be seen the dependence on the optical properties described by the metal dielectric function and the optical response of the medium, described by ε_0, the medium dielectric constant. On a first analysis, it is hard to visualize the local field enhancement from Eq. 10.14, since it is heavily dependent on the dielectric function of the metal. Some of the tabulated [34, 35] values of the real and imaginary parts of the dielectric function for some metals as function of the wavelength is shown in Fig. 10.4.

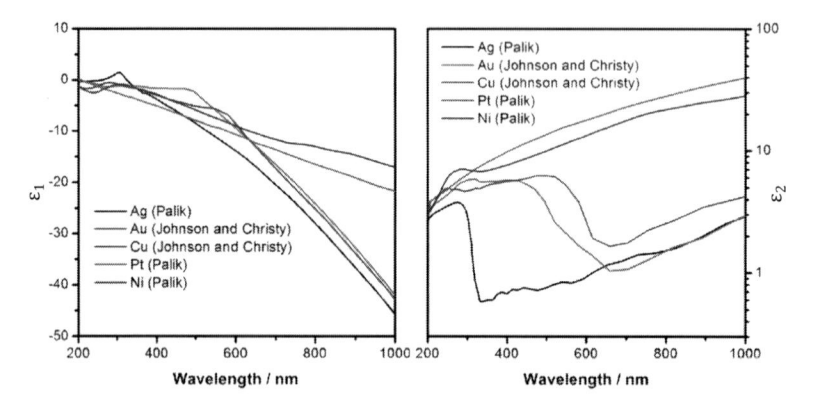

Figure 10.4 Real (left, ε_1) and imaginary (right, ε_2) parts of the dielectric function of some metals, as indicated in the figure, extracted from the compilations of experimental values by Palik [34] and Johnson and Christy [35].

One of the most important observations from Fig. 10.4 is that the real part is negative in the visible-near infrared spectral range, and it presents a strong wavelength dependence in that region. The fact that $\varepsilon_1(\omega)$ is negative permits the following resonance condition in Eq. 10.14: $\varepsilon_1(\omega) = -2\varepsilon_0$; provided that the nanosphere is small compared to the wavelength of light in the visible region (electrostatic approximation) and that the imaginary part, $\varepsilon_2(\omega)$, also present in the denominator as a relaxation factor, is relatively small. Among the listed metals in Fig. 10.4, silver, gold, and copper have the smallest values for ε_2 when compared to platinum and nickel, which justifies the use of Ag, Au and (in a less extent) Cu nanoparticles for SERS studies.

The relevant quantity to evaluate the EF_{EM} resulted from the coupling of light and surface plasmon modes is the local field enhancement factor (M_{Loc}), defined as

$$M_{Loc} = \frac{|E_{Loc}|^2}{|E_0|^2}, \tag{10.15}$$

where $|E_{Loc}|$ is the local (on the surface) electric field amplitude. In Fig. 10.5, the surface average local field enhancement factor ($<M_{Loc}>$) is presented as function of wavelength for silver (black line) and gold (red line) nanospheres of 12 nm radius in water as the dielectric medium ($\varepsilon_0 \approx 1.78$). The nanoparticles dimensions are much smaller than the wavelength of visible light and, therefore, the electrostatic approximation should be valid. The values of $<M_{Loc}>$ were calculated within the Mie theory formalism (which constitutes the exact solution of Maxwell equations for a sphere) using the program SPLac made available by Le Ru and Etchegoin [12].

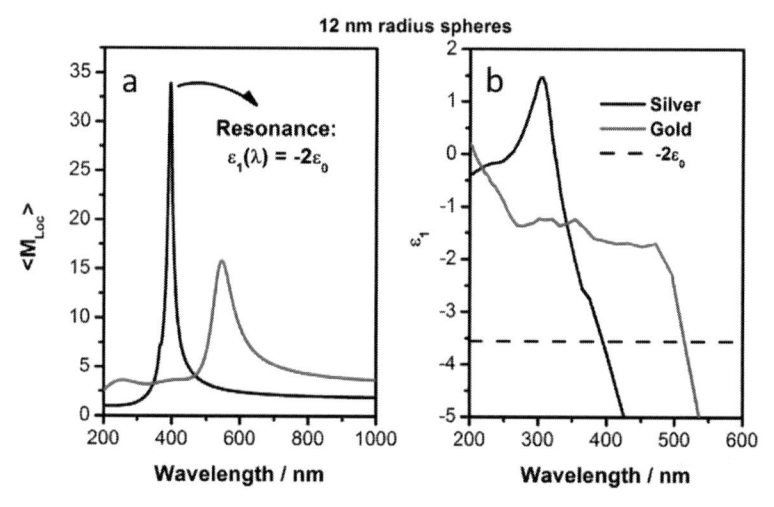

Figure 10.5 (a) Wavelength dependence of the surface average value of the local field enhancement factor for 12 nm radius spheres of silver (black line) and gold (red line) embedded in water as dielectric medium ($\varepsilon_0 \approx 1.78$). (b) Real part of the dielectric function of silver and gold in comparison to the value of $-2\varepsilon_0$, where the resonance wavelength is expected according to the electrostatic approximation.

It is possible to observe in Fig. 10.5a that for the silver nanosphere the resonance in M_{Loc} (resonance in E_{Loc}) occurs at 395 nm, where the value of ε_1 for silver equals $-2\varepsilon_0 \approx 3.56$, as indicated in Fig. 10.5b. This happens because $\varepsilon_2(\omega)$ is small in the visible for silver. For gold, on the other hand, $\varepsilon_2(\omega)$ is not small around 500 nm, where it is possible to observe a spectral structure that is associated to interband absorption [12, 20]. This leads to two effects on the $<M_{Loc}>$ profile: (I) the resonance condition occurs red-shifted from the $\varepsilon_1 = -2\varepsilon_0$ and (II) the resonance band for Au is broader than for Ag, which leads to a smaller $<M_{Loc}>$ for an Au nanosphere. For wavelengths above 650 nm, Fig. 10.4 shows that ε_2 for these two metals are very similar and, therefore, they present similar SERS performances, which justifies the broad use of Au nanoparticles in near IR SERS studies due to its much better chemical stability compared to Ag and Cu.

The local field enhancement factor is the most important consequence of the EM mechanism, since it leads to a stronger field experienced by the adsorbed molecule, and a higher Raman scattering. However, the origin of Raman scattering, discussed earlier in this chapter, involves the formation of an induced dipole, which acts as a source of electromagnetic radiation. The dipole emission can be greatly modified if the dipole is in a heterogeneous medium [12, 36, 37]. For example, a modification in the emission pattern is expected when the dipole inside a dielectric with constant ε_0 is close to a surface of a metal with dielectric function $\varepsilon(\omega)$. The modification in the dipole emission can be observed in the power extracted from the dipole, which can be characterized by the enhancement factor in the dipole radiation, $<M_{Rad}>$. In order to demonstrate this behavior, the system described by a dipole on the surface of a silver nanosphere of 25 nm radius, which is the typical size of silver nanoparticles synthesized for SERS studies [38–40], was evaluated. The results in Fig. 10.6 show the application of Mie theory for the problem of a dipole oscillating at a certain frequency (ω_R), close to a silver nanosphere, acting as source of electromagnetic radiation. Two dipole orientations, parallel (//) and perpendicular ($\perp$) were considered in this case, which translate, in the case of SERS, to a parallel and a perpendicular (relative to the surface, described in the figure by a gray sphere) normal mode of vibration, respectively.

Figure 10.6a shows the comparison of $<M_{Loc}>$ (solid line) and M_{Rad} for perpendicular (dashed) and parallel (dotted) dipoles as function of wavelength. In all cases, the radiation (from the incident laser or from the dipole) couples to the surface plasmon modes of the metal, leading to an enhancement in the Raman scattering. Therefore, besides the local electric field experienced by the adsorbed molecule, enhanced by the surface plasmon resonance, the Raman scattering is also enhanced by the coupling of the induced dipole with the plasmon resonance. Because the scattered light by the molecule has frequency close to the incident light (only shifted by a vibrational quantum), both the incident and the scattered electromagnetic radiation can couple to the same plasmon mode. Therefore, EF_{EM} contains contributions from these two effects.

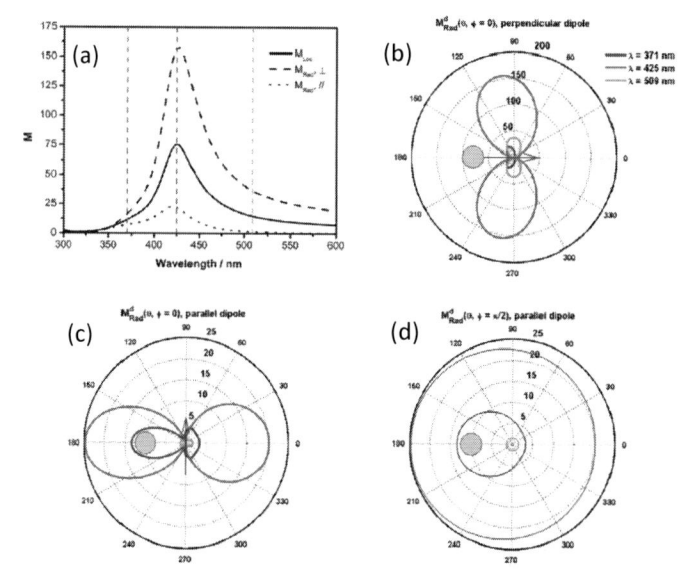

Figure 10.6 (a) Wavelength dependence of the enhancement factor for the dipole radiation, M_{Rad}, in two configurations: parallel and perpendicular orientations in respect to the nanoparticle surface. These enhancement factor profiles are compared to the local field enhancement factor, M_{Loc}. (b) Directional profile for the dipole radiation enhancement in the plane parallel ($\phi = 0$ rad) to a dipole that is oriented perpendicular to the nanoparticle surface (notice that the dipole (molecule) and the nanoparticle are not in scale). (c) Same as (b), but for a dipole oriented parallel to nanoparticle surface. (d) Directional profile for the dipole radiation enhancement in the plane perpendicular ($\phi = \pi/2$ rad) to a dipole (represented by a point at the center of the figure) that is oriented parallel to the nanoparticle surface.

One of the most interesting characteristics of Fig. 10.6a is that the magnitude of the enhancement for the perpendicularly oriented dipole is higher (~150) than that for the parallel dipole (~20), which maps into the "SERS world" as the statement that vibrational modes perpendicular to the surface are preferentially enhanced (surface selection rules) [41–43].

From the experimental point of view, the scattering geometry also needs to be considered. In other words, the direction of the detector with respect to the system is very important, i.e., the direction of the enhancement factor for the dipole radiation, which is the important parameter for SERS.

Figure 10.6b shows the scattering profile (M_{Rad}^d) as function of the angle (θ) for a dipole (represented in the figure by a vector) oriented perpendicularly to the surface of the sphere. This corresponds to a scattering profile in the plane that contains the dipole and the metal sphere ($\phi = 0°$) as indicated in the figure. Figure 10.6c is equivalent to Fig. 10.6b for a parallel dipole (relative to the surface). Figure 10.6d corresponds to the scattering profile for a parallel dipole (relative to the surface) in a plane perpendicular to that of Fig. 10.6c ($\phi = 90°$). In all cases, the scattering profiles are shown for a dipole oscillating with three different characteristic wavelengths: 371 nm (blue), 425 nm (surface plasmon resonance) and 509 nm (green). Figures 10.6b,c clearly show the directionality of the dipole emission, and so, of the SERS intensity. It is important to emphasize the emission not only in the direction perpendicular to the dipole, but Figs. 10.6b,c also show that the dipole radiates more power towards the supporting plasmonic nanostructure.

Considering the discussion above, EF_{EM} can be described as

$$EF_{EM} \approx \langle M_{Loc}(\omega) \rangle \langle M_{Rad}^d(\omega_R) \rangle, \qquad (10.16)$$

where the brackets correspond to surface average. The first factor, the local field enhancement can be calculated by using a variety of numerical tools like Mie theory in the case of spheroids, Discrete Dipole Approximation (DDA) [44–46], Finite Element Method (FEM) [47, 48], Finite-difference time-domain method (FDTD) [48], and others that can be used to describe nanostructures of different shapes [12]. The dipole radiation enhancement, on the other hand, is not a simple parameter to calculate, once it needs the handling of singularities at the dipole position [12]. Therefore, for the theoretical evaluation of EF_{EM} a good approximation for M_{Rad}^d is sometimes required. It can be observed from Fig. 10.6a that M_{Rad}^d has similar magnitude as M_{Loc}, suggesting a possible approximation to M_{Rad}^d that is much easier to calculate:

$$EF_{EM} \approx \langle M_{Loc}(\omega)\rangle\langle M_{Loc}(\omega_R)\rangle = \left(\frac{|E_{Loc}(\omega)|^2}{|E_0(\omega)|^2}\right)\left(\frac{|E_{Loc}(\omega_R)|^2}{|E_0|(\omega_R)^2}\right) \quad (10.17)$$

The so-called E^4-approximation expression refers to the fact that at the limit of low-frequency vibrations ($\omega_R \to \omega$) the local field enhancement at the laser and scattered light frequencies are approximately the same, i.e., $M_{Loc}(\omega) \to M_{Loc}(\omega_R)$, which leads to a very common approximate expression to the enhancement factor of the form

$$EF_{EM,E^4} \approx \langle M_{Loc}(\omega)\rangle^2 = \left(\frac{|E_{Loc}(\omega)|}{|E_0(\omega)|}\right)^4,$$

where the subscript E^4 is used to characterize the approximation for low-frequency vibrations. According to this approximation, the enhancement factor can be calculated through the fourth power of the local electric field probed by the molecule at frequency ω, $E_{Loc}(\omega)$.

Figure 10.7 shows the comparison between the E^4-approximation and the calculation of EF_{EM} by Eq. 10.16, where $<M_{Rad}>$ was taken as the average between the parallel and the perpendicular results. This shows that the E^4-approximation is a very good approximation to describe the average between the parallel and perpendicular dipole emission. In fact, it can be shown by the optical reciprocity theorem [12, 49] that for a given set of conditions, the E^4-approximation is actually the exact result for EF_{EM} [49].

From the above considerations, and in analogy to Eq. 10.9, the SERS intensity can be described by

$$I_{SERS} \propto N_{SERS}(\omega \pm \omega_j)^4 M_{Loc}(\omega) M_{Loc}(\omega_R) |\vec{E}_0|^2 \sum_{\rho,\sigma}(\hat{\alpha}_{SERS})_{\rho\sigma}^2, \quad (10.18)$$

where, N_{SERS} is the number of molecules responsible for the observed SERS intensity. Note that the term $M_{Loc}(\omega)M_{Loc}(\omega_R)$ in Eq. 10.18 corresponds to the electromagnetic enhancement factor, whose wavelength profile is presented in Fig. 10.7. Therefore, it can be seen from Eq. 10.18 and 10.7 that the SERS intensity can be orders of magnitude higher than the Raman scattering of the probe, as characterized by Fig. 10.7. The enhancement in the SERS intensity can be even larger if the molecule is located in the vicinity between two nanoparticles separated by a distance of the order of 1 nm. In this condition, the surface plasmon modes of the two nanoparticles can couple and the electric field can be greatly enhanced and localized in the junction. Figure 10.8 shows a model example of two silver nanospheres separated by a distance of 1 nm.

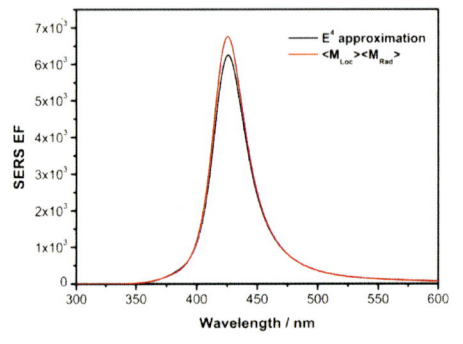

Figure 10.7 SERS enhancement factor (considering only the electromagnetic contribution) for a 25 nm radius silver sphere as calculated by E^4-approximation (black line) compared to the SERS enhancement factor calculated by considering the dipole radiation enhancement (red line).

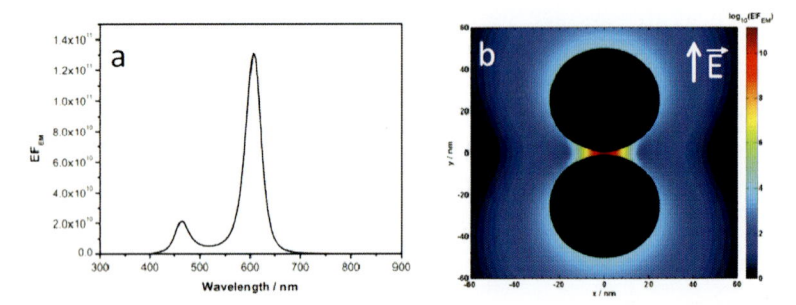

Figure 10.8 (a) Wavelength dependence of the electromagnetic enhancement factor (EF_{EM}) calculated in the vicinity of two silver nanospheres of 25 nm radius separated by 1 nm distance. (b) Map of EF_{EM} around the nanosphere dimer calculated at 608 nm (wavelength of maximum enhancement, as indicated in (a)).

Figure 10.8a shows the values of EF_{EM} (calculated using the E^4-approximation) at the vicinity between the two nanospheres (junction) as function of the wavelength. The results were obtained using the DDA method implemented by the DDSCAT program [44, 50], and the generalized Mie theory through the GMM-FIELD program [51–55]. The simulations were done for incident polarization parallel to the interparticle axis, as indicated in Fig. 10.8b. The resonance observed at 608 nm corresponds to the coupling between the dipolar plasmon modes within each nanosphere. The other resonance mode involves higher order plasmon modes. The important aspect from Fig. 10.8a is the magnitude of the enhancement, which can be as high as 10–11 orders of magnitude. Notice that the EF at the junction is much larger than expected from a single nanoparticle, which was presented in Fig. 10.7. This huge SERS enhancement factor is highly localized in a very small region between the two nanospheres, as it can be seen in the EF_{EM} map of Fig. 10.8b. This region of high field localization is called a "SERS hot spot." The magnitude of the enhancement obtained from the hot spots are of fundamental importance in SERS, once it enables the acquisition of spectra from a very small number of adsorbed molecules, and ultimately the acquisition of Raman spectra from single-molecules.

10.4 SERS Substrates

Much of the SERS studies concentrate on the development of very active (signal enhancing) substrates that not only present the power to observe high SERS intensities but also can be fabricated with a high degree of reproducibility [56–61]. There exist in the literature some review articles that explore this aspect of SERS research. In this chapter, this very important part of SERS will not be discussed but the interested reader is referred to the references [62, 63]. This chapter will instead focus in the single-molecule SERS (SM-SERS) statistics and the importance of nanoparticle local field properties. Therefore, one very important aspect for such discussion is that the nanoparticles must support very strong local field enhancements for the observation of single-molecule events, which can be obtained by nanoparticle aggregation. We want to caution the reader that the aggregation of metal nanoparticles not always leads to SERS enhancements, which depends on the format of the nanoparticles and the geometry of aggregation. One example described in the literature is for the case of gold nanorods in the end-to-end and side-by-side aggregation geometries, presented in Figs. 10.9a,b, respectively.

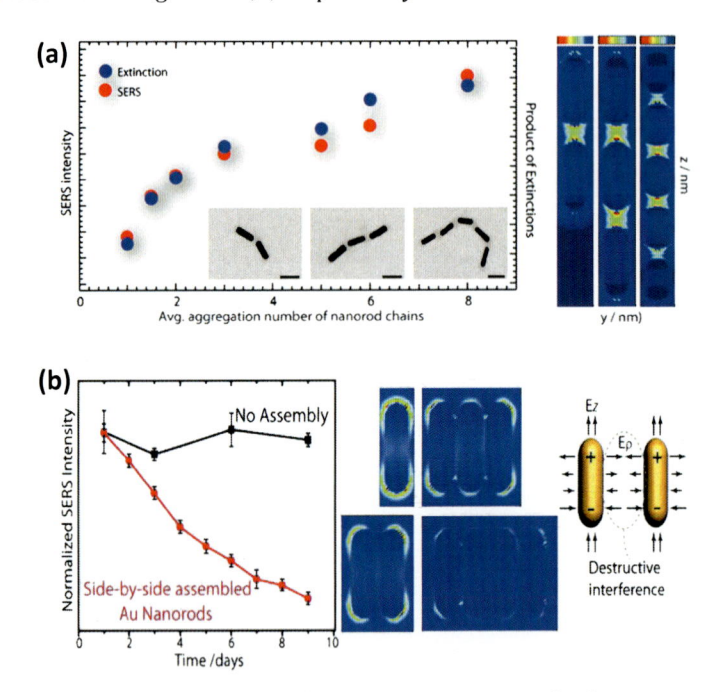

Figure 10.9 Dependence of the SERS intensity and electromagnetic enhancement factor simulations (by FDTD method) with the aggregation state of gold nanorods in (a) end-to-end geometry (reprinted with permission from Ref. [64], Copyright 2011, American Chemical Society) and (b) side-by-side geometry (reprinted with permission from Ref. [65], Copyright 2012, American Chemical Society), which also shows a schematic representation of the destructive interference in such geometry.

The end-to-end configuration leads to an increase in the SERS intensity, as shown in Fig. 10.9a, when the longitudinal surface plasmon mode of the nanorods (along the principal axis) is excited, which is mainly due to formation of hot spots. The side-by-side configuration, on the other hand, leads to a decrease of the SERS intensity, when the same surface plasmon mode is excited. The reason for such a decrease in the SERS intensity is due to destructive interferences that cause the electric field to cancel out in the region between nanoparticles, as shown in the scheme of Fig. 10.9b. This example illustrates an important aspect of SERS, which is its dependence on the local properties of nanoparticles, especially in the single-molecule limit, which will be discussed from now on.

10.5 Single-Molecule SERS

The SM problem

Much of the current interest in SERS has its origin from the first reports on the detection of single-molecules in 1997 [66, 67]. Such experimental observations, together with advances in nanofabrication and computational methods for electromagnetic calculations, opened up a whole new world of investigations in SERS and revitalized the field. The SM-SERS spectra not only contain a lot of information of vibrational sort (fingerprint) but also offer the possibility of investigating the environment probed by the single-molecule. These very attractive features have turned SERS in one of the most investigated optical spectroscopic method in nanotechnology.

The initial reports in 1997 on the single-molecule limit of detection of SERS were based on two main experimental observations: (I) SERS signal was obtained from solutions at ultra-low concentrations (as low as 10^{-14} M). Considerations about the confocal volume probed in Raman microscopes, in conjunction with the ultra-low concentration of the analyte in solution, led to the conclusion that, on average, the observed SERS originated by single-molecules within the probed volume; (II) the experimental SERS from ultra-low concentrations presented very strong intensity fluctuations, characterized by the observation of very strong and very weak (inclusive null) spectra in a given time sequence. To illustrate the format of such intensity fluctuations, we presented in Fig. 10.10 an example of a time-series of SERS spectra of rhodamine 6G (R6G), concentration 10 nM, using the Lee and Meisel [40] silver colloid as SERS substrate. In Fig. 10.10, strong intensity fluctuations can be clearly identified, with the appearance of very intense spectra, as well as the observation of null intensity spectra.

These fluctuations in SERS intensities were initially called "blinking," in analogy to similar observations from single-molecule fluorescence emission experiments. However, it was later clarified that the physics behind the fluctuations in SERS intensities are very different than in fluorescence and the term "blinking" should then be avoided in the context of single-molecule SERS [39, 68, 69].

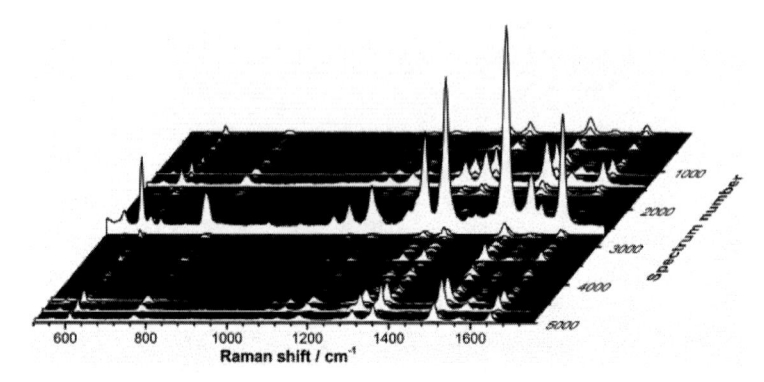

Figure 10.10 Time-series of single-molecule SERS spectra of rhodamine 6G at concentration of 10 nM obtained using the Lee and Meisel [40] silver colloid aggregated with KBr (10 mM).

The important feature behind the single-molecule observation in SERS is the presence of hot spots (Fig. 10.8), which is a region of strong field localization. Therefore, the characteristics of SERS hot spots drive the statistics of intensity fluctuations observed in a given experiment. The objective in this chapter, from now on, is to analyze the single-molecule SERS statistics, which is usually presented in the form of histograms that describe the intensity distributions, i.e., a graph that shows the number of observed events for a given range (bin) of intensities. Figure 10.11 shows histograms of the normalized intensities for the band at 1503 cm^{-1} of R6G. The term normalized intensity means that the intensities in a given series of spectra are normalized by the average intensity in that series, which is a very important normalization if we want to compare the intensity distributions in different contexts, for instance, different concentrations, as shown in Fig. 10.11.

The histograms shown in Fig. 10.11 were obtained for time-series experiments using a silver electrode submitted to an electrochemical roughening procedure to generate nanostructures that support surface plasmon resonance [70]. The results of

Fig. 10.11 clearly show that as the analyte concentration is decreased, i.e., as the system changes from "average SERS" (where a large number of molecules are responsible for the overall SERS signal at given time) to "single-molecule SERS" conditions, the shape of the intensity distribution histogram changes from a normal to an exponential-like distribution, with a high number of null or very weak intensity events and a small number of events with strong SERS signal (see Fig. 10.10). The tailed (exponential) distribution is then a characteristic shape obtained in single-molecule SERS conditions.

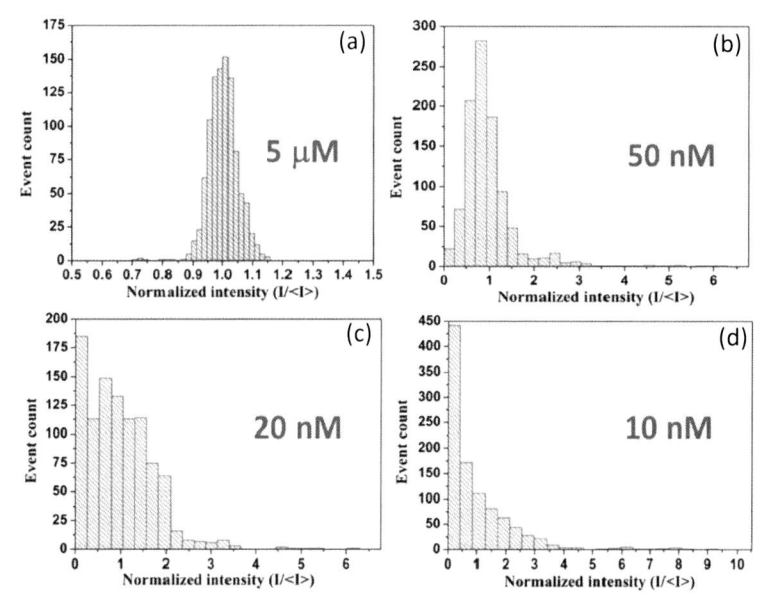

Figure 10.11 SERS intensity histograms for rhodamine 6G adsorbed on an electrochemically roughened (by oxidation-reduction cycles [ORCs]) silver electrode at different concentrations, as indicated in the figure: (a) –5 µM; (b) –50 nM; (c) –20 nM; (d) –10 nM. Adapted with permission from Ref. [70]. Copyright 2009, American Chemical Society.

The single-molecule interpretation for these experimental observations, reported in 1997, was met by a lot of skepticism from the literature. In fact, those experiments could be interpreted using other arguments; for instance, photodecomposition could lead to the fluctuations described in Fig. 10.10. A significant amount of experimental efforts to prove the single-molecule capability of

SERS and to provide a better understanding of the fluctuations followed the original papers from 1997 [66, 67]. One fundamental experiment that settled this dispute, and it is considered nowadays as a definitive prove of the single-molecule detection limit of the SERS method, is the bi-analyte technique first introduced by Le Ru et al. [71, 72], which is outlined in Fig. 10.12.

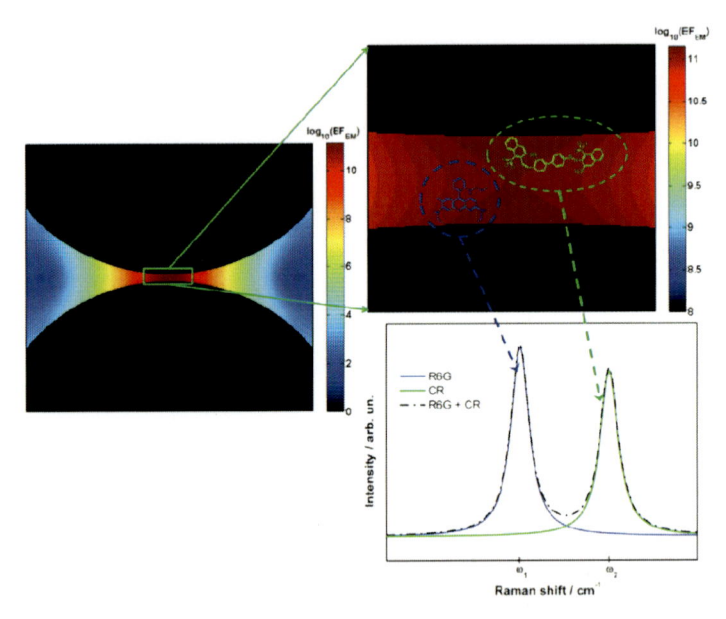

Figure 10.12 Illustration of the bi-analyte concept with the dyes rhodamine 6G (blue) and congo red (green). Each molecule inside a given hot spot gives rise to a SERS spectrum, as represented in the figure. If only one of the two types of molecules is in the hot spot (single-molecule condition), only vibrational signatures of that molecule is observed (blue or green spectrum). In the average SERS condition the SERS spectrum contains the vibrational signatures of both molecules (black dashed line).

The bi-analyte experiment is based on the following argument: If the solution consists of a mixture of two different molecules (with detectable differences in the SERS spectrum) at low concentration, then the probability of two molecules to adsorb on the same SERS hot spot is small. This means that the single-molecule regime can be characterized when in a given set of spectra in a time series one observes events in which the SERS features contains the vibrational characteristics of only one of the species in solution. This can better be visualized in Fig. 10.12. If the single-molecule regime is achieved, there is a higher probability of seeing only one of the two molecules in a given hot spot. In the scheme of Fig. 10.12, the two molecules in the bi-analyte experiment were rhodamine 6G and congo red (CR, an anionic dye [70]). These two dyes will be later used in an actual experimental example in this chapter. Therefore, in the single-molecule limit, there is a high probability of detecting, for instance, unique signatures from R6G only (SERS spectrum represented by the blue line in Fig. 10.12). On the other hand, when only CR adsorbs from the mixture in the hot spot region, then the observed SERS spectrum would be described by the green line. When the two molecules visit the hot spot at the same time, the resulting spectrum would be the black dashed line. This event, however, has low probability at low solution concentrations.

In the single-molecule limit, the majority of the collected spectra are of the type represented by the blue and green lines in Fig. 10.12 (single component spectrum), although some spectra with the vibrational features of the two dyes (like the black dashed line in Fig. 10.12) might be observed as a minor component of the statistics. However, in the average SERS condition (normal distribution of Fig. 10.11a—high concentration of the dyes in the mixture), every spectrum in a time series will all look like the black dashed line in Fig. 10.12, with the bands attributed to each of the dyes from the mixture being observed at the same time. This summarizes the overall concept behind the bi-analyte experiment as a proof of single-molecule detection in SERS. In what follows, these arguments will be analyzed for a given experimental conditions through a Monte Carlo simulation, that will clearly illustrate the dynamics related to the single-molecule statistics. In this simulation, a model SERS hot spot formed by a dimer of 25 nm radius spheres, as discussed in Figs. 10.8 and 10.12, is used. The simulation procedure is similar to that described by Etchegoin et al. [73].

As described in Fig. 10.8, the SERS hot spot formed by two particles is responsible for the high enhancement factors that allow single-molecule detection. Several experiments suggest that an enhancement factor (EF) of about 10^8 is enough for single-molecule detection [73, 74]. It is clearly seen in Fig. 10.8 that the

SERS hot spot is much smaller than the dimensions of the nanoparticle dimer. This can be further visualized in Fig. 10.13 where the angular dependence of EF_{EM} about the hot spot is presented (see the definition in Fig. 10.13a).

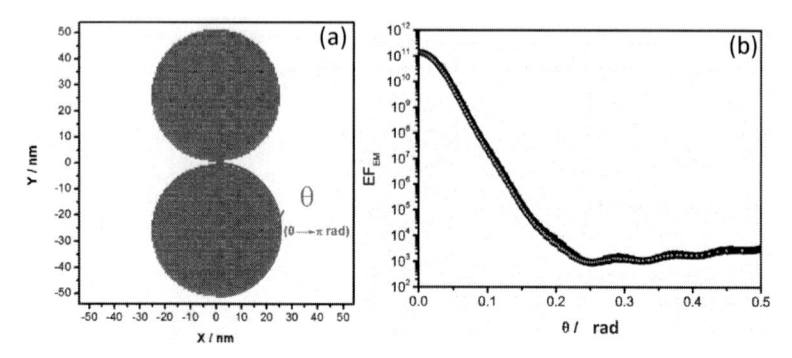

Figure 10.13 Model of a hot spot consisting of a dimer of two nanospheres of 25 nm radius and the angle dependence (defined in the figure) of EF_{EM} for such dimer.

As indicated in Fig. 10.13b, the enhancement factor drops by a factor of almost 1000 times at about 5° from the hot spot center. This result reinforces the strong field localization that gives rise to giant SERS enhancements. Therefore, in a very dilute solution of the molecule of interest, its surface concentration is also expected to be small. Since the SERS signal is only detectable when molecules populate the hot spot region (in the model in Fig. 10.13a an angular range of $-5° \leq \theta \leq 5°$), the number of events in a time series with strong SERS will be small. In order to illustrate this even further, if only one hot spot is considered, the intensity fluctuations can be explained by a combination of small hot spot volume and low surface concentrations when diluted solutions are utilized. The experimental histograms can then be obtained through simulations that consider a hot spot area that is much smaller than the total area of the nanostructures. For instance, considering a Monte Carlo simulation in which the system dynamics is described in each step by choosing a set of angles θ for N adsorbed molecules (remember that the surface coverage is tuned by the solution concentration), according to the following probability distribution [73]:

$$p(\theta) = \frac{1}{2}\sin(\theta) \qquad (10.19)$$

Such probability distribution is taken into account to recover the 3D characteristic of the problem out of the 2D simulation. In other words, the very small probability that a given molecule will diffuse from the solution to adsorb at the hot spot should be taken into account. This is done by imagining that each value of θ describes a circumference about the interparticle axis where the molecule can adsorb. As this imaginary circumference increases (by changing θ), more possibilities a molecule will have to adsorb on the surface. But because this circumference reaches its maxim value for $\theta = 90°$, for this value of θ there is a higher probability of adsorption and a small probability close to the hot spot ($\theta = 0°$) or in its opposite side ($\theta = 180°$). The function that describes such relationship is precisely the probability distribution of Eq. 10.19. The results of the simulation are presented in Fig. 10.14.

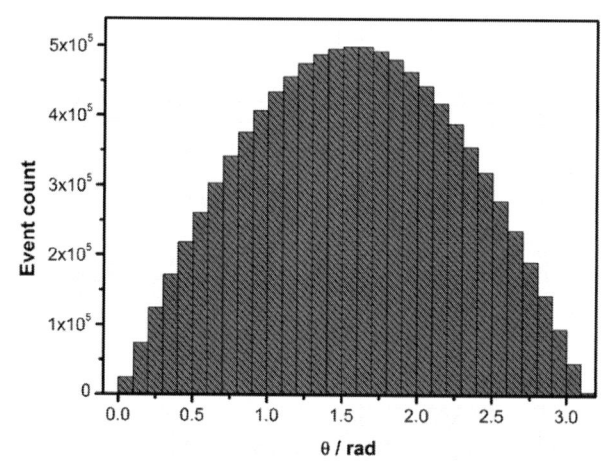

Figure 10.14 Histogram of sorted angles according to the probability distribution of Eq. 10.19.

As time evolves during an experiment, so does the system configuration, i.e., the distribution of the N adsorbed molecules around the angle θ (adsorption position in the nanoparticles). EF_{EM} for each molecule can be calculated for each value of the angle θ, and the individual results are summed up. The total SERS intensity is therefore proportional to the total EF_{EM}. A threshold of $EF = 10^8$ is used, since this has been suggested as the minimum amount of enhancement required for a single-molecule SERS to be observable above the noise of a typical Raman microscope [73, 74].

It is possible to further refine the simulation by considering a mixture of two distinct adsorbed molecules that contribute to the total signal. In this model, the two molecules have the same adsorption characteristics, SERS cross sections, and the adsorption energy anywhere in the surface is the same (Langmuir hypothesis). Figure 10.15 shows the results obtained from several Monte Carlo simulations for different number of molecules adsorbed (indicated in the figure) on the surface of the dimer of Figs. 10.8b and 10.12.

Two types of histograms are presented in Fig. 10.15 for each surface coverage (number of adsorbed molecules). The histograms in the left-hand side correspond to the overall intensity distribution normalized by the average intensity (SERS intensity fluctuation histograms, analogous to the experimental ones shown in Fig. 10.11). The histograms in the right-hand side of Fig. 10.15 were calculated by considering the percentage contribution to the overall intensity from molecule 1 (mol1). Since this simulation was done for a mixture of two species, the fractional contribution to the total SERS intensity from molecule 2 can also be easily extracted. This latter histogram permits the direct visualization of the number of events in which only one species contributed to the signal. For a low surface coverage (for instance, 200 adsorbed molecules), the majority of the events (SERS spectrum in a time series) allows the observation of SERS signal exclusively from one or the other molecule ($p_{mol1} = 0$ and $p_{mol1} = 1$) in the mixture. This type of histogram has been realized experimentally and characterizes the single-molecule regime in a SERS experiment. As the surface concentration is increased, by changing, for instance, the solution concentration, the number of events containing SERS signal from the two molecules simultaneously ($p_{mol1} = 0.5$) increases as well, and the single-molecule regime is lost (for example, see the histograms for 1500 adsorbed molecules in Fig. 10.15). Note that the requirements in this analysis that the two molecules present the same SERS cross section and adsorption properties have been achieved experimentally by the isotopic editing technique [72, 75], although it is important to emphasize that the principle of the bi-analyte technique can also be applied using distinct molecules (different adsorption properties and Raman cross sections). In those cases, however, the histograms of p_{mol1} are not expected to be symmetric as the simulated in Fig. 10.15.

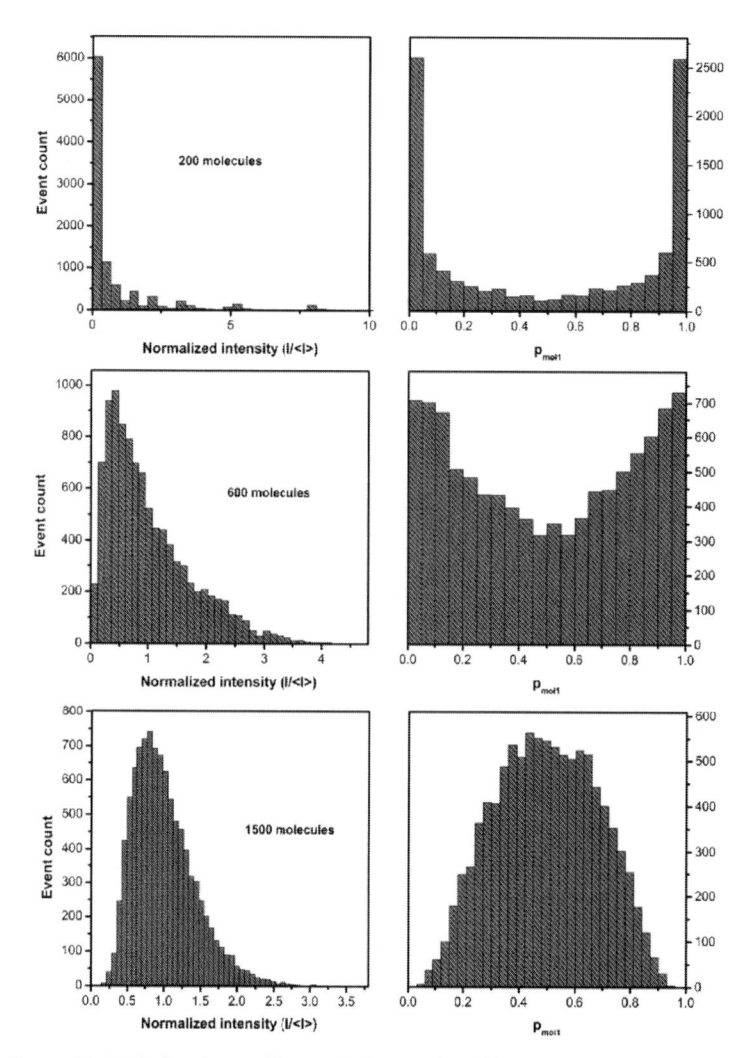

Figure 10.15 Left column: Monte Carlo simulated histograms of normalized intensity (with respect to the average EF_{EM} for all the steps in the simulation) for different number of molecules adsorbed on the nanoparticle surface. Right column: Monte Carlo simulated histograms of fraction of signal for which one of the two types of molecule (bi-analyte) in the simulation is responsible. Each line in the column corresponds to different surface concentrations as in the left column histograms.

The histograms of p_{mol1} (histograms in the right side of Fig. 10.15, particularly the one at the top of the figure) are then usually used as a proof of single-molecule detection, since it reflects the most intimate physical property of SERS: only molecules in the hot spot region can be detected. On the other hand, the normalized intensity distributions (histograms in the left side in Fig. 10.15) shows a significant number of null events at the single-molecule condition, which is related to the low surface coverage. Therefore, long-tailed distributions dominated by null events can also be used to indirectly infer the single-molecule regime (as it was done in the discussion of the shape of the histograms of Fig. 10.11).

It is important to emphasize that the shape of the intensity distribution histograms is a complicated function of a number of properties that include surface concentration, adsorption properties (adsorption constant), electrochemical potential (when applicable), density of hot spots, and so on. In order to provide an idea of the effect of one of these parameters, the discussion will now concentrate only on the effect of the number of hot spots in the intensity distribution histograms under the single-molecule regime (low surface coverage). The same simulation, as discussed in Fig. 10.15, will be presented, but a different number of model hot spots contributing to the overall SERS signal will be used. Figure 10.16 shows results for simulations containing either 5 or 50 hot spots (all with the same characteristics of Fig. 10.13), but keeping the same surface coverage of 100 molecules per nanostructure, i.e., 500 and 5000 molecules in the systems of 5 and 50 hot spots, respectively. According to Fig. 10.15, this level of surface coverage is certain to produce a characteristic single-molecule response from a single hot spot.

Figure 10.16 shows strong deviations in the normalized intensity distribution histograms for systems with different number of hot spots, even though the surface coverage was kept the same. This result can be interpreted by considering that the overall SERS signal is now originated by contributions from a distribution of hot spots. Therefore, molecules at different hot spots contribute to the total SERS intensity, which leads to a decrease in the number of null events. Besides the change in the shape of the intensity distribution, the increase in the number of hot spots led to a decrease in the number of events associated to the single-molecule limit. This result suggests that in different SERS substrates, where the difference is only the number of hot spots, the substrate with the highest number of hot spots being illuminated by the incident radiation will require a smallest surface concentration to exhibit single-molecule behavior. This can be observed in Fig. 10.17, where two model substrates with different number of hot spots presented the same single-molecule statistics (same shape for the histogram of p_{mol1}) when the surface coverage was adjusted. For practical reasons, the substrate with 50 hot spots was compared to a substrate with one hot spot in Fig. 10.17. The observation of strong SERS spectrum in a time series for the substrate with one hot spot is a very rare event (several null events are seen in the intensity distribution), even when the surface coverage is set to 200 molecules per nanostructure (see Fig. 10.17).

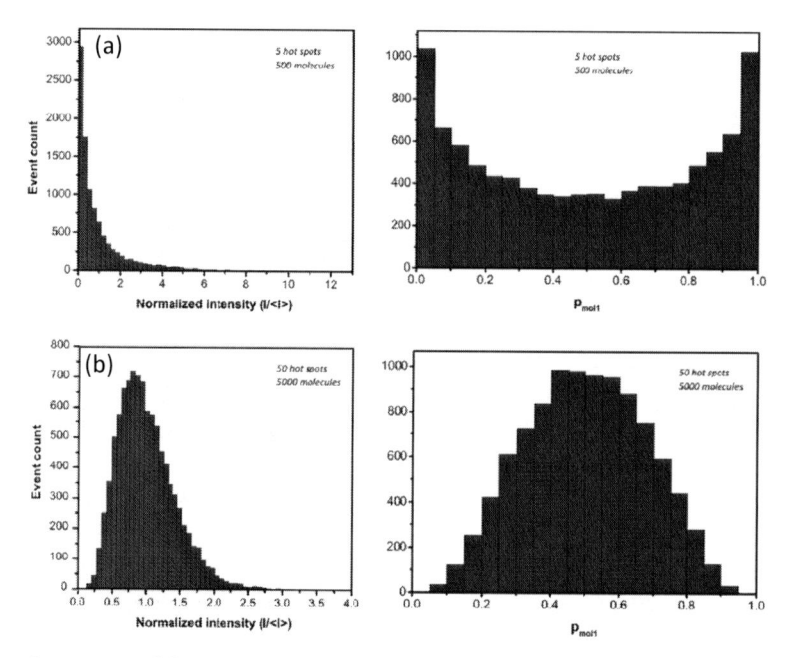

Figure 10.16 (a) Monte Carlo simulation for histograms of normalized intensity (left) and fraction of signal for which one of the two types molecules (bi-analyte) is responsible. The simulation was done for 500 molecules randomly distributed over 5 nanosphere dimers as in Fig. 10.13. (b) Same as in (a), but increasing the number of hot spots and the number of molecules by a factor of 10.

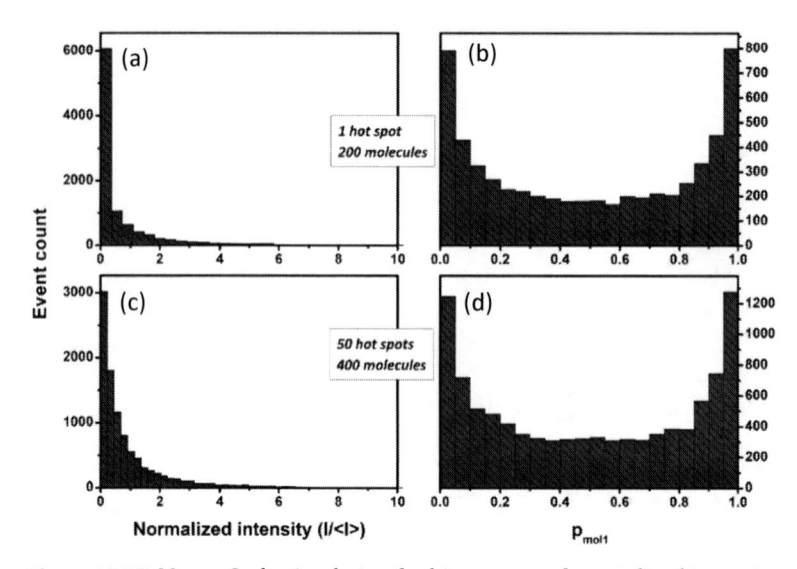

Figure 10.17 Monte Carlo simulation for histograms of normalized intensity (a) and fraction of signal for which one of the two types molecules (bi-analyte) is responsible (b), for a system of 200 molecules randomly distributed over one nanosphere dimer. (c and d) Same as (a) and (b), but for 400 molecules randomly distributed over the surface of 50 dimers.

The substrate with more hot spots (50 hot spots) yields the same single-molecule statistics for the p_{mol1} histogram as the substrate with one hot spot only when the surface coverage is smaller. The surface concentration for the substrate with 50 hot spots was 8 molecules per nanostructure, compared to 200 molecules per nanostructure for the substrate with one hot spot. Figures 10.17c,d show that the single-molecule statistics for the two substrates are very similar for those coverage values. The same reasoning can be extended to other systems with different number of hot spots, and the results are summarized in Fig. 10.18.

Figure 10.18 clearly demonstrates the necessity of decreasing surface (or solution) concentration to observe single-molecule statistics as the number of hot spots in the illuminated confocal volume increases.

In real systems, the strength of each hot spot and the number density of hot spots both contribute to the substrate SERS efficiency in the sense of average SERS. Therefore, the differences in efficiency of distinct SERS substrates can be interpreted based on those parameters. For instance, the higher SERS efficiency reported for aggregated Ag colloids relative to roughened electrodes activated by ORC, is due to a higher number of hot spots with strong resonances around the laser excitation. Experimentally, in the case of the two aforementioned substrates, it should be expected that single-molecule statistics should be then observed at very distinct concentrations of the same dye (considering an equilibrium between the solution and the surface concentration). This prediction is confirmed by a number of papers in the literature, where the typical concentrations for single-molecule detection in colloidal systems are very low (ultra low concentrations), as the ones used in the first reports of single-molecule detection by Kneipp et al. [66] and Emory et al. [67], whereas more recent publications show the possibility of single-molecule detection using the ORC electrode but at dye concentrations in the nanomolar range [70].

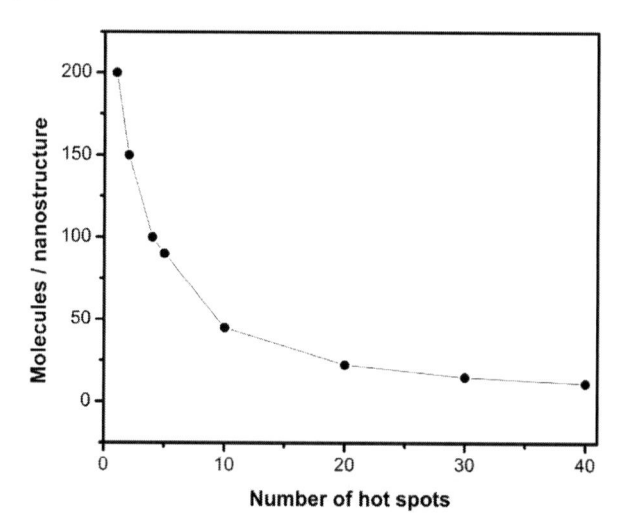

Figure 10.18 Dependence of the molecule concentration on the number of hot spots in the system. As the number of hot spots increases, the molecule concentration (surface concentration, and therefore solution concentration) has to decrease to keep the single-molecule condition (Figs. 10.17b,d).

Figure 10.19 is an example of single-molecule SERS detection on Ag ORC electrode. The results were obtained for a bi-analyte experiment involving rhodamine 6G (R6G, a cationic species) and congo red (CR, an anionic species) at 10 nM each. Figure 10.19 presents results for two applied potentials to the electrode (−0.1 and −0.2 V, left and right column, respectively). A number of features discussed up to now can be identified in the experimental results presented in Fig. 10.19. For instance, the possibility of using two very distinct molecules (a cationic and an anionic species in this case) in the bi-analyte technique is illustrated. These two chemical species have very different SERS cross sections and distinct adsorption properties. These differences are reflected in the histograms in Figs. 10.19c,f. As already discussed, these histograms are not symmetric as in the case of the Monte Carlo simulations in Figs. 10.15 and 10.16. Another important aspect is that it is possible to observe single-molecule fluctuations for this SERS substrate (Ag ORC) at the concentrations of the order of 20 nM, which is not as low as reported in the literature for some colloidal systems (in the order of 10^{-14} M), which reflects the smaller SERS efficiency of the ORC electrode.

One very important advantage of using ORC electrode as SERS substrate is the possibility of following single-molecule SERS fluctuations as a function of the applied potential. The control over the applied potential allows the tuning of the surface concentration of the dye adsorbed to the metal surface. In the example in Fig. 10.19, the two dyes have opposite charges, and their interaction with the polarized electrode surface will be different at different potentials. Since the applied potential modulate the charge at the metallic electrode surface, the direct coulombic interaction controls the surface concentration of dye at a given potential on the nanostructured surface. In order to illustrate this behavior, Fig. 10.20 shows the dynamics of two kinds of molecules with opposite charges (positively and negatively charged) around a negatively charged metal nanoparticle. In this particular example, the charge on the nanoparticle surface arises from a negative potential bias. Figure 10.20 shows the trajectories of different molecules, simulated by Monte Carlo, in which the probability for diffusion in a given direction is controlled by the coulombic potential energy (ϕ) in the system. This probability is high when the diffusion for that particular direction leads to a decrease in the potential energy of the system, and it is low otherwise:

$$p_{dif} = \exp\left(-\frac{\Delta\phi}{k_B T}\right), \tag{10.20}$$

where $\Delta\phi$ is the change in the potential energy upon diffusion to a particular direction.

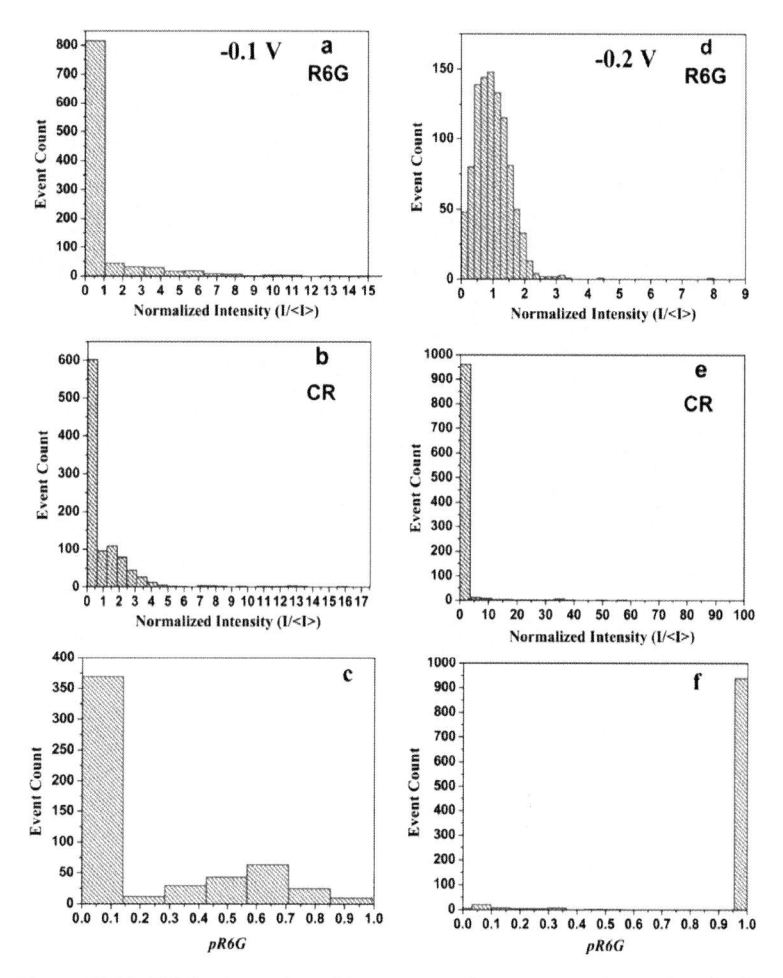

Figure 10.19 SERS intensity histograms from an electrochemically roughened silver electrode immersed in an equimolar mixture of 20 nM rhodamine 6G (R6G) and congo red (CR) in 0.1 M KCl for different potentials. −0.1 V Vs Ag|AgCl$_{(sat)}$|KCl 0.1 M: (a) Normalized intensity histogram for the 1503 cm^{-1} band of R6G; (b) normalized intensity histogram for the 1591 cm^{-1} band of CR. (c) Histogram of fraction of signal for which R6G is responsible. The same types of histograms is presented in the right column but for a potential of −0.2 V Vs Ag|AgCl$_{(sat)}$|KCl 0.1 M. Reprinted with permission from Ref. [70]. Copyright 2009, American Chemical Society.

Figure 10.20 shows, as expected, that the negative molecule has a small probability of being found close to the surface of a negatively charged nanoparticle, especially close to the SERS hot spot. Therefore, according to this interpretation, the application of negative potential leads to a decrease in the probability to have a molecule like CR at a given hot spot, which is observed experimentally in Fig. 10.19 by the decrease in the number of events for pR6G = 0. This is indeed the case of Fig. 10.19, when more negative potentials were applied to the electrode the number of events associated to observation of SERS from only CR (pR6G = 0) decreased.

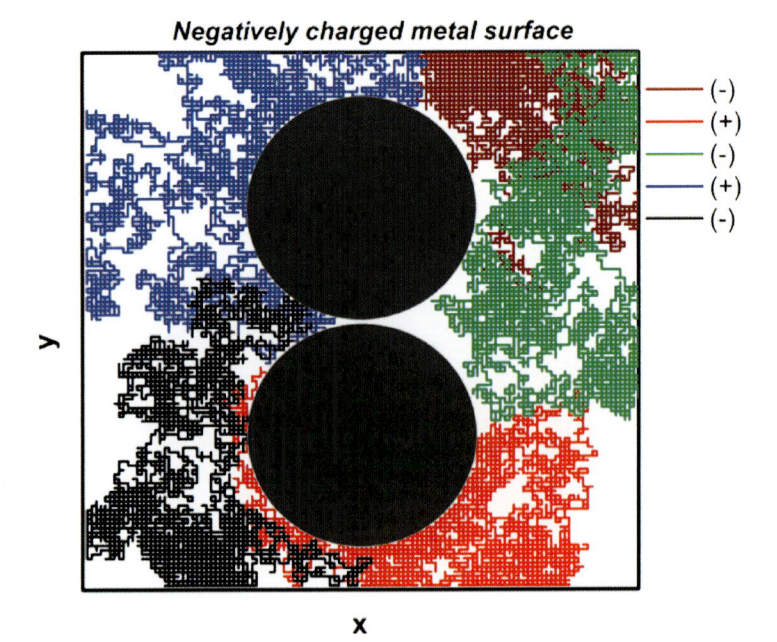

Figure 10.20 Monte Carlo simulation of the trajectory of negatively (−) and positively (+) charged molecules (see figure) subjected to an electrostatic potential interaction with the negatively charged nanoparticle spheres. In this example, the charge on metal is due to a negative potential bias.

This result also shows the possibility of utilization of SERS to study electrochemical process at single-molecule level, which has already been demonstrated in the literature [76, 77].

10.6 Conclusion

In this chapter, the fundamentals of SERS were discussed. The main contributions (chemical and electromagnetic) to the overall enhancement in the Raman intensity of molecules adsorbed on metallic (mainly Ag, Au, and Cu) nanostructures were presented. Special emphasis was given to the electromagnetic (plasmonic) mechanism, since it shows the highest contribution to the enhancement in the majority of systems. The EM contribution is especially important when the nanoparticles are aggregated, a situation that gives rise to the possibility of highly efficient hot spots and single-molecule detection by SERS.

Single-molecule SERS is a subject of fundamental importance to plasmonics. Not only due to the obvious analytical possibilities, but also because the understanding of the single-molecule behavior can provide very important fundamental information. In fact, the dynamics of adsorbed molecules subjected to such strong electric field enhancements and information about site-specific adsorption behavior is encrypted in the SERS spectra. Once the electromagnetic mechanism is of vital importance to SERS (especially single-molecule), the understanding of such system in its more fundamental aspect may provide valuable information about local properties of plasmonic nanostructures, like electric field distribution around hot spots, resonance energy of the SERS hot spots [15], etc. These more advanced aspects of single-molecule SERS are still evolving, but they have as fundamental basis the proper understanding of what single-molecule SERS exactly means. This main aspect was treated in this chapter, including the statistics behind the single-molecule SERS phenomenon. This fundamental understanding is crucial as a basic starting point to the further development of more advanced applications of SM-SERS.

The ultimate goal for SPR biosensing technology is to effectively compete with other popular labeled techniques in particular fluorescence methods. Unlike fluorescent-based assays, SPR as a label-free method eliminates errors caused by the unlikely but possible binding of the fluorescent label itself to proteins being screened. The increasing support and interest from industry to investigate different aspects of SPR makes the future of SPR biosensing very promising.

References

1. Itoh, T., Kikkawa, Y., Yoshida, K., Hashimoto, K., Biju, V., Ishikawa, M., Ozaki, Y. (2006). *J. Photochem. Photobiol. A*, **183**, 322.
2. Itoh, T., Yoshikawa, H., Yoshida, K.-I., Biju, V., Ishikawa, M. (2010). *J. Chem. Phys.*, **133**, 124704.
3. Itoh, T., Kikkawa, Y., Biju, V., Ishikawa, M., Ikehata, A., Ozaki, Y. (2006). *J. Phys. Chem. B*, **110**, 21536.
4. Perassi, E. M., Coronado, E. A. (2013). *J. Phys. Chem. C*, **117**, 7744.
5. Schatz, G. C., Young, M. A., Van Duyne, R. P. (2006). *Surface-Enhanced Raman Scattering: Physics and Applications*, Springer.
6. Masiello, D. J., Schatz, G. C. (2008). *Phys. Rev. A*, **78**, 042505.
7. Lombardi, J. R., Birke, R. L. (2012). *J. Chem. Phys.*, **136**, 144704.
8. Kitahama, Y., Tanaka, Y., Itoh, T., Ozaki, Y. (2011). *Phys. Chem. Chem. Phys.*, **13**, 7439.
9. Kitahama, Y., Tanaka, Y., Itoh, T., Ozaki, Y. (2011). *Chem. Commun.*, **47**, 3888.
10. Perassi, E. M., Canali, L. R., Coronado, E. A. (2009). *J. Phys. Chem. C*, **113**, 6315.
11. Aroca, R. (2006). *Surface Enhanced Vibrational Spectroscopy*. Wiley: Chichester.
12. Ru, E. C. L., Etchegoin, P. G. (2008). *Principles of Surface-Enhanced Raman Spectroscopy and Related Plasmonic Effects*. Elsevier: Amsterdam.
13. Long, D. A. (2002). *The Raman Effect: A Unified Treatment of the Theory of Raman Scattering by Molecules*. Wiley.
14. dos Santos, D. P., Andrade, G. F. S., Brolo, A. G., Temperini, M. L. A. (2011). *Chem. Commun.*, **47**, 7158.
15. dos Santos, D. P., Temperini, M. L. A., Brolo, A. G. (2012). *J. Am. Chem. Soc.*, **134**, 13492.
16. Brolo, A. G., Sanderson, A. C., Smith, A. P. (2004). *Phys. Rev. B*, **69**, 045424.
17. Löwen, H., Allahyarov, E., Likos, C. N., Blaak, R., Dzubiella, J., Jusufi, A., Hoffmann, N., Harreis, H. M. (2003). *J. Phys. A*, **36**, 5827.
18. Stevens, M. J., Falk, M. L., Robbins, M. O. (1996). *J. Chem. Phys.*, **104**, 5209.

19. Iberi, V., Mirsaleh-Kohan, N., Camden, J. P. (2013). *J. Phys. Chem. Lett.*, **4**, 1070.
20. Etchegoin, P. G., Le Ru, E. C. (2010). Basic electromagnetic theory of SERS. In *Surface Enhanced Raman Spectroscopy*, Wiley-VCH Verlag GmbH & Co. KGaA, pp. 1.
21. Bockris, J. O. M., Khan, S. U. M., *Surface Electrochemistry: A Molecular Level Approach*. Plenum: 1993.
22. Jeanmaire, D. L., Van Duyne, R. P. (1977). *J. Electroanal. Chem. Interfacial Electrochem.*, **84**, 1.
23. Hale, G. M., Querry, M. R. (1973). *Appl. Opt.*, **12**, 555.
24. Le Ru, E. C., Blackie, E., Meyer, M., Etchegoin, P. G. (2007). *J. Phys. Chem. C*, **111**, 13794.
25. Frens, G. (1973). *Nature*, **241**, 20.
26. Crocker, J. C., Grier, D. G. (1996). *Phys. Rev. Lett.*, **77**, 1897.
27. Lombardi, J. R., Birke, R. L., Lu, T. H., Xu, J. (1986). *J. Chem. Phys.*, **84**, 4174.
28. Lombardi, J. R., Birke, R. L. (2008). *J. Phys. Chem. C*, **112**, 5605.
29. Lombardi, J. R., Birke, R. L. (2009). *Acc. Chem. Res.*, **42**, 734.
30. Wu, D.-Y., Li, J.-F., Ren, B., Tian, Z.-Q. (2008). *Chem. Soc. Rev.*, **37**, 1025.
31. Mullin, J. M., Autschbach, J., Schatz, G. C. (2012). *Comput. Theor. Chem.*, **987**, 32.
32. Mullin, J., Schatz, G. C. (2012). *J. Phys. Chem. A*, **116**, 1931.
33. Kelly, K. L., Coronado, E., Zhao, L. L., Schatz, G. C. (2003). *J. Phys. Chem. B*, **107**, 668.
34. Palik, E. D. (1985). *Handbook of Optical Constants of Solids*. Academic Press.
35. Zuloaga, J., Nordlander, P. (2011). *Nano Lett.*, **11**, 1280.
36. Galloway, C. M., Etchegoin, P. G., Le Ru, E. C. (2009). *Phys. Rev. Lett.*, **103**, 063003.
37. Etchegoin, P. G., Le Ru, E. C. (2006). *J. Phys. Condens. Matter.*, **18**, 1175.
38. Kleinman, S. L., Ringe, E., Valley, N., Wustholz, K. L., Phillips, E., Scheidt, K. A., Schatz, G. C., Van Duyne, R. P. (2011). *J. Am. Chem. Soc.*, **133**, 4115.
39. Maher, R. C., Dalley, M., Le Ru, E. C., Cohen, L. F., Etchegoin, P. G., Hartigan, H., Brown, R. J. C., Milton, M. J. T. (2004). *J. Chem. Phys.*, **121**, 8901.

40. Lee, P. C., Meisel, D. (1982). *J. Phys. Chem.*, **17**, 3391.
41. Moskovits, M. (2013). *Phys. Chem. Chem. Phys.*, **15**, 5301.
42. Le Ru, E. C., Meyer, S. A., Artur, C., Etchegoin, P. G., Grand, J., Lang, P., Maurel, F. (2011). *Chem. Comm.*, **47**, 3903.
43. Moskovits, M. (1982). *J. Chem Phys.*, **77**, 4408.
44. Draine, B. T., Flatau, P. J. (1994). *J. Opt. Soc. Am. A*, **11**, 1491.
45. Yurkin, M. A., Hoekstra, A. G. (2007). *J. Quant. Spectrosc. Radiat. Transfer*, **106**, 558.
46. Yang, W. H., Schatz, G. C., Vanduyne, R. P. (1995). *J. Chem. Phys.*, **103**, 869.
47. McMahon, J. M., Li, S. Z., Ausman, L. K., Schatz, G. C. (2012). *J. Phys. Chem. C*, **116**, 1627.
48. Zhao, J., Pinchuk, A. O., McMahon, J. M., Li, S., Ausman, L. K., Atkinson, A. L., Schatz, G. C. (2008). *Acc. Chem. Res.*, **41**, 1710.
49. Le Ru, E. C., Etchegoin, P. G. (2006). *Chem. Phys. Lett.*, **423**, 63.
50. Draine, B. T., Flatau, P. J. *User Guide to the Discrete Dipole Appoximation Code DDSCAT 7.1*, 2010.
51. Kleinman, S. L., Sharma, B., Blaber, M. G., Henry, A.-I., Valley, N., Freeman, R. G., Natan, M. J., Schatz, G. C., Van Duyne, R. P. (2012). *J. Am. Chem. Soc.*, **135**, 301.
52. Meyer, M., Le Ru, E. C., Etchegoin, P. G. (2006). *J. Phys. Chem. B*, **110**, 6040.
53. Gray, S. K. (2012). *J. Phys. Chem. C*, **117**, 1983.
54. Cang, H., Labno, A., Lu, C., Yin, X., Liu, M., Gladden, C., Liu, Y., Zhang, X. (2011). *Nature*, **469**, 385.
55. Cortés, E., Etchegoin, P. G., Le Ru, E. C., Fainstein, A., Vela, M. E., Salvarezza, R. C. (2013). *J. Am. Chem. Soc.*, **135**, 2809.
56. Willets, K. A. (2013). *Phys. Chem. Chem. Phys.*, **15**, 5345.
57. Yang, Z., Chen, S., Fang, P., Ren, B., Girault, H. H., Tian, Z. (2013). *Phys. Chem. Chem. Phys.*, **15**, 5374.
58. Le Ru, E. C., Somerville, W. R. C., Auguié, B. (2013). *Phys. Rev. A*, **87**, 012504.
59. Cortes, E., Etchegoin, P. G., Le Ru, E. C., Fainstein, A., Vela, M. E., Salvarezza, R. C. (2013). *J. Am. Chem. Soc.*, **135**, 2809.
60. Pazos-Perez, N., Wagner, C. S., Romo-Herrera, J. M., Liz-Marzán, L. M., García de Abajo, F. J., Wittemann, A., Fery, A., Alvarez-Puebla, R. A. (2012). *Angew. Chem. Int. Ed.*, **51**, 12688.

61. Batista, E. A., dos Santos, D. P., Andrade, G. F. S., Sant'Ana, A. C., Brolo, A. G., Temperini, M. L. A. (2009). *J. Nanosci. Nanotech.*, **9**, 3233.
62. Fan, M., Andrade, G. F. S., Brolo, A. G. (2011). *Anal. Chim. Acta*, **693**, 7.
63. Laurence, T. A., Braun, G. B., Reich, N. O., Moskovits, M. (2012). *Nano Lett.*, **12**, 2912.
64. Lee, A., Andrade, G. F. S., Ahmed, A., Souza, M. L., Coombs, N., Tumarkin, E., Liu, K., Gordon, R., Brolo, A. G., Kumacheva, E. (2011). *J. Am. Chem. Soc.*, **133**, 7563.
65. Lee, A., Ahmed, A., dos Santos, D. P., Coombs, N., Park, J. I., Gordon, R., Brolo, A. G., Kumacheva, E. (2012). *J. Phys. Chem. C*, **116**, 5538.
66. Kneipp, K., Wang, Y., Kneipp, H., Perelman, L. T., Itzkan, I. (1997). *Phys. Rev. Lett.*, **78**, 1667.
67. Nie, S. M., Emory, S. R. (1997). *Science*, **275**, 1102.
68. Etchegoin, P. G., Meyer, M., Le Ru, E. C. (2007). *Phys. Chem. Chem. Phys.*, **9**, 3006.
69. Wang, Z., Rothberg, L. J. (2005). *J. Phys. Chem. B*, **109**, 3387.
70. dos Santos, D. P., Andrade, G. F. S., Temperini, M. L. A., Brolo, A. G. (2009). *J. Phys. Chem. C*, **113**, 17737.
71. Le Ru, E. C., Meyer, M., Etchegoin, P. G. (2006). *J. Phys. Chem. B*, **110**, 1944.
72. Dieringer, J. A., Lettan, R. B., Scheidt, K. A., Van Duyne, R. P. (2007). *J. Am. Chem. Soc.*, **129**, 16249.
73. Le Ru, E. C., Etchegoin, P. G., Meyer, M. (2006). *J. Chem. Phys.*, **125**, 204701.
74. Etchegoin, P. G., Lacharmoise, P. D., Le Ru, E. C. (2009). *Anal. Chem.*, **81**, 682.
75. Blackie, E., Le Ru, E. C., Meyer, M., Timmer, M., Burkett, B., Northcote, P., Etchegoin, P. G. (2008). *Phys. Chem. Chem. Phys.*, **10**, 4147.
76. Cortes, E., Etchegoin, P. G., Le Ru, E. C., Fainstein, A., Vela, M. E., Salvarezza, R. C. (2010). *Anal. Chem.*, **82**, 6919.
77. Cortes, E., Etchegoin, P. G., Le Ru, E. C., Fainstein, A., Vela, M. E., Salvarezza, R. C. (2010). *J. Am. Chem. Soc.*, **132**, 18034.

Chapter 11

Graphene-Based Plasmonics

Sinan Balci,[a] Emre Ozan Polat,[b] and Coskun Kocabas[b]

[a]*University of Turkish Aeronautical Association, Department of Astronautical Engineering, 06790 Ankara, Turkey*
[b]*Bilkent University Department of Physics, 06800 Ankara, Turkey*

ckocabas@fen.bilkent.edu.tr

The unique electrical, optical, and chemical properties of the two-dimensional crystal of carbon, graphene, stimulated great interest for plasmonics in reduced dimensions. Notably, controlled electrostatic doping on graphene opens new perspectives for gate-tunable active plasmonics. In this chapter, we review the fundamental aspects of plasmons on graphene and their applications ranging from surface plasmon sensors to active plasmonic devices.

11.1 Introduction: Plasmons in Reduced Dimensions

Graphene is a two-dimensional (2D) crystal. Studying plasma oscillation on 2D materials sometimes causes confusions between surface plasmons and surface plasmon polaritons (SPPs). First, we would like to clarify these definitions. Plasmons are quasiparticles

---

Introduction to Plasmonics: Advances and Applications
Edited by Sabine Szunerits and Rabah Boukherroub
Copyright © 2015 Pan Stanford Publishing Pte. Ltd.
ISBN 978-981-4613-12-5 (Hardcover), 978-981-4613-13-2 (eBook)
www.panstanford.com

of quantized plasma oscillations. Polaritons are also quasiparticles, but they are formed by strong coupling of electromagnetic waves and excitations on a material. For example, (SPPs) are the coupled state of plasmons and electromagnetic waves that propagate on the surface of 3D material. Plasmons on 2D materials are usually named as surface plasmons because they live on a surface of 2D material but they are not polaritons. When we say surface plasmons of graphene we talk about bulk plasma oscillations on graphene surface. To understand the physics of plasmons, we should start with the comparison of plasma oscillations in different dimensions such as 3D bulk materials, i.e., gold and silver; 2D materials, i.e., graphene and inversion layer on silicon; 1D materials, i.e., carbon nanotubes or atomic nanowires. Depending on the dimensionality of the material, dispersion characteristics of the plasma oscillations differ. For 3D materials, the plasma oscillations are dispersionless with a plasma frequency of

$$\omega_{3D} = \sqrt{\frac{4\pi n e^2}{m}}, \tag{11.1}$$

where e is the elementary charge, m is the mass of an electron, and n is the charge density, which is the only material-dependent parameter that defines the plasma frequency. In reduced dimensions, however, plasma oscillations yield dispersion characteristics because the electric field of the plasma oscillations penetrates out of the material resulting in less restoring forces. This causes momentum-dependent plasma frequency. The early experiments on 2D plasma oscillations have been performed on liquid helium surface by Gregory Adams in 1976 [6]. The image-potential-induced surface states on liquid helium traps electrons in close proximity to the surfaces. These trapped electrons form a sheet of 2D electron gas with extremely large carrier mobilities of 10^7 cm^2/Vs. However, the charge densities on liquid helium are limited to 10^8 cm^{-2}. Later, metal-oxide-semiconductor (MOS) structures have been used to generate tunable 2D electron gas in the inversion layer on silicon surface [3]. A grating-shaped gate electrode was used to generate the inversion layer and to couple light to plasmons. With Si-MOS devices, carrier densities of 4×10^{12} cm^{-2} with plasma frequency of 30 cm^{-1} can be achieved. Very recently, graphene has been used to reveal the physics of 2D-plasmons [4, 5]. Graphene in backgatedtransistor geometry provides a unique system to study plasma oscillations. Various techniques have been used to probe the plasmons on graphene sheet. Because of large momentum mismatch, plasmon on graphene cannot be excited by free propagating light. Therefore, a sharp atomic force microscope tip attached with an infrared spectrometer has been used to probe the tunable graphene plasmon in a back-gated transistor.

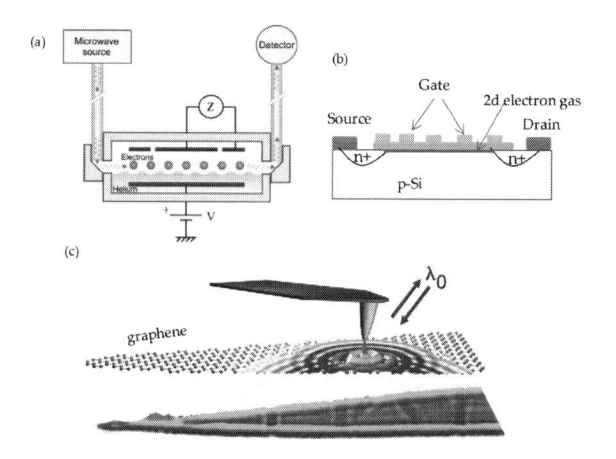

Figure 11.1 Platforms used for studying plasmons in two dimensions. (a) 2D electron gas on liquid helium probed by microwave excitations [1]. (b) Plasmons on silicon surface generated by a MOS structure with grating shaped gate electrode. Gate electrode is used to couple light to plasmons on the inversion layer [3]. (c) Plasmons on graphene probed by infrared nanoimaging. A sharp atomic force microscope tip is used for excitation of plasmons on graphene [4, 5].

Figure 11.1 summarizes various platforms used to study physics of 2D-plasmons. These experiments proved the dispersive characteristics of plasma frequency as

$$\omega_{2D}(q) = \sqrt{\frac{2\pi n e^2 q}{\varepsilon m}}, \qquad (11.2)$$

where ε is the dielectric constant of the 2D material and q is the momentum of plasma oscillations. In 1D, i.e., atomic chains, carbon nanotubes, plasma oscillations are called wire plasmons [9]. Significant portion of the restoring electric field is out of the material and strongly reflects the confined nature of the electron motion. Experiments on atomic chain of gold provide dispersion relation of

$$\omega_{1d}(q) = q\sqrt{\frac{2en\ln\left(\dfrac{a}{q}\right)}{m}}, \qquad (11.3)$$

where n is the linear charge density and a is the geometric parameter of the 1D system. Figure 11.2 illustrates the dispersion relation of the plasma oscillations in reduced dimensions.

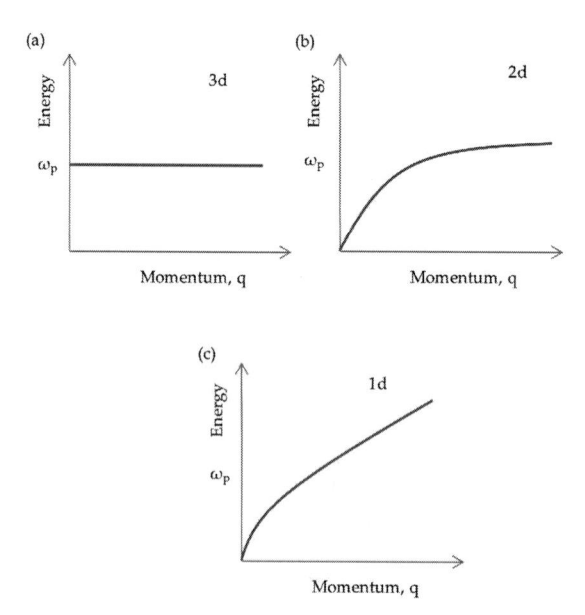

Figure 11.2 Dispersion relation of plasmons in reduced dimensions.

11.2 Optical Properties of Graphene

Graphene has interesting optical properties. Monoatomic thickness of graphene together with optical transparency and broadband absorption provide unique platform for electro-optical devices [10–13]. The interband and intraband electronic transitions are main mechanisms behind the broadband optical absorption of graphene [13]. Figure 11.3 shows the band structure of graphene and the dominant electronic transitions. Depending on the energy of light, the contributions of inter- and intraband transitions vary. In the visible spectrum, interband transition is dominant because the momentum of light is not sufficient to create electron–hole pair in the same band. Due to the linear band structure of the graphene, interband transitions yield broadband optical response with constant optical conductivity. By electrostatic doping one can tune the Fermi energy of graphene. The interband transition with energies less than the $2E_f$ can be blocked due to Pauli blocking resulting in transparent graphene (see Fig. 11.3). A recent work by Liu et al. demonstrated an electro-absorption modulator using graphene printed on an optical waveguide [2]. Their device operates at near-infrared (NIR) wavelengths. They achieved Fermi energy of up to 0.5 eV through a dielectric capacitor. Similarly, back-gated transistors have been implemented to make active photonic and plasmonic cavities operating at NIR [7, 14–16]. The common drawback of these devices is the dielectric breakdown limits the operation in the visible and NIR spectra. Different gating schemes such as electrolyte gating or ferroelectric gating, can reduce the operation wavelength down to 700 nm. In the electrolyte gating schemes, application of a gate voltage polarizes the electrolyte and forms an electrical double layer (EDL) near the graphene surface. Since the thickness of EDL is around a few nanometers, EDL generates very large electric fields and associated Fermi energies. The electrochemical potential window of the electrolyte limits the maximum bias voltage and the carrier concentration.

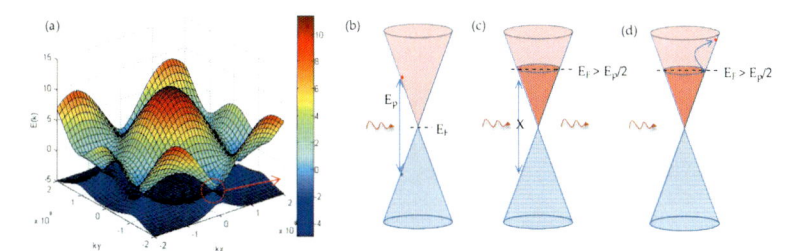

Figure 11.3 (a) Electronic band structure of graphene calculated using tight-binding method. (b) Interband electronic transition on the Dirac cone of graphene. (c) For doped graphene when the Fermi energy is larger than half of the excitation energy, interband transitions are blocked due to Pauli blocking. (d) Intraband electronic transitions, which define the optical properties of graphene at far-infrared spectrum. Graphene behaves like a Drude metal.

More recently, we have demonstrated another type of gating scheme using graphene supercapacitors to control optical properties of graphene in the visible spectrum [8]. Supercapacitors, ultracapacitors, or electric double-layer capacitors, store charges at the electric double-layers formed at the electrolyte-electrode interface. We used graphene supercapacitors as broadband optical modulators. The demonstrated device consists of two parallel graphene electrodes and electrolyte between them. The modulator has simple parallel plate geometry, yet, with a very efficient device operation. The graphene supercapacitors operate as electro-absorption modulators over a broad range of wavelengths from 450 nm to 2 μm under ambient conditions. We also studied various device geometries to increase the modulation amplitude. We were able to obtain modulation of 35% using a few-layer graphene with an ionic liquid electrolyte.

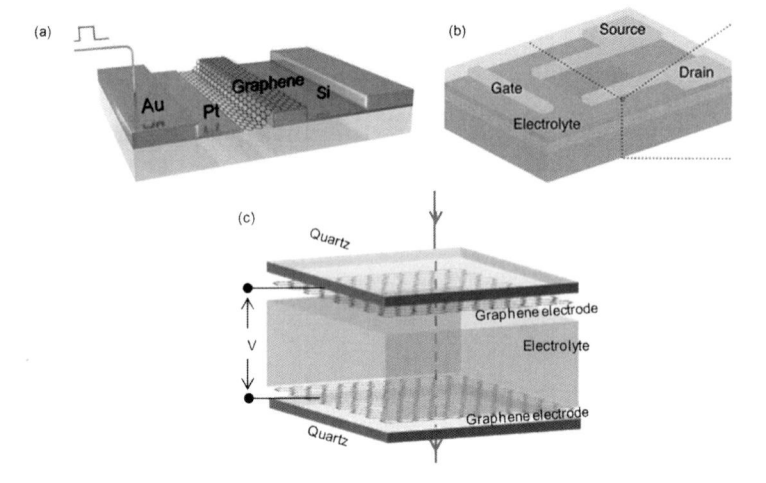

Figure 11.4 Tunable graphene-based optical devices: (a) Electro-optical modulator using a Si waveguide integrated with graphene electrode [2]. (b) Electrolyte gated graphene integrated on a photonic crystal [7]. (c) Broadband graphene modulator using a graphene supercapacitor [8].

At long wavelengths (far infrared and terahertz frequencies), the interband transitions are blocked due to unintentional doping. Therefore, the optical response is dominated by the low-energy intraband transition. These transitions yield gate-tunable Drude-like optical conductivities of

$$\sigma(\omega) = \frac{\sigma_{DC}}{1-i\omega\tau},\qquad(11.4)$$

where σ_{DC} is the low frequency conductivity, ω is the angular frequency of light, and τ is the electron scattering time. Active optical devices such as detectors, modulators, tunable antennas, and meta-materials working at far infrared and terahertz wavelengths have been demonstrated based on gate-tunable intraband transitions. A recent review paper by Wang et al. summarizes the current understanding of the optical properties of graphene [12].

11.3 Synthesis of Graphene

Recent progress on the synthesis of graphene on large-area substrates fosters new applications of the 2D crystal [17, 18]. Catalytic decomposition of hydrocarbons on transition metals provides a scalable route for graphene synthesis. Various methods such as chemical vapor deposition, surface segregation, and solid carbon source and ion implantation are alternative techniques for synthesis of graphene layers on metal surfaces. Nickel and copper are the most commonly used metal substrates for the graphene growth [18, 19]. Other metals such as Ru, Ir, Pt, Pd, and Co have been also used as a substrate for the chemical vapor deposition of graphene [20–23]. For plasmonic applications, ability to grow graphene on gold and silver are important. We developed a method to directly grow graphene layers on gold surface in various forms such as thin films, foils, and wires via chemical vapor deposition technique [24].

Formation of graphitic carbon on metal surfaces has been known for many years. For example, Ru and Pt are the earliest metals on which epitaxial graphitic carbon due to the segregation of carbon impurities has been observed [20, 22]. The inert nature of gold hinders the uses of gold surface for catalytic purposes. However, atomic step edges show some degree of catalytic activity. Adsorption of graphene-like polycyclic hydrocarbons across gold step edges has been reported. Gold nanoparticles, however, show some diameter-dependent catalytic activity for oxidation of carbon monoxide and alcohols. It is found that the smaller particles are more active for oxidation. Growing other types of graphitic materials such as single-walled (SWNT) and multi-walled carbon nanotubes (MWNT) over gold nanoparticles also shows catalytic activity of the gold nanoparticles. The growth mechanism of catalytic activity of Au nanoparticles during CVD process still remains unclear.

Figure 11.5 shows the schematic representation of the steps to be followed for the preparation of the gold foils, deposition, and transfer processes of the graphene layers on dielectric substrates. We used 25 µm-thick gold foils obtained by pressing high-purity gold plates (99.99%). In order to remove impurities and reconstruct the single crystalline gold surface, the gold foils were annealed with a hydrogen flame for 20 min before use. Owing to the fast heating and cooling rates, hydrogen flame annealing provides more complete crystallization of the gold foil than the furnace annealing. The result of this treatment was a polycrystalline gold surface partially (111) oriented, with a roughness lower than a few nanometers. After the annealing step, the gold foil is placed in a quartz chamber and the chamber is flushed with Ar gas for 5 min. The foil is heated up to 975°C under Ar and H_2 flow (240 sccm and 8 sccm, respectively). Methane gas with a flow rate of 10 sccm is sent to the chamber for 10 min. After stopping the methane flow, we cooled the chamber with a natural cooling rate of around 10°C/s. We have also developed a transfer printing method to transfer the graphene from the gold foil to insulating dielectric substrates. After the deposition, an elastomeric stamp PDMS (polydimethylsiloxane) is applied on the graphene-coated gold foil. The gold layer is etched by diluted gold etchant (type TFA,

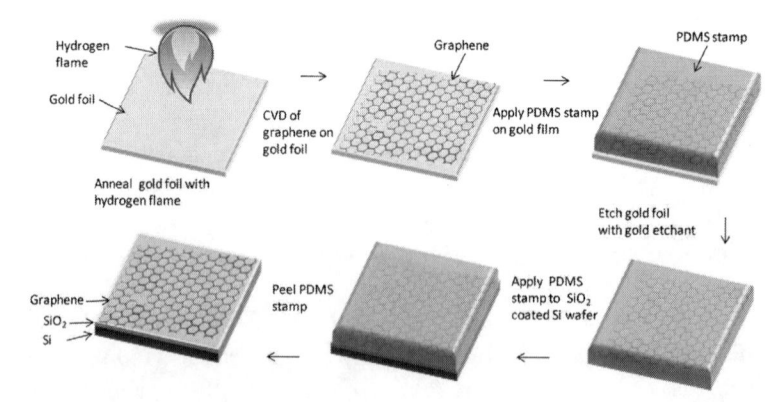

Figure 11.5 Synthesis and transfer printing steps of graphene on gold [24].

Transene Company Inc.). After complete etching of the gold foil, the graphene layer on PDMS is applied on a 100 nm SiO_2-coated silicon wafers. Peeling the PDMS releases the graphene on the dielectric surface.

Figure 11.6 shows Raman spectra of the graphene on gold foils. The inset in Fig. 11.6a shows the optical micrograph of the graphene on the gold foil. The grain boundaries and wrinkles on the polycrystalline gold surface are clearly seen.

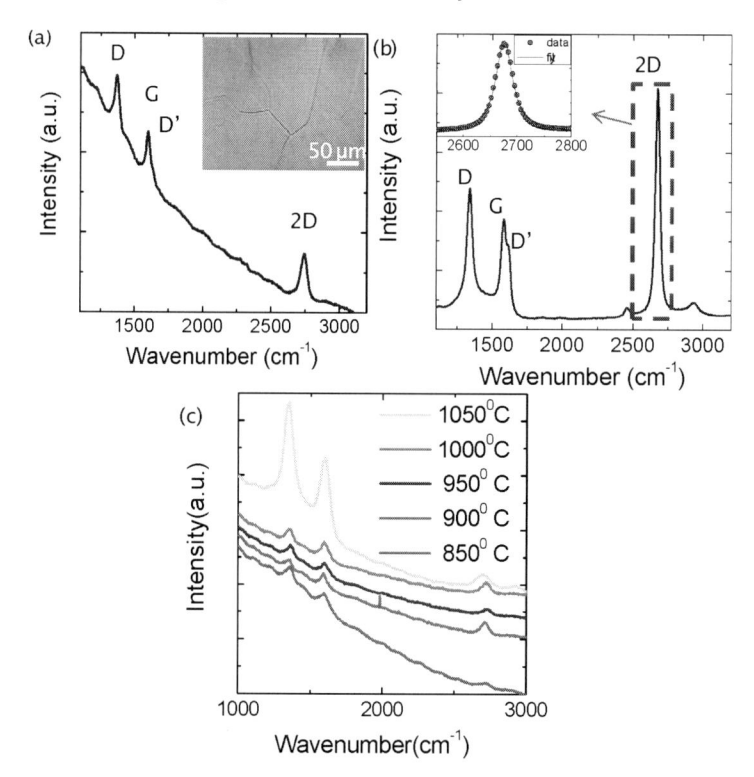

Figure 10.6 (a) Raman spectrum of the graphene as grown on gold surface. The inset shows the optical micrograph of the same area. (b) Raman spectrum of the graphene on SiO_2/Si substrate. The inset shows the zoomed spectrum of the 2D peak and a Lorentzian fit. The width of the Lorentzian is around 37 cm^{-1}. (c) Raman spectra of the graphene samples grown at a range of temperatures between 850 and 1050°C [24].

Raman spectroscopy provides clear fingerprints of graphene layers. Raman spectrum of the samples was measured using a confocal microRaman system (Horiba Jobin Yvon) in back-scattering geometry. A 532 nm diode laser is used as an excitation source and the Raman signal is collected by a cooled CCD camera. Figures 11.6b,c show the Raman spectra of the graphene as grown on gold and after its transfer onto silicon wafer. The expected Raman peaks of D, G, D′, and 2D can be seen on the graph. The position of D, G, D′, and 2D peaks are 1371, 1600, 1640, and 2743 cm^{-1} on gold and 1339, 1584, 1620, and 2676 cm^{-1} on SiO$_2$/Si substrate, respectively. There is a significant red shift after the transfer process, likely due to the release of compressive strain on the graphene film. The inset in Fig. 11.6c shows the zoomed 2D peak and the Lorentzian fit. The fit has a symmetric Lorentzian shape with a width of 37 cm^{-1}. The shape and the width of the 2D peak are a good indication of single layer graphene layer or noninteracting a few layers of graphene. The Raman D-band signal of graphene on gold is relatively more intense than the graphene grown on copper. This intense D-band is likely due to the large lattice mismatch between gold and graphene. The lattice constant of bulk graphite and hexagonally closed-packed gold surface are 2.46 Å and 2.88 Å, respectively. In order to find the optimum growth conditions, we performed Raman spectroscopy on samples grown at temperatures between 850 and 1050°C. At low temperatures, around 850°C, we do not observe a significant 2D peak. On the other hand, at very high temperatures, the D and G peaks are present, which indicate the formation of multilayer defective graphene layers. For the samples grown at temperatures between 850 and 1000°C, the Raman spectra look relatively the same. We have also studied the effect of the cooling process on the grown graphene layers. However, we do not observe any significant variation in the Raman spectra for cooling rates between 10 and 0.5°C/s.

11.4 Plasma Oscillations on Graphene–Metal Surface

In this section, we will summarize the application of graphene in plasmonics as an example for the functional coating [25, 26]. We studied the excitation SPPs on 50 nm-thick gold and silver surfaces coated with monolayer and multilayer graphene [27]. Dispersion curves of SPPs on graphene-coated metal surfaces reveal the essential feature of the SPP. This configuration provides an interesting platform to study adsorption of molecules on graphene layer from an aqueous solution. Particularly, we studied nonspecific adsorption of a serum albumin protein (BSA, bovine serum albumin) on a graphene layer using a microfluidic device integrated with a graphene functionalized SPR sensor.

The graphene layers grown on copper foils were transfer-printed on metal surfaces using a photoresist layer as a mechanical support. After coating the graphene layers with photoresist, we etched the metal substrate and place the polymer layer on the target metal surface. After reflowing the photoresist on the metal at 80°C, we removed the photoresist in acetone. Figure 11.7 shows Raman spectra of as grown graphene on copper foils and transfer printed graphene on silver and gold surfaces. The G and 2D Raman signals are clearly seen from the spectra. The intensity of D-band is very small, indicating high-quality graphene layer even after the transfer printing process. The intensity ratio of 2D to G band

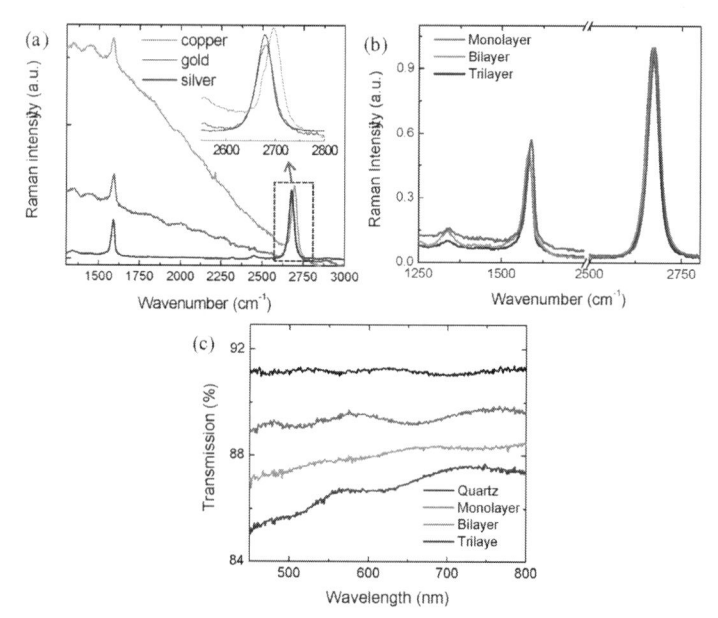

Figure 10.7 (a) Raman spectra of the graphene as grown on copper foil and transfer-printed on gold and silver surfaces. (b) Raman spectra and optical absorption of multilayer graphene layers transferred on quartz substrate. Each layer introduces around 2% optical absorption [27].

is around 1.7 and the Lorentzian linewidth of 2D band is 37 cm^{-1}. There is a slight red shift (~43 cm^{-1}) in the frequency of 2D peak after the transfer process likely because of release of the stress on the graphene. Raman spectra of multilayer graphene are shown in Fig. 11.7b. We do not observe any significant variation in the spectra, indicating that the graphene layers are not interacting. Furthermore, we obtained the optical transmission spectra of the multilayer films on a transparent quartz substrate (see Fig. 11.7c). The transmission spectra show 2% absorption for each graphene layer. This further supports the transfer process of single layer graphene.

After transferring graphene layers on metal surfaces, we characterized the resonance condition for excitation of surface SPPs on the metal surface coated by graphene layers. Figure 11.8a shows the schematic drawing of the experimental setup. The SPPs were exited through a prism in the Kretschmann configuration. This configuration overcomes the momentum mismatch between the excitation source and SPPs. A supercontinuum laser (Koheras-SuperK Versa) with an acousto-optic tunable filter was used as a tunable light source with a spectral width of ~1 nm. The incidence angle was controlled with a motorized rotary stage with an accuracy of ~0.01°. The polarization-dependent reflection from the metal surface is detected with a photodiode (Newport 818) connected to an amplifier. Figure 11.8b shows the reflectivity of the gold surface as a function of incidence angle before and after transfer printing graphene. The reflection goes to minima where the phase matching condition between the incidence light and SPP is satisfied. The phase matching condition is satisfied when the horizontal component of momentum of light matches the real part of the momentum of SPPs. This condition can be written as

$$k_0 n_p \sin(\theta_r) = \mathrm{Re}(k_{sp}) = \mathrm{Re}\left(\frac{2\pi}{\lambda}\sqrt{\frac{\varepsilon_m \varepsilon_d}{\varepsilon_m + \varepsilon_d}}\right), \quad (11.5)$$

where λ is the wavelength and k_0 is the free space wave vector of the excitation light, k_{sp} is the wave vector of SPPs, n_p is the refractive index of the prism, θ_r is the resonance angle, and ε_m and ε_d are the dielectric constants of graphene-coated metal surface and dielectric medium on the surface, respectively. The resonance angles for the bare gold surface and graphene functionalized gold surface are 44° and 45°, respectively. There is around one degree shift in the plasmon resonance angle. The area of transferred graphene layers is around 1 cm². The inset in Fig. 11.8b shows the reflectivity curves for these points. The resonance angle remains constant at around 45°, indicating a good level of coverage and homogeneity. Figures 11.8c,d show angular dispersion curves for graphene–gold and graphene–silver surfaces obtained by TM-polarized reflection measurements, respectively.

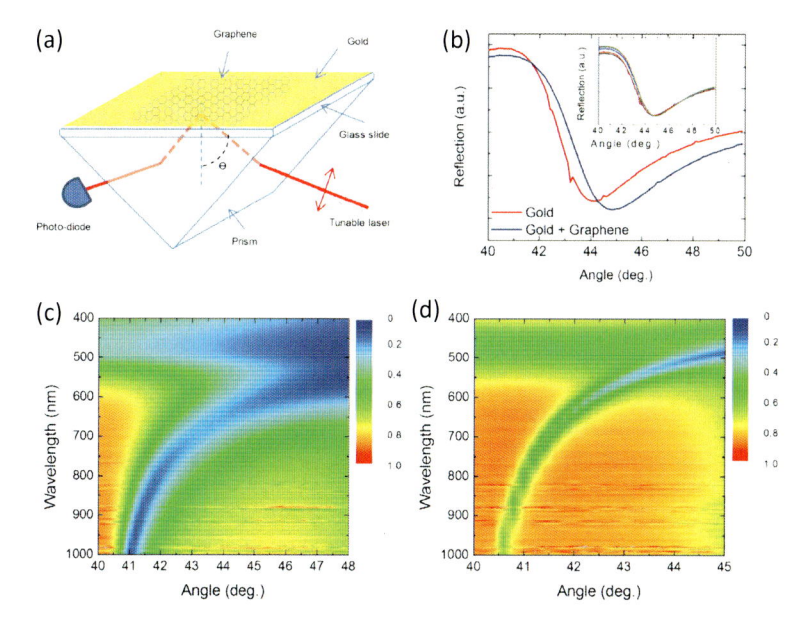

Figure 11.8 (a) Schematic representation of the Kretschmann configuration used to excite surface plasmon polariton on graphene–gold surface. (b) Surface plasmon resonance curves for gold surface before (red) and after (blue) transfer printing graphene. The wavelength of the incident light is 632 nm. The presence of the graphene introduces one degree shift in the resonance angle. The inset shows the SPR curves from six different points on the sample. (c, d) Angular dispersion curves for graphene–gold and graphene–silver surfaces obtained by TM-polarized reflection measurements [27].

11.5 Graphene Functionalized SPR Sensors

Coating gold and silver surfaces with graphene yields an atomically thin functionalization for plasmon sensors. Having the basic characterization of the SPPs on graphene–metal surface, we would like to demonstrate a plasmon resonance sensor based on graphene–metal surface. The fabricated sensor provides a unique setup to study the adsorption of organic molecules on graphene surface [28–31]. Understanding the adsorption of organic molecules from aqueous solution on carbon materials has significant importance for wide spectrum of applications such as analysis of drinking water, waste water treatment, and pharmaceutical industry [29]. Activated carbon is the most commonly used material for these types of applications. Due to the large surface area, graphene-based materials could be used for similar applications. The fabricated graphene based surface plasmon sensor is used to elucidate the adsorption of organic molecules on graphene surface. Owing to the evanescent nature of the electric field of the surface plasmons, SPR sensors provide surface specific information. The sensitivity of the sensor decreases due to the lower quality factor of the plasmon resonance. The presence of graphene coating enhances the nonspecific binding of organic molecules due to pi-stacking interactions. We do not expect any specificity; however, the surface can be functionalized for specific binding. Figure 11.9a shows the fabrication steps of the sensor. We fabricated an elastomeric microfluidic device to control the flow on the graphene-coated gold surface. The microfluidic device uses a 200 μm-thick spacer and an elastomeric cover on the top as a flow chamber. The fluid is injected into the flow chamber via polymer pipes by a gravity driven pressure difference. The rate of flow is kept constant at a rate of 1 mL/min adjusted by a flow regulator. The variation of SPR signal due to temperature and laser is reduced by a reference photodetector. The wave vector of the SP increases when the chamber is filled with water. In order to excite SP at water–gold interface, we used a prism with a high index of refraction (n_p = 1.78). Figure 11.9b shows the optical image of the fabricated plasmon resonance sensor. For water–gold interface, the shift of SPR angle due to the graphene layer is less than the case for air-gold interface due to the large effective index of plasmons on water–gold interface.

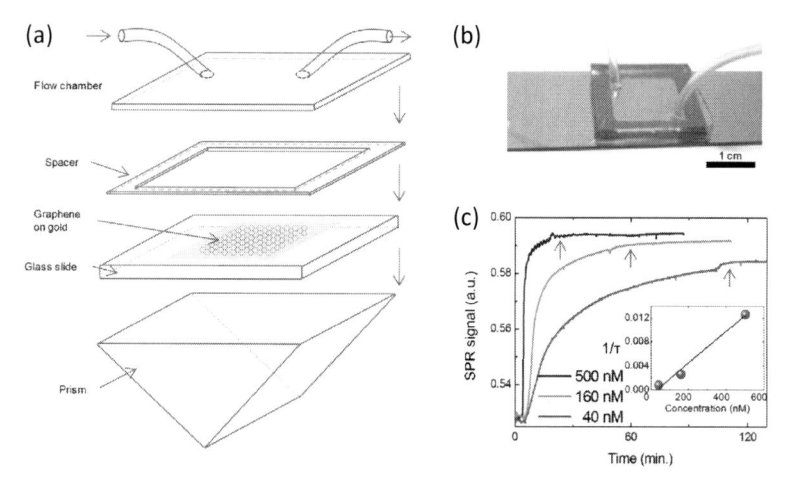

Figure 11.9 (a) Fabrication steps of the microfluidic flow chamber integrated with the graphene functionalized SPR sensor. (b) Optical image of the fabricated sensor. (c) Overlaid binding interaction plot for BSA (Bovine serum albumin) for concentration from 40 to 500 nM interacting with graphene layer. The inset shows the calculated time constant of the exponential saturation curves. The slop of the curve provides association constant k_a of 2.4 × 10^{-5} $M^{-1}s^{-1}$. The calculated dissociation constant is zero, indicating that BSA is kinetically stable on graphene layer [27].

The microfluidic flow chamber allows us to study the adsorption of molecules on graphene surface from an aqueous solution. Binding of molecules on a surface can be modeled with a single association-dissociation step. The change of surface concentration of the adsorbent can be described with the following differential equation:

$$\frac{dC}{dt} = k_a A(B_0 - C) - k_d C, \tag{11.6}$$

where C is the surface concentration of the adsorbent, k_a and k_d are the association and dissociation constants, A is the analyte concentration, and B_0 is the surface concentration of binding side. During association and dissociation phases, the solution of the differential equation gives exponential saturation and decay curves, respectively. Fitting the curves provides quantitative kinetic parameters. Using the fabricated SPR sensor we first examined adsorption of a serum albumin protein (BSA, bovine serum albumin) on graphene surface. The incidence angle was set to 49°, which provides the steepest slope in the reflectivity curve. First, we flowed deionized water (DI water) for 20 min and then introduced 500 nM BSA aqueous solution. The real-time reflection is shown in Fig. 11.9c. The SPR signal increases as the graphene surface is covered by the proteins. After 20 min. when the SPR signal becomes saturated, we stopped the flow of BSA solution and washed the chamber with DI water. We recorded the SPR signal for 60 min in order to see desorption of the protein from the graphene surface. We do not observe any significant change in the reflection for 60 min. This observation indicates a very small dissociation rate of BSA on graphene surface. To determine the association constant, we repeated the experiments for three different BSA concentrations. The inset in Fig. 11.9d shows the inverse of the time constant as a function of analyte concentration. The slope of the curve determines the association (k_a) constant of 2.4×10^{-5} $M^{-1}s^{-1}$. The intersection provides the dissociation constant (k_d), which is found to be very small indicating that BSA is kinetically stable on graphene surface. The association constant of BSA on graphene layer is three times smaller than the anti-BSA-coated surface (7.4×10^{-5} $M^{-1}s^{-1}$).

11.6 Graphene Passivation for SPR Sensors

Silver is an ideal material for plasmonic applications. Detrimental effects of chemicals on the silver surfaces dramatically reduce the performance of the SPR sensors. Passivation of the silver surface with inert dielectric layers can reduce the tarnishing. However, the dielectric coating significantly reduces the sensitivity of the sensor. An ultrathin passivation layer would be an ideal coating for SPR sensors. Therefore, atomically thin graphene coating is an ideal candidate and can prevent chemicals such as sulfur to degrade the silver surface. Reed et al. used CVD grown graphene to passivate the surface of silver nanostructures [32]. Figure 11.10 shows the schematic drawing and SEM images of silver nanostructures passivated by the graphene coating. With single layer graphene, atmospheric sulfur containing compounds were unable to degrade the surface of the silver. They showed that, after 30 days, graphene-passivated silver nanoantennas exhibit a much higher sensitivity over that of the bare Ag nanoantennas and two orders of magnitude improvement in peak width endurance.

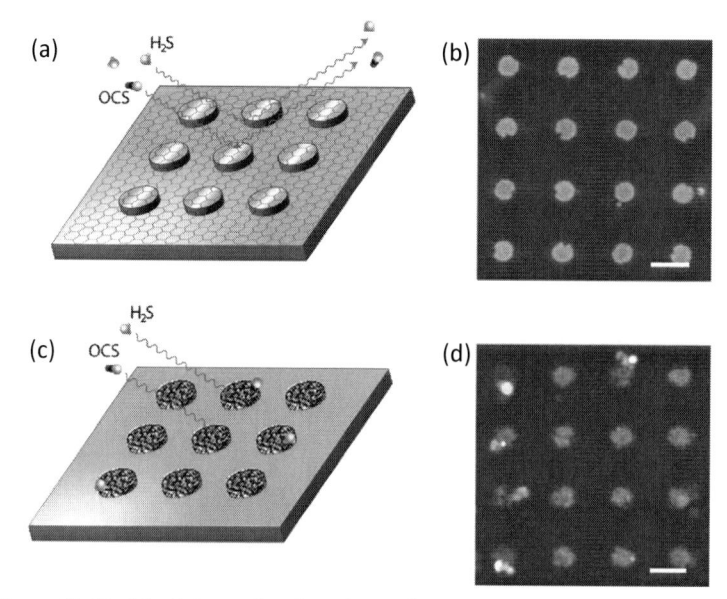

Figure 11.10 (a) Schematic drawing of Ag nanoantennas passivated by graphene. The graphene layer blocks the penetration of atmospheric H_2S and OCS. (b) Illustration of bare Ag nanoantennas. SEM images of (c) graphene-passivated Ag nanoantennas, and (d) bare Ag nanoantennas after 30 days; scale bars are 200 nm [31].

Graphene can also function as anticorrosion and antioxidation coating. Single-atom-thick graphene layer has impressive impermeability to gases. A recent work by Schriver et al. studied the long-term and short-term performance of graphene passivation on copper and Si surfaces [33]. They observed that graphene can protect Cu surface from thermal oxidation reasonably well at high temperatures over short time scales; however, graphene promotes galvanic corrosion on Cu surface over long-time scales like months. Figure 11.11 shows aging of graphene-coated copper surface for long time scale. They observed that within a month the graphene-coated Cu surface starts to tarnish nonuniformly. Within a few months, most regions of the graphene-covered Cu appear highly oxidized, with an oxide layer thickness of tens or hundreds of nanometers.

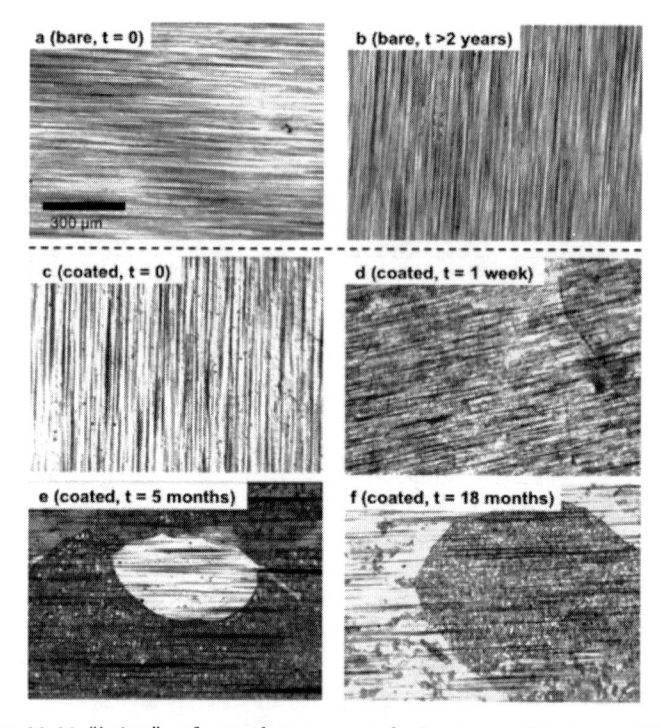

Figure 11.11 "Aging" of graphene-covered Cu in ambient conditions. Optical micrographs of "aged" bare and graphene-covered Cu foils, captured with the same lighting conditions and at the same magnification (scale bar in part (a) applies to all photos) [32].

11.7 Biomolecular Detection Using Graphene Functionalized SPR Sensors

The functionalized metal surfaces provide label-free detection of organic and biological molecules from aqueous solution with improved sensitivity and binding sites. Determining the concentration of organic or biological molecules in aqueous solution is crucial for pharmaceutical, waste water treatment, and drinking water applications. Graphene functionalized SPR sensors provide a unique device configuration to study adsorption of organic molecules from the aqueous solution on graphene surface [34–36]. Figure 11.12a shows detection of varying concentration of BSA protein in an aqueous solution. Bovine serum albumin proteins adsorb on the graphene surface and thereby affecting the plasmon resonance angle and wavelength. Therefore, the concentration of BSA can be easily determined. The graphene layer on the metal surface boosts the nonspecific binding of organic molecules because of the π-stacking interactions [37]. Since the graphene layer does not contain any functional group, no specific binding is expected. However, chemical functionalization of graphene surface has been extensively studied and binding sites to graphene surface can be added. In this way, the specificity of the adsorption can be controlled. For example, 1-pyrenebutanoic acid succinimidyl ester can be irreversibly adsorbed on the graphene surface, because highly aromatic pyrenyl group of 1-pyrenebutanoic acid succinimidyl ester interacts strongly with the basal plane of the graphene via π-stacking interaction [38].

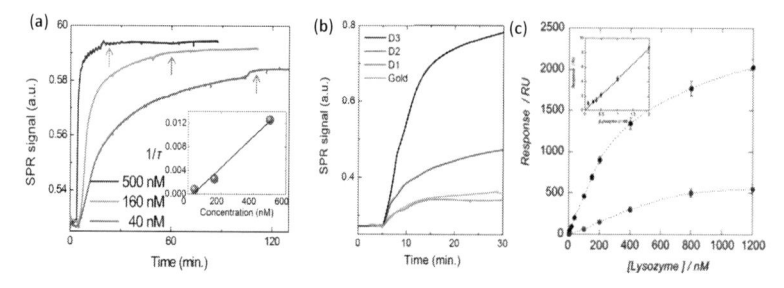

Figure 11.12 Responses of graphene, SWNT, and GO functionalized SPR sensors. (a) Graphene functionalized SPR sensor. Binding interaction plot for BSA proteins on the graphene functionalized gold surface. The inset shows the calculated time constant of the exponential saturation curve; (b) SPR sensor coated by SWNTs; (c) GO functionalized SPR sensor. Calibration curve for different concentrations of lysozyme on gold surface (blue) and gold/rGO (black). The inset shows the standard curve obtained at lower concentration [34–36].

11.8 Graphene Oxide Functionalization

Another carbon containing nanomaterial that has been recently used to functionalize SPR sensors are graphene oxide (GO), oxidized form of graphene [39]. Different from graphene, GO contains hydroxyl, carboxylic, epoxy, and even five- and six-membered ring lactol functional groups [40]. Therefore, GO is water soluble and additional functional groups like amine can be introduced by chemical modification of the graphene oxide. Since GO is water soluble, there are various ways of depositing it on metal surfaces. For instance, Subramanian et al. have recently shown electrophoretic deposition of GO on gold surfaces [39], Fig. 11.13. In order to electrophoretically deposit GO on metal surfaces, first, gold-based SPR surfaces are fabricated by depositing 5 nm-thick Ti film and followed by deposition of 50 nm-thick gold film. GO is synthesized from graphite powder by a modified Hummer's method [39]. GO water solution with a concentration of 0.5 mg/mL is prepared. The electrophoretic deposition is performed by using two electrode cell containing the aqueous GO solution. A DC voltage of 150 V is applied to these electrodes for a duration of 20 s. Platinum electrode acts as the cathode and gold electrode as the anode. The electrodes are separated by 1 cm apart and placed parallel to each other in the GO solution. After completing the deposition process, the gold electrode is washed with deionized water and blow-dried.

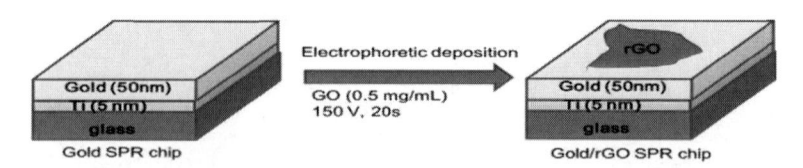

Figure 11.13 Schematic diagram showing electrophoretic deposition of graphene oxide flakes on a gold surface used for sensing applications [39].

11.9 Gate-Tunable Graphene Plasmonics

Many features of graphene can be tuned by electrostatic doping. For example, optical absorption, plasma frequency, and electron–phonon coupling in graphene can be tuned by electrostatic gating [4, 8, 14, 41–44]. These tunable physical properties provide possibilities for new optical devices based on tunable light–matter interactions. Graphene interacts with light through interband and intraband electronic transitions. The gate dependence of the optical response of graphene varies with the wavelength of light, for example, in the visible region; the interband electronic transitions are dominant and yield a constant optical conductivity.

By shifting the Fermi energy, one can block the interband transition via Pauli blocking and reduce the optical conductivity. Ability to tune the optical conductivity yields new optical modulators [2]. At far-infrared regime, however, graphene behaves like a Drude metal with tunable charge density [45]. By tuning the charge density on graphene, one can tune the plasma frequency. Two recent studies have shown the gate-tunable plasma oscillations on graphene. In the first study, Chen et al. launched and detected propagating optical plasmons in tapered graphene nanostructures using near-field scattering microscopy with infrared excitation light [5]. They used infrared laser and sharp AFM tip to lunch plasmons on graphene and detected the standing wave patterns. They also tuned the charge density on graphene with a back-gate electrode. They found that the extracted plasmon wavelength is very short—more than 40 times smaller than the wavelength of illumination. In the second work, Fei et al. measured the gate-tuning of graphene plasmons using infrared nanoimaging [4]. They integrated an infrared spectrometer with a sharp AFM tip and measured the scattering cross section of the tip. The scattering cross section provides the amplitude of the excited plasma waves. They have succeeded in altering both the amplitude and the wavelength of these plasmons by varying the gate voltage.

Active plasmonic devices operating in NIR spectrum are also possible. Kim et al. demonstrated a gate-tunable graphene–nanorod hybrid device by integrating gold nanorods in a graphene transistor [46]. Figure 11.14 shows the used device geometry and the SEM image of the gold-nanorods. They used ionic liquid as an electrolyte to tune the electronic transition of graphene. Ionic liquids provide an efficient gating with charge densities larger than 10^{13} $1/cm^2$ and Fermi energies of 1 eV. By electrostatic gating of graphene, they blocked interband optical transitions of graphene; they are able to significantly modulate both the resonance frequency and quality factor of gold nanorod plasmon. Their analysis showed that the plasmon–graphene coupling for graphene–nanorod structure is unexpectedly very high and they speculate that even a single electron in graphene at the plasmonic hotspot could yield an observable effect on plasmon resonance. Such hybrid graphene–nanometallic structure provides a powerful way for active plasmonic devices.

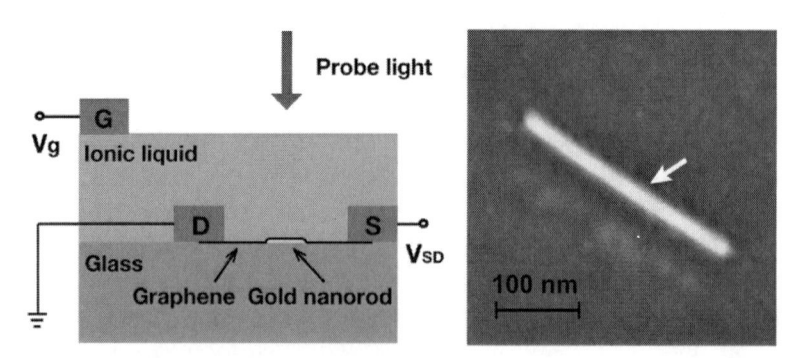

Figure 11.14 (a) Electrolyte-gated graphene–gold nanorod hybrid structure. A top electrolyte gate with ionic liquid is used to control the charge density and optical transitions on graphene. (b) A SEM image of the single gold nanorod covered by graphene [46].

Patterning of graphene into nanometer sized ribbons or discs provide an additional level of control on the plasmonic response of graphene for new types of tunable metamaterials. Figure 11.15 shows various approaches to control the plasma resonance of graphene by patterning. Ju et al. explored plasmon excitations in engineered graphene micro-ribbon arrays [42]. They showed that graphene plasmon resonances can be tuned over a broad terahertz frequency range by varying the micro-ribbon width and in situ electrostatic doping. Patterning graphene into ribbons yields remarkably large oscillator strengths, and enhances room-temperature optical absorption peaks. Their results are the first for realization of tunable graphene-based terahertz metamaterials. Another approach to control plasma resonance of graphene is to use graphene/insulator stacks, which are formed by depositing alternating wafer-scale graphene sheets and thin insulating layers [47]. Yan et al. have patterned the graphene/insulator stacks into photonic-crystal-like structures. They showed that the plasmon in such stacks is non-classical due to direct consequence of the unique carrier density scaling law of the plasmonic resonance of Dirac fermions. Their work could lead to the development of new types of photonic devices such as detectors, modulators, and three-dimensional metamaterial systems. However, graphene has inherently weak optical absorption at visible and infrared wavelengths, which limits realistic applications. To overcome this challenge Fang et al. introduced nanopatterning of a graphene layer into an array of closely packed graphene nanodisks [48]. The optical absorption of these arrays can be increased from less than 3% to 30% in the infrared region of the spectrum. Furthermore, they studied the dependence of the enhanced absorption on nanodisk size and interparticle spacing. By integrating the patterned disk in transistor geometry, they were able to tune plasma frequency with the gate voltage. These active plasmonic devices highlight-possible ways for realization of novel devices using gate-tunable nature of graphene.

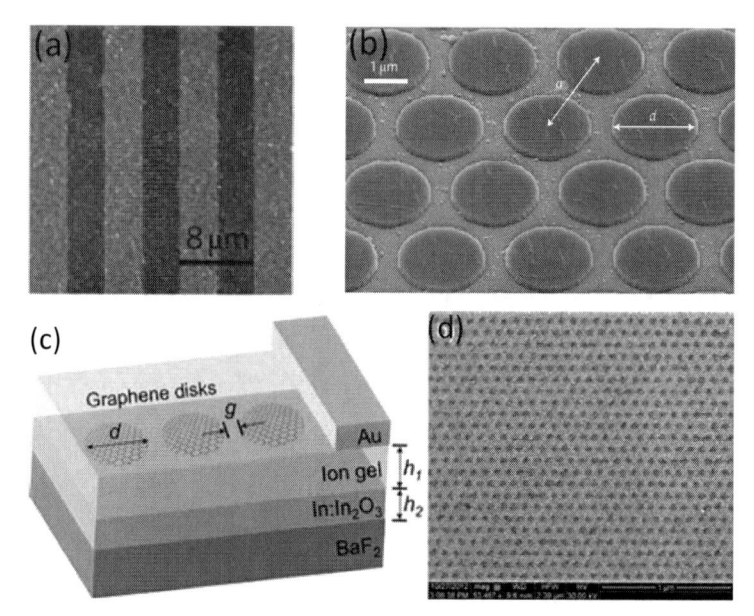

Figure 11.15 Controlling the plasma frequency of graphene by patterning graphene into (a) ribbons [42], (b) graphene/insulator stacks [47], or (c, d) nanodisks [48]. Electrostatic doping of these patterned graphene introduces an additional control on the plasma frequency.

11.10 Conclusion

As a conclusion, the discovery of the two-dimensional crystal of carbon stimulated enormous interest to fabricate new functional devices ranging from electronics to plasmonics. The unique band structure and atomic thickness of graphene yield novel electrical, optical, and chemical properties that are suitable for new applications. For plasmonic applications, graphene provides three unique advantages: (1) a tunable platform to study plasma oscillations in 2D, (2) an active material for tunable plasmonics, and (3) ultrathin functional coating for plasmon sensors. In this chapter, we summarized the fundamental aspects of plasmons on graphene and their applications ranging from surface plasmon sensors to active plasmonic devices. We anticipate that the integration of gate-tunable nature of graphene with the existing plasmonic devices will generate new perspectives in the field of plasmonics.

References

1. Konstantinov, D., Kono, K. (2010). Photon-induced vanishing of magnetoconductance in 2D electrons on liquid helium. *Phys. Rev. Lett.*, **105**, 226801.
2. Zhang, X., et al. (2011). A graphene-based broadband optical modulator. *Nature*, **474**, 64.
3. Allen, S. J., Tsui, D. C., Logan, R. A. (1977). Observation of the two-dimensional plasmon in silicon inversion layers. *Phys. Rev. Lett.*, **38**, 980.
4. Fei, Z., et al. (2012). Gate-tuning of graphene plasmons revealed by infrared nano-imaging. *Nature*, **487**, 82.
5. Chen, J. N., et al. (2012). Optical nano-imaging of gate-tunable graphene plasmons. *Nature*, **487**, 77.
6. Grimes, C. C., Adams, G. (1976). Observation of two-dimensional plasmons and electron-ripplon scattering in a sheet of electrons on liquid helium. *Phys. Rev. Lett.*, **36**, 145.
7. Gan, X. T., et al. (2013). High-contrast electrooptic modulation of a photonic crystal nanocavity by electrical gating of graphene. *Nano Lett.*, **13**, 691.
8. Polat, E. O., Kocabas, C. (2013). Broadband optical modulators based on graphene supercapacitors. *Nano Lett.*, **13**, 5851.
9. Goni, A. R., et al. (1991). One-dimensional plasmon dispersion and dispersionless intersubband excitations in GaAs quantum wires. *Phys. Rev. Lett.*, **67**, 3298.
10. Bonaccorso, F., Sun, Z., Hasan, T., Ferrari, A. C. (2010). Graphene photonics and optoelectronics. *Nature Photon.*, **4**, 611.

11. Li, Z. Q., et al. (2008). Dirac charge dynamics in graphene by infrared spectroscopy. *Nat. Phys.*, **4**, 532.
12. Wang, F., et al. (2008). Gate-variable optical transitions in graphene. *Science*, **320**, 206.
13. Mak, K. F., Ju, L., Wang, F., Heinz, T. F. (2012). Optical spectroscopy of graphene: From the far infrared to the ultraviolet. *Solid State Commun.*, **152**, 1341.
14. Koppens, F. H. L., Chang, D. E., de Abajo, F. J. G. (2011). Graphene plasmonics: A platform for strong light–matter interactions. *Nano Lett.*, **11**, 3370.
15. Liu, M., Yin, X. B., Zhang, X. (2012). Double-layer graphene optical modulator. *Nano Lett.*, **12**, 1482.
16. Yao, Y., et al. (2013). Broad electrical tuning of graphene-loaded plasmonic antennas. *Nano Lett.*, **13**, 1257.
17. Nandamuri, G., Roumimov, S., Solanki, R. Chemical vapor deposition of graphene films. *Nanotechnology*, **21**, 145604.
18. Li, X. S., et al. (2009). Large-area synthesis of high-quality and uniform graphene films on copper foils. *Science*, **324**, 1312.
19. Reina, A., et al. (2009). Large area, few-layer graphene films on arbitrary substrates by chemical vapor deposition. *Nano Lett.*, **9**, 30.
20. Sutter, E., Albrecht, P., Sutter, P. (2009). Graphene growth on polycrystalline Ru thin films. *Appl. Phys. Lett.*, **95**, 133109.
21. Coraux, J., et al. (2009). Growth of graphene on Ir(111). *New J. Phys.*, **11**, 023006.
22. Sutter, P., Sadowski, J. T., Sutter, E. (2009). Graphene on Pt(111): Growth and substrate interaction. *Phys. Rev. B*, **80**, 245411.
23. Varykhalov, A., Rader, O. (2009). Graphene grown on Co(0001). films and islands: Electronic structure and its precise magnetization dependence. *Phys. Rev. B*, **80**, 035437.
24. Oznuluer, T., et al. (2011). Synthesis of graphene on gold. *Appl. Phys. Lett.*, **98**, 183101.
25. Wu, L., Chu, H. S., Koh, W. S., Li, E. P. Highly sensitive graphene biosensors based on surface plasmon resonance. *Opt. Express*, **18**, 14395 (2010).
26. Choi, S. H., Kim, Y. L., Byun, K. M. (2011). Graphene-on-silver substrates for sensitive surface plasmon resonance imaging biosensors. *Opt. Express*, **19**, 458.

27. Salihoglu, O., Balci, S., Kocabas, C. (2012). Plasmon-polaritons on graphene-metal surface and their use in biosensors. *Appl. Phys. Lett.*, **100**, 213110.

28. Chen, J. Y., Chen, W., Zhu, D. (2008). Adsorption of nonionic aromatic compounds to single-walled carbon nanotubes: Effects of aqueous solution chemistry. *Environ. Sci. Technol.*, **42**, 7225.

29. Moreno-Castilla, C. (2004). Adsorption of organic molecules from aqueous solutions on carbon materials. *Carbon*, **42**, 83.

30. Horing, N. J. M. (2009). Coupling of graphene and surface plasmons. *Phys. Rev. B*, **80**, 193401.

31. Oznuluer, T., et al. (2011). Synthesis of graphene on gold. *Appl. Phys. Lett.*, **98**, 183101.

32. Reed, J. C., Zhu, H., Zhu, A. Y., Li, C., Cubukcu, E. (2012). Graphene-enabled silver nanoantenna sensors. *Nano Lett.*, **12**, 4090.

33. Schriver, M., et al. (2013). Graphene as a long-term metal oxidation barrier: Worse than nothing. *ACS Nano*, **7**, 5763.

34. Reckinger, N., Vlad, A., Melinte, S., Colomer, J. F., Sarrazin, M. (2013). Graphene-coated holey metal films: Tunable molecular sensing by surface plasmon resonance. *Appl. Phys. Lett.*, **102**, 211108.

35. Wang, P., Liang, O., Zhang, W., Schroeder, T., Xie, Y. H. (2013). Ultra-sensitive graphene-plasmonic hybrid platform for label-free detection. *Adv. Mater.*, **25**, 4918.

36. Kakenov, N., Balci, O., Balci, S., Kocabas, C. (2012). Probing molecular interactions on carbon nanotube surfaces using surface plasmon resonance sensors. *Appl. Phys. Lett.*, **101**, 223114.

37. Yang, J. W., Chen, X. R., Song, B. (2013). Interaction of graphene-on-Al(111). composite with D-glucopyranose and its application in biodetection. *J. Phys. Chem. C*, **117**, 8475.

38. Chen, R. J., Zhang, Y. G., Wang, D. W., Dai, H. J. (2001). Noncovalent sidewall functionalization of single-walled carbon nanotubes for protein immobilization. *J. Am. Chem. Soc.*, **123**, 3838.

39. Subramanian, P., et al. (2013). Lysozyme detection on aptamer functionalized graphene-coated SPR interfaces. *Biosens. Bioelectrons*, **50**, 239.

40. Gao, W., Alemany, L. B., Ci, L. J., Ajayan, P. M. (2009). New insights into the structure and reduction of graphite oxide. *Nat. Chem.*, **1**, 403.

41. Grigorenko, A. N., Polini, M., Novoselov, K. S. (2012). Graphene plasmonics. *Nature Photon.*, **6**, 749.

42. Ju, L., et al. (2011). Graphene plasmonics for tunable terahertz metamaterials. *Nature Nanotechnol.*, **6**, 630.
43. Fei, Z., et al. (2011). Infrared nanoscopy of Dirac plasmons at the graphene–SiO$_2$ interface. *Nano Lett.*, **11**, 4701.
44. Das, A., et al. (2008). Monitoring dopants by Raman scattering in an electrochemically top-gated graphene transistor. *Nat. Nanotechnol*, **3**, 210.
45. Horng, J., et al. (2011). Drude conductivity of Dirac fermions in graphene. *Phys. Rev. B*, **83**, 165113.
46. Kim, J., et al. (2012). Electrical control of optical plasmon resonance with graphene. *Nano Lett.*, **12**, 5598.
47. Yan, H. G., et al. (2012). Tunable infrared plasmonic devices using graphene/insulator stacks. *Nature Nanotechnol.*, **7**, 330.
48. Fang, Z. Y., et al. (2014). Active tunable 002. Absorption enhancement with graphene nanodisk arrays. *Nano Lett.*, **14**, 299.

Chapter 12

SPR: An Industrial Point of View

Iban Larroulet

SENSIA, Poligono Aranguren 9, Apartado Correos 171,
20180 OIARTZUN, Gipuzkoa, Spain

ilarroulet@seimcc.com

12.1 Introduction

Characterization of therapeutic drugs, monoclonal antibodies, lectins, and other biological components is one of the required steps during the regulatory approval process. The manner and strength in which antibodies interact and bind with target molecules for example is of primary importance and binding assays are used to define biological activity and immunological properties. The surface plasmon resonance (SPR) technology has become a leading technology in the field of real-time observation of biomolecular interactions (Fig. 12.1). SPR allows efficient selection, optimization, and characterization of biological components and helps in an easy and rapid manner in the comprehension of molecular mechanisms and structures-function relationships. The advantage of SPR is that it is a label-free analysis technique. However, its acknowledgment as a technique for some types of analysis remains controversial, since it lacks recognition regulatory-wise, in the legal environment

---

Introduction to Plasmonics: Advances and Applications
Edited by Sabine Szunerits and Rabah Boukherroub
Copyright © 2015 Pan Stanford Publishing Pte. Ltd.
ISBN 978-981-4613-12-5 (Hardcover), 978-981-4613-13-2 (eBook)
www.panstanford.com

of the homologation of assays. The developed SPR instruments have generally been designed to target the research market. The SPR market is a small niche market of around €200 million per year, corresponding to an installed park of approximately 3000 machines worldwide, to be compared with the world biochip market, which weighs $3.6 billion (source: BCC Research—Global biochip Markets 2009. SDI Strategic Direction International 2009). SPR is thus a nano droplet in a wide ocean of other analysis approaches.

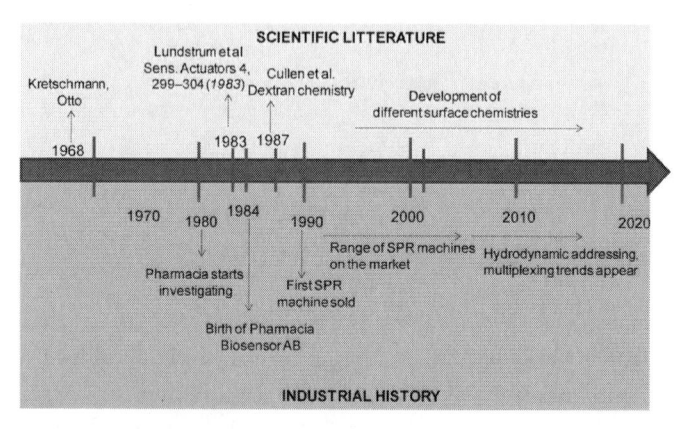

Figure 12.1 History of SPR at a glance.

This market did not exist 30 years ago. The first commercial device introduced in the market was in 1990, by Pharmacia Biosensor AB. Few companies have been able to master all the technologies included in a SPR device, and one company was able to get into a monopolistic position until some years ago, with a market share superior to 85% at some stages. Nowadays, a dozen of actors are present in this market, with three/four of them having more presence than the others. One of the most important factors that have contributed to the growth of the SPR market has been the development of the market of antibodies: the technique could find applications thanks to them.

12.2 Companies

After the first SPR instrument was introduced 1990 by Pharmacia Biosensor AB under the name of Biacore, a range of analytical systems conceived around the same principle have been built. The commercial market for SPR is competitive and SPR instruments are available from various manufacturers.

Among all the relevant actors of the market (if I forget any, I ask to be forgiven, the purpose here being to show, although partially, the diversity of solutions that have been developed), Biacore dominates the SPR market and has set the standards serving as reference for all other instrument manufacturers. The BI-SPR of Biosensing Instruments (BI, Temple, AZ, USA) uses an innovative method to detect the SPR angle, key to the high performance of the instrument. The proteomic processor of Lumera (Bohell, WA, USA) uses SPR microscopy in which a beam of light is directed onto a spot of a microarray allowing the monitoring of thousands of spots simultaneously.

The optical setup of the GenOptics SPRi-Plex and SPRi-Lab system (now part of HORIBA Scientific) allows the spotting of several hundreds of different molecules on the biochip and makes advantage of the multiplexing capabilities for a rapid screening (>100 sensograms in parallel) of the molecules.

Graffinity Pharmaceuticals GmbH (Heidelberg, Germany) has developed the Plasmon Imager for the discovering of small molecular hit compounds. The platform is used to find novel small molecules in drug discovery and chemical genomics approaches. In the 1990s, the IBIS I and II systems by IBIs Technologies (Hengelo, The Netherlands) were introduced in the market. In 2007, the IBIs Technologies launched a new advanced imaging SPR instrument with a patent pending for angle scanning imaging technology.

The production of the electrochemical SPR instrument by Metrohm Autolab, unique of its kind, was stopped in October 2011. The decision was made due to an ongoing effort to refocus the activities of the company on its core activity, which is the development and production of electrochemical instrumentation.

The first Basque initiative to establish a leading company in the field of analytical instrumentation based on SPR for life science laboratories and environmental measurements is SENSIA. Initially SENSIA spun-off in 2004 from the technologies developed at the Biosensors Group within the National Microelectronic Center (CNM) of the Spanish National Research Council (CSIC). Since 2008, SENSIA has been following an innovative process of machine development and is present on the market with a fully automated, bi-channel SPR device, delivering high sensitivity, transportability, robustness, and extreme stability; this device is called the Indicator (Fig. 12.2). One of its utmost innovative features is its versatility in terms of surfaces for biosensing that can be used in it, with no manual goniometer, and no modification of its optical platform: golden prisms coated with several substances such as graphene, silica gel, for instance, can be used within the same system, indifferently. This is how and why we started working with Sabine. MONDRAGON Componentes acquired share in SENSIA in fall 2007 and took total control of the company in 2012.

Figure 12.2 Indicator, SPR instrumentation from SENSIA.

As someone living in the industrial world, far from academics, I want to underline that living the adventure from a lab prototype to an industrial product that has gone through all the field tests through to technical maturity, the birth of a new SPR machine, bringing new features in the market is a total adventure. In a mature market where standards of quality are well established, the "law of the plane" is a strict marketing law you are submitted to. It says: "*If a product does not take off well, it will never fly high.*" It is, thus, better to postpone the introduction of a product to the market if you are not totally sure of its success. The market leader faces other problems and is in less need to submit to this strict rule.

The adventure of SENSIA in the development of a new SPR machine has been based upon the pedagogy of failure. In any well-established and structured industrial group, when the technological portfolio of the product is mastered, a new complex product requires at least two years of time from the decision to move forward to its launching on the market. If the technological portfolio is not mastered, this time soars up to more than five years. These are long cycles that everybody has to be aware of before making a decision, with all its consequences.

Research and development is an investment in which crucial decisions that condition the future of a company are being made: miss your investment and your company will be in danger. A friend of mine told me once, "Iban, if you think of something that seems new to you, please be aware that someone else has thought of it before you." A lot of competition takes place between the different R&D programs worldwide, rather than organizing where possible to avoid money spill and wasted resources. The world is full of good thinkers, and our educational systems build them: A good brain is like a V12 engine. But from there on, how many actually build up the mechanical systems to transmit the generated power from the engine to the wheels, put the wheels on the course, build up the car, find the road, and pilot it all?

Move from the idea to the product, from the lab to the market: that is one evolution in which the R&D Realm has to enter. It means moving into new dimensions, opening them, giving a sense of utility to what money making can be, finding a function developed by a technology that will meet the customer's need. Research and develop, and if possible do it in the bending and ex-centered downstream attitude toward application for market purposes.

12.3 Future Trends

SPR is like cooking: Dear reader, if you are reading this book, it is because you are a cooker. And to cook, you have the kitchen, i.e., the machine, and the receipt, i.e., the chemistry. In terms of "kitchen," future trends of the SPR market aim at more sensitivity, since the commonly reached refractive indexes are a limit to applications in some cases. They also aim at the multiplication of the number of simultaneously performed trials, using multiplexing techniques. Miniaturization and the consequent reliability of the developed solutions is still a challenge for the devices to come.

In terms of "receipt," future trends of the SPR market aim at new antibodies allowing new applications, and new surfaces delivering specific binding properties and/or more sensitivity. The ultimate goal is the ready-to-use chip, with the antibody fixed on it, and all the parameters of conservation totally mastered, for the non-denaturation of the protein. There is still some work in front of us.

The future of SPR companies depends on the development of the market of SPR: if this development is not sufficient to make this market more widespread and bigger, in a word: wealthier, the SPR companies will be integrated in bigger conglomerates in a strategy of market share or a strategy of vertical integration. It already happened with the leader (Biacore) and will happen again.

Sir Winston Churchill said, "Success in the ability to go from one failure to another with no loss of enthusiasm." Wisdom is universal.

Index

anisotropy 97–98
annealing, thermal 220, 232
anti-Stokes scattering 278–280
AZO, see Al-doped ZnO

bacteria 54–55
BCL, see block copolymer lithography
biomolecules 41, 52, 214, 220, 256, 260
bioreceptors 36–37, 41, 43, 47
biosensing 22, 43, 91, 119–120, 129, 350
biosensors, optical 22–23, 103
block copolymer lithography (BCL) 208–210
block copolymers 231
bovine serum albumin (BSA) 47, 329, 333–334, 336–337
BSA, see bovine serum albumin

carboxymethylated dextran layers 37, 39
cell membranes 89, 100, 112
cell-penetrating peptides (CPPs) 103–104, 158
chemical vapor deposition 325
CMOs, see conducting metal oxides
colloidal gold 252–253
colorimetric sensors 253, 256
conducting metal oxides (CMOs) 144, 146–147, 149, 151–152, 157, 159–160, 162

conductivity, optical 323–324, 338–339
copper 226, 251, 282, 288, 325, 328, 335
  graphene-covered 335–336
copper foils 329
core–shell nanoparticles 224
coupled plasmonic systems 183, 185
CPPs, see cell-penetrating peptides

denaturant molecules 70
dielectric films 186–187, 189, 191
dielectric medium 2, 4, 6, 9–10, 21, 24, 120–122, 126, 128–129, 221, 289, 331
dipole 183, 290–292
dipole emission 287, 290, 292
dipole radiation enhancement 291–292, 294
DNA 41, 67–70, 72, 75, 84–85, 260, 265
DNA microarrays 62
Al-doped ZnO (AZO) 160–161
Drude model 6–7, 144–145, 151–152, 157, 163, 174–175, 178

EBL, see electron beam lithography
EDL, see electrical double layer
EDSA, see evaporation-driven self-assembly

electrical double layer (EDL) 323
electrochemistry 43, 62, 67
electromagnetic field intensities 93–94
electromagnetic fields 5–6, 9, 12, 14, 22, 31, 92–93, 147
electromagnetic mechanism 30, 91, 146, 284–285, 287, 313
electron beam lithography (EBL) 203–204, 207, 219
electronic mobility 153, 155–156
electronic states 279–280, 284, 286
electronic transitions 282, 286, 322–323, 338–339
electrophoretic deposition 53, 225, 338
electrostatic approximation 287–289
ENFOL, see evanescent near-field optical lithography
evanescent near-field optical lithography (ENFOL) 203
evaporation-driven self-assembly (EDSA) 228–230
extinction coefficient 92, 96–98

FDTD, see finite difference time domain
FIB, see focused ion beam
finite difference time domain (FDTD) 186, 292
fluorescence 23, 31, 61–62, 122–124, 132, 298
fluorescence intensities 124, 131–132, 135
focused ion beam (FIB) 203, 210–211, 219
formamide 70–72
free electron conductors 149–150

full width half maximum (FWHM) 27, 84, 161, 194
FWHM, see full width half maximum

G-protein coupled receptors (GPCRs) 89–90, 100, 104–105, 107–108
G-proteins 89, 104–105, 107–109
  subtypes 107–108
gate-tunable graphene plasmonics 338–339
glycans 41, 49, 53–55
gold evaporation 220
gold films 43, 127–128, 130–131, 185, 216
gold foils 326–327
gold nanoparticles 121, 189, 227, 232, 253, 255, 258, 260–267, 290, 326
  synthesis 252, 254
gold nanorods 188, 204, 214, 254, 296, 339
gold nanostars 268–269
gold salts 221, 223, 252, 254
gold surface 37, 40, 67, 69, 325, 327, 329–331, 337–338
  polycrystalline 326–327
gold surface atoms 257
GPCRs, see G-protein coupled receptors
graphene 52, 236, 319–323, 325–333, 335–342, 350
  multilayer 328, 330
  optical properties of 323–325
  synthesis of 325, 327
graphene-based plasmonics 319–342
graphene coating 332, 334
graphene functionalized SPR sensors 332–333

graphene layers 52, 325–327, 329–330, 332–334, 337
graphene oxide 53, 337–338
graphene oxide functionalization 337
graphene plasmonics 236
graphene plasmons 339
graphene supercapacitors 324
graphene surface 320, 323, 332–334, 336–337
graphene–metal surface 328–329, 331–332
graphene–silver surfaces 331

high mobility conducting metal oxides 161
hydrogel 120, 132–134
hydrogel film 133–134
hydrogel matrix 133, 136
hydrogen peroxide 268–269

IMI, see insulator-metal-insulator
indium tin oxide (ITO) 52, 113, 144–147, 149, 151, 153–154, 157–160, 194, 236
infrared plasmonic sensors 207
  high-throughput fabrication of 207
infrared surface plasmon resonance 143–164
insulator-metal-insulator (IMI) 17
internal reflecting element (IRE) 146–150
IRE, see internal reflecting element
ITO, see indium tin oxide

Kretschmann configuration 20–21, 63, 123, 147, 330–331

lamellar SPR structures 47, 49, 51, 53
Langmuir–Blodgett technique 226–227, 231
laser ablation 212
layer-by-layer (LbL) 227
LbL, see layer-by-layer
LbL technique 227–228
lectins 49, 54, 347
Lens culinaris 49–50, 54
ligand binding 106, 109
lipid bilayer membranes 95, 97
lipid membranes 90, 96, 99, 102, 104–105, 111, 113
lipid model systems 89–90
lipids 96, 100, 103–104, 107–108, 111–113, 263
  anionic 111–112
liquid helium 320–321
lithography 170, 200, 203, 214
  block copolymer 208–210
  multilevel interference 205–206
  nanosphere 208–209
  nanostencil 206–208
lithography techniques 201, 205, 209
localized surface plasmon (LSP) 31, 170, 174, 176–177, 186–187, 189, 191, 193, 195, 209
localized surface plasmon resonance (LSPR) 31, 144–146, 151–152, 157, 169–182, 184, 186, 188, 190–192, 194, 196, 220, 225, 254, 287
long-range surface plasmon (LRSP) 120, 126–131, 134–135, 137
long-range surface plasmon fluorescence spectroscopy 126–131
LRSP, see long-range surface plasmon

LRSP excitation 127, 129, 131
LRSP modes 128–130, 133, 136–137
LSP, *see* localized surface plasmon
LSP resonances 220–222, 234–235
LSPR, *see* localized surface plasmon resonance

melting curves analysis 62, 78, 85
metal deposition 201, 210, 214, 218
metal dielectric function 287–288
metal-insulator-metal (MIM) 17, 19
metal nanoparticles 171, 174–175, 200, 210, 287, 295
metal-oxide-semiconductor (MOS) 320
metallic gold 252
metallic nanoparticles 187, 230
metallic nanostructures 236, 276, 282, 284
mid-infrared surface plasmon polaritons 159
MIM, *see* metal-insulator-metal
molecular polarizability 97–98, 277–279
molecular recognition 263, 265, 267
molecules
 bio-recognition 23
 biological 52, 336
 thiolated 41–42
monolayer protected clusters (MPCs) 254
monolayers
 ordered nanoparticle 229
 self-assembled 37, 40, 214, 225, 256

Monte Carlo simulation 301–302, 304, 307–308, 310, 312
MOS, *see* metal-oxide-semiconductor
MPCs, *see* monolayer protected clusters

nano-imprint lithography (NIL) 217–219
nanoparticles
 hollow 212
 spherical 178, 254
nanospheres 208–209, 221, 288–289, 295, 302
nanostencils 207–208
nanostructures 31, 157, 192, 203, 208, 213–214, 285, 292, 298, 302, 306–308, 313
NIL, *see* nano-imprint lithography
noble metals 6–7, 144, 162, 200

oligonucleotides 80–81, 259–260, 267

particle beam lithography 201, 204
PCR, *see* polymer chain reaction
PECVD, *see* plasma-enhanced chemical vapor deposition
penetratin 104, 111
peptides 111–112, 261–262, 264
 amyloid 89, 111–112
periodic nanostructure 186–187, 189, 191, 193, 195
photolithography 201–204, 207, 219, 227
plasma-enhanced chemical vapor deposition (PECVD) 48

plasmon dispersion 24–25
plasmon modes 126, 176–177, 288, 291
  localized 182–183
plasmon spectroscopy 90–99
plasmon-waveguide resonance 90, 94, 104, 108
plasmon-waveguide resonance (PWR) 90, 93–95, 104–106, 108, 111–113
plasmon waveguide resonance spectroscopy 89–90, 92, 94, 96, 98, 102, 104, 106, 108, 110, 112
plasmonic interfaces 35–36, 38, 40, 42, 44, 46, 48, 50, 52, 54, 56, 191
plasmonic materials 147, 163, 211, 226, 236
plasmonic metal deposition 207–208
plasmonic metal nanoparticles 251
plasmonic nanoparticles 201, 213, 219, 221, 225–227, 230
  self-organization of 201, 225
plasmonic nanostructures 7, 214, 225, 268, 292, 313
  fabrication of 199–230
plasmons 14–16, 19, 24, 27, 48, 52, 91, 93–94, 144–145, 149, 151, 176–177, 319–322, 339–340, 342
  hybrid 157
point mutation detection 84–85
polymer chain reaction (PCR) 62, 80, 83, 85
polymers 43, 53–54, 129, 209, 214, 225
polypeptides 261–263
polypyrrole-3 44–46
prostate specific antigen (PSA) 125–126, 269
proteases 261, 264

proteins 40, 85, 96, 103–105, 108, 112, 121, 253, 261–262, 334, 337, 352
PSA, *see* prostate specific antigen
PWR, *see* plasmon-waveguide resonance
pyrrole 44, 67, 69

quantum dots 227, 230
quasi-static approximation 172–173, 175–179

Raman scattering 30, 276–282, 287, 290–291, 294
Raman spectra 295, 327–329
reactive ion etching (RIE) 209–210, 233
replica molding 214–216
rhodopsin 105, 108
RIE, *see* reactive ion etching

SA, *see* sialic acid
SAMs, *see* self-assembled monolayers
self-assembled monolayers (SAMs) 37, 40–42, 68, 214, 225–226, 256
SERS, *see* surface enhanced Raman scattering
SERS enhancement factor 30, 294–295
SERS substrates 276, 295, 297, 306, 310
short-range surface plasmon (SRSP) 120
sialic acid (SA) 104, 112
silica 47, 93, 95–96, 103, 106, 174–175, 320

silica substrate 183–185
silver 7–8, 10, 15, 51–53, 93, 95, 175–177, 226, 234, 282, 288–290, 320, 325, 329, 334
  optical properties of 51–52
silver colloid 297–298
silver electrode 283, 298–299
silver nanocrystals 268
silver nanoparticles 223, 252–253, 290
silver nanospheres 290, 294
silver nanostructures 334
silver surfaces 283, 328–329, 332, 334
single-molecule detection 275, 283, 301, 306, 309, 313
single-molecule SERS 295, 297–299, 301, 303, 305, 307, 309, 311, 313
single nucleotide point (SNP) 62, 79, 85
single point mutation 76–84
SNP, see single nucleotide point
SPFS, see surface plasmon fluorescence spectroscopy
SPP, see surface plasmon polariton
SPR, see surface plasmon resonance
SPR imaging (SPRi) 61–63, 65–67
SPR sensors 23, 27, 29, 36, 224, 332, 334, 337
SPRi, see SPR imaging
SPWs, see surface plasmon waves

SRSP, see short-range surface plasmon
surface enhanced Raman scattering (SERS) 30–31, 275–276, 282–285, 287, 289–293, 295, 297–298, 300–302, 304, 306, 310, 312–313
surface plasmon excitation 63, 91, 93, 137, 284
surface plasmon fluorescence spectroscopy (SPFS) 120, 122, 124–125, 131–132
surface plasmon modes 120, 122, 137, 182, 186, 287, 289, 291, 294, 296
surface plasmon polariton (SPP) 2–3, 10, 19–24, 29–32, 119, 144, 146–147, 149–152, 156, 158–159, 161, 213, 236, 319–320, 328–332
surface plasmon resonance (SPR) 20, 28–29, 35–37, 54–56, 61–62, 91–94, 119, 143–144, 148–149, 151–152, 162–163, 200, 291–292, 313, 347–352
surface plasmon waves (SPWs) 2, 22–23, 27
surface-wave enhanced biosensing 119–120, 122, 124, 126, 128, 130, 132, 134, 136, 138

two-photon polymerization 212–213